100 Years in Maintenance and Reliability

Practical Lessons from Three Lifetimes at Process Plants

100 Years in Maintenance and Reliability

Practical Lessons from Three Lifetimes at Process Plants

V. Narayan
J.W. Wardhaugh
M.C. Das

Foreword by
Charles J. Latino

Illustrations by
Steven van Els

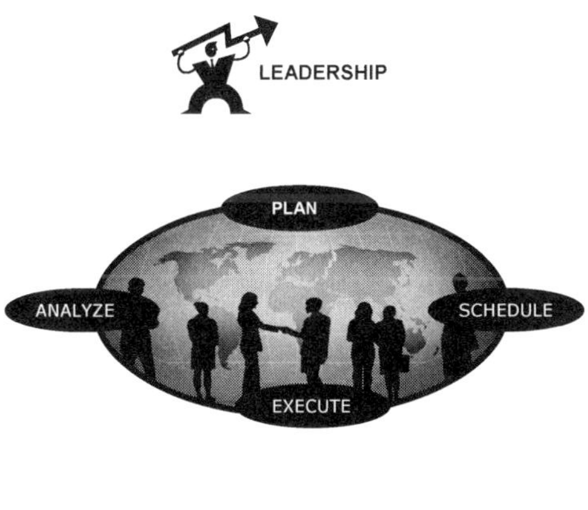

Industrial Press Inc.
New York

Library of Congress Cataloging-in-Publication Data

Narayan, V.
100 years in maintenance: practical lessons from three lifetimes / V. Narayan, M.C. Das, J.W. Wardhaugh.
p. cm.
ISBN 978-0-8311-3323-8
1. Maintenance. 2. Plant engineering. 3. Facility management. I. Das, M.C. II. Wardhaugh, J.W. III. Title.IV. Title: 100 years of maintenance.

TS192.N353 2007
658.2'02--dc22

2006053263

Industrial Press, Inc.
989 Avenue of the Americas
New York, NY 10018

Sponsoring Editor: John Carleo
Interior Text and Cover Design: Janet Romano
Developmental Editor: Robert Weinstein

10 9 8 7 6 5 4 3 2 1

Table of Contents

Foreword

This book is written by three engineers who have had exceptional experiences in industry, particularly the hydrocarbon process industry. All of them have held positions of authority in the maintenance and reliability of the companies they worked for. A reader will quickly grasp that they were self starters and still are, as evidenced by the creation of this book. These are not men who needed to be pushed. Indeed, I suspect that that would have dampened their motivation to create and install the myriad of solutions and systems that they introduced.

The book is a selection of work problems that these men had to struggle with and solve. The chapters penetrate every aspect of field engineering, maintenance, and field management. Each author was able to make contributions to each section. They were able to do this because of their remarkable breadth of experience, which readers will appreciate as they read and assimilate the various sections.

In my own career as a field engineer and manager in the process industry, I learned to listen and even enjoy the experience of others. Everything you read in this book will not be directly applicable to your particular job at the moment. However, as you read and enjoy their related experiences you will be storing their experiences in your mind. You will build connections to your own experiences that will make the text memorable. Finally, you will form ideas about how to approach problems that will make your respective jobs easier and more fulfilling.

The three authors provide their experiences in facilities in the Middle East, Far East, Europe, and Central America. As a reader, you may want to put yourself in the place of the writers as you study each episode in their long litany of experiences. In this way, you will taste the cultures that formed their experience.

I was gratified that the book did not linger on mechanical "How To" ideas, but got to the heart of what makes a refinery, or a machine manufacturing facility, or any production facility really work well. These men are really pointing out, although they do not specifically say it, that the greatest impetus to successful operations is how people manage themselves and set up the procedures that provide rapid and accurate work products.

At the end of each of their related experiences, the writer delineates the lessons he learned from that encounter and the principles that emanated from the experience. In reading the draft, I found that the lessons and principles capped the learning; they made the narration of the depicted experiences complete.

In summary, this is a must read for people who have to struggle with the day-to-day problems of plant life. If you have a subordinate field position in a manufacturing facility, this book will reveal why bosses do the things they do. If you are in a supervisory or management role, this book will help you steer your career.

Charles J. Latino
CEO and President
Reliability Center, Inc.

About the Authors

V. Narayan is a leading authority on maintenance and reliability engineering. He is a graduate mechanical engineer from Pune University in India and has over 40 years of experience in maintenance and project management. He has worked in the automobile, pharmaceutical, liquefied natural gas, oil & gas production, and petroleum refining industries. In his long career he has trained, consulted and worked in many countries, including spending eight years as the head of the Maintenance Strategy group in Shell UK Exploration and Production. He headed the Maintenance and Reliability Center of Excellence in Shell.

At Shell, he developed refinery performance measurement methods in the 1990's that are still effectively used today. For the last 16 years, he has been teaching Reliability Engineering, Maintenance Management, Reliability-Centered Maintenance and Root Cause Analysis to engineers in the USA, Europe, Middle East and Far East. He is an advisor to the Robert Gordon University in Aberdeen, on their Asset Management MSc program development. He has published many articles and presented papers at international conferences. His book, "Effective Maintenance Management–Risk and Reliability Strategies for Optimizing Performance" is also published by Industrial Press.

Jim Wardhaugh graduated from the University of Liverpool and is a Chartered Engineer. In a 30-year career in Shell he demonstrated success in many different roles (projects, construction, maintenance, technical, inspection, warehousing, transport, quality, and training) in a number of different countries. Within Shell's Technical Head Office, he guided refineries worldwide on best maintenance practices and on "Computerized Management and Information Systems". He was a founding member of the Shell MERIT consultancy group. He is an external faculty member of the Robert Gordon University, Aberdeen, for their MSc program in Asset Management. He provides consultancy services that target performance improvement, particularly in the fields of asset management, operations, and maintenance.

Mahen Das has a mechanical engineering degree from Benares University, India, and is a Chartered Mechanical Engineer. He retired from Shell International in 2002 after 42 years in optimization of maintenance and operational reliability of petroleum refineries and gas plants. His learning and experience has been drawn from hands-on work, at all levels of process plant asset management, at 40 sites in 22 countries. After working for 32 years in the "field", the last 10 years of his career were spent in transferring this learning to Shell, as well to as third-party clients in the form of consultancy services. During this period he helped establish Shell's MERIT initiative. As a leader of MERIT, he visited and reviewed the business processes of more than 30 operating plants – Shell's as well as third-parties' -- and helped achieve significant improvements in their maintenance and reliability performance. He is currently a free-lance consultant in the field of maintenance and reliability of process plants.

LEADERSHIP

PART 1: INTRODUCTION

Chapter 1

Introduction and Navigation Guide

Author: V. Narayan

1.1 Authors' Background

The school of hard knocks taught me most of the really useful things I learned about reliability and maintenance. I worked with many talented people during my career in industry, who were my best teachers. My co-author Mahen Das and I are both mechanical engineers and worked in a small petroleum refinery in the early stages of our careers. This company had a "can-do" attitude and dynamic culture. We could make occasional mistakes without fear of reprisal. Innovative ideas and creativity thrived at every level. Mahen and I took full advantage of this wonderful social laboratory. But, as the saying goes, all good things eventually come to an end. When this happened, we left the company within a week of each other and went our separate ways.

A few years later both of us happened to rejoin the parent company of this refinery in Europe. Some years later, both of us began working in their corporate headquarters. That was when we met Jim Wardhaugh, the third author of this book. Jim is an electrical engineer who worked in the power generation and distribution industry before coming to the oil and gas industry. The three of us got along brilliantly, and were members of the maintenance and reliability advisory team in the parent company.

The parent company had global operations in the upstream and downstream oil and gas business. It had responsibility for day-to-day management for many operating companies distributed around the world. Corporate headquarters provided technical support and governance. The parent company wholly owned or had a significant stake in the operating companies. In most cases, it was responsible for the design of the facilities in the operating company, as well as for commissioning and initial operations. The support included providing the operating companies with skilled staff and technical advice.

During the first few years of any new venture, key positions were held by staff assigned from the parent company. They were responsible for operating the facilities safely and efficiently. They also trained local employees to take over these positions within the first few years. Most of the assigned staff stayed in any one location for three-to-five years. Mahen, Jim, and I were among this group of gypsies working in different parts of the world. Later, during our tenure in the corporate headquarters, we traveled widely in a consulting role.

All three of us have worked in other industries, e.g., engineering, pharmaceuticals, textile machinery, chemicals, power distribution and manufacturing. Some of the events described in this book are from these industries. Together, the three of us have more than 100 years of experience in this field—which explains the title of the book.

1.2 Impetus for this Book

Books dealing with maintenance subjects seem to focus on answering these questions: What should be done? How should it be done? And sometimes, when or why should it be done? Books on reliability seem to focus on mathematical aspects; the average maintenance manager or supervisor finds it hard to relate their content to the reality they face in their work.

We decided we would write about learning experiences from our working lives. We describe the hand we were dealt and how our team handled the situation in those circumstances. In hindsight, we found some underlying truths or principles in these experiences which we believe may be applicable in other situations. Based on these descriptions, readers can decide whether they should consider a different approach from the ones they currently follow.

1.3 The Shewhart Cycle

Edward Deming[i] describes Shewhart's continuous improvement cycle with the Plan-Do-Check-Act sequence. We use an adaptation of this, with the Plan, Schedule, Execute, and Analyze phases, shown in Figure 1.1. We have grouped our chapters under these four headings. Although some of the chapters could be placed under two or more headings, we chose the heading that seemed appropriate from our perspective. To these four, we added two more headings: Leadership and People. The subjects covered in the various chapters fall under one of these six headings. We hope that the grouping helps readers to find what they are looking for easily.

1.4 Chapter Contents

In each chapter, we describe an event or situation that one of us experienced personally. We have tried to relate the events factually, at least as far as we could remember them. In order to protect the identities of those in-

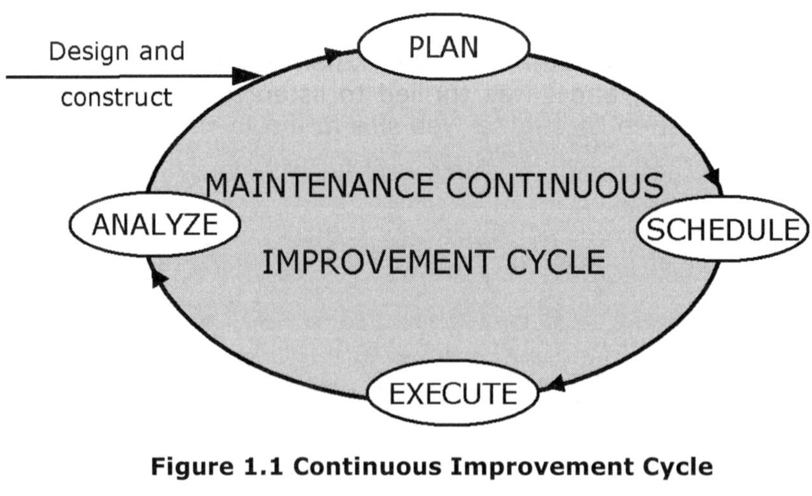

Figure 1.1 Continuous Improvement Cycle

volved, we have not revealed the names of the locations or of the individuals. The key issues are 1) how the people involved handled the event and 2) the results or outcome of their effort. We have summarized our own learning from each situation. We also included underlying principles we feel are relevant: these are stated at the end of each chapter.

At the time of the events described in the book, we did not know many of the relevant underlying theories or philosophies. We picked up most of the concepts subsequently, sometimes many years later. Had we known them at the time, we may have found the solutions with less effort. For the benefit of readers who may need them, we have described the relevant theory or methodology in appendices.

1.5 Locations

We have, between the three of us, worked in about 20 locations. In a consulting role, we have advised at least 30 locations around the world. Each site had its own way of doing things. So that readers can appreciate the different pressures the locations faced, we have provided an overview in Chapter 2. Because there may be more than one chapter about a given location, some of the common information is described in this chapter. Hopefully, this will avoid needless repetition, but it does mean that before reading a chapter, readers may have to go back to Chapter 2 to get the overview.

Where relevant, we have given some information about the cultural and social climate that prevailed in that location. So that the events described are placed in context, we have provided additional background material at the beginning of each chapter.

1.6 Glossary, Acronyms and Abbreviations

Please refer to these if certain words or acronyms are not clear.

1.7 Acknowledgements

I had the pleasure of meeting Charles Latino in Chicago at the Maintenance and Reliability Technology Summit in 2005. Charles is a well-known reliability and maintenance guru, and I was thrilled to listen to his brilliant lecture. I knew his son Bob Latino through a web site forum in which we both participate, and from his excellent book Root Cause Analysis[ii]. Hearing Charles' talk spawned the idea for this book. I consider it a great honor that he has written the foreword.

A former fellow student and friend from my University days, Satish Shirke, agreed to do the illustrations for the book. Satish lives in California, but we managed the trans-Atlantic communications quite well. He did a great job, but because of his workload, he could not continue. I was in a spot, desperately seeking a good illustrator to replace Satish. When Steven van Els, based in Suriname in South America, offered to help, I was delighted. I 'met' Steven on the reliability.com web site and have a great deal of respect for his knowledge and experience. Steven has done an absolutely superb job, converting our crude sketches and charts into excellent figures or tables. He also added value by creating drawings to illustrate the text at his own initiative. As a real world practicing maintenance manager, his comments on the chapters were invaluable. My friend Narmada Guruswamy helped design the cover pages separating the six main parts of the book.

The International Labour Office in Geneva gave us permission to use two tables and three charts from an excellent reference book entitled Introduction to Work Study[iii] published by them. Mr. Peter Morgon of Lithgow & Associates and MPI Publications, publishers of Fitz's Atlas[iv], very kindly agreed to our reproducing graphical coating breakdown standards from their book.

Earlier I have had the pleasure of working with the team from Industrial Press Inc., the publishers of this book. Janet Romano designed the cover and provided much needed support with the publication and printing, and Suzanne Remore kept us on our toes in meeting schedules. Patrick Hansard is a pleasure to work with and a great person to handle sales and marketing. I have known John Carleo, the Director of Publications, for over three years. He has been a friend, philosopher and guide, and an enthusiastic supporter. In practical terms, this meant fast responses to my queries and requests, and guidance in all aspects of publication.

Christine Wardhaugh, Madhu Das, and my wife Lata have been ever so patient and tolerant with the three of us. Both Christine and Madhu accused me of being a slave driver. Lata came to my support, saying I was both a slave and a slave driver!

Mahen, Jim, and I are grateful to all of these wonderful people.

References

i Deming, W. Edwards. 2000. *Out of Crisis.* Cambridge: MIT Press. ISBN:026 254 1157.

ii Latino, R.J. and K.Latino. 2002. *Root Cause Analysis: Improving Performance for Bottom-Line Results.* Boca Raton: CRC Press. ISBN: 084 931 318X.

iii Kanawaty, G., ed. 1992. *Introduction to Work Study.* 4th (revised) ed. Geneva: ILO Publications. ISBN: 92-2-107108-1.

iv Weatherhead, Roger and Peter Morgan, Lithgow & Associates, ed. *Fitz's Atlas™ of Coating Defects.* Surrey: MPI Publications. ISBN: 0 9513940 2 9. URL: http://www.mpigroup.co.uk/fitzs-atlas.asp

Chapter 2

The Locations

Author: V. Narayan

We, the three authors—Jim Wardhaugh, Mahen Das, and Vee Narayan—have worked in a number of locations around the world. In this chapter, we will describe each location briefly so that you have an overview of the sites and get an idea of the facilities and prevailing culture. In the chapters that follow, we will refer to these locations by their reference number. Please see the relevant section here before proceeding to the chapter you wish to read.

2.1 Locations in the Middle East and South Asia

2.1.1 Facility: Pharmaceutical Plant

This small company made a range of over-the-counter drugs. The main products were throat lozenges, pain-relief balms, and tablets for relief from colds and headaches. They also produced menthol crystals from oil extracted from the menthol plant.

The facilities included ointment blending vessels, tablet forming, coating and packaging machines, bottling machines, packaging lines, and a refrigeration plant. The research and development facilities were located at the factory site; this group was also responsible for product quality management.

2.1.2 Facility: Automobile Parts Manufacturer

This large company made fuel injection pumps for diesel engines and spark plugs for petrol engines. The factory had about 3000 machine tools, many of which were of high precision and cost. They had a European principal who provided technical expertise and governance. The principal operated similar factories in five other countries. In this site, they had about 8000 employees, working six days a week. About 2500 production staff, mainly machinists, worked in each of the first and second shifts. About 1500 production staff worked in the third shift. There were about 1500 employees in the 'day' shift.

Company employees had a strong work ethic; people were disciplined, kept to schedules, and worked to high quality standards.

2.1.3 Facility: Petroleum Refinery

This facility was a semi-complex petroleum refinery, with process plants, utilities, product packaging, and storage facilities. The process plants were grouped into two main sections. The primary processing units included crude distillation, high-vacuum distillation, and bitumen blowing units. The secondary processing plants consisted of a fluid catalytic cracker and a reformer (or platformer) unit, grouped with the utilities.

There were two other operational sections, responsible for the storage and handling of crude oil and products. One of them managed bitumen and liquefied petroleum gas storage, packaging, handling, and dispatch. Another managed the storage and handling of crude oil and products. The maintenance areas were aligned to these sections, with a supervisor in charge of each area.

Breakdowns and trips of equipment were common, resulting in excessive downtime and costs. Maintenance in the refinery had become a firefighting activity. Craftsmen were constantly being moved from job to job, resulting in low productivity and quality. As a result, morale was low, both in Maintenance and in Operations.

2.1.4 Facility: A Large Petroleum Refinery

At the time of the events described in this book, this refinery was fairly new. Two large distillation units and a high vacuum unit provided primary distillation capacity. Secondary processing included thermal and hydro-cracking units. There was a large benzene unit and hydro-treaters for kerosene, naphtha, and gas-oil. Electricity, potable water, and sea cooling water were provided by public utility companies. Product-to-feed heat exchangers and air-cooled heat exchangers were used for cooling, with some limited final cooling with sea water exchangers.

Most of the maintenance work was reactive, but the condition monitoring program and minor preventive maintenance work (lubrication, alignment checks, etc.) were planned and executed satisfactorily. Local craftsmen were being trained, and the bulk of the maintenance work was done by expatriate contract workers. Skill levels were reasonable, but the company's approach was that it was employing 'hands' not people who could use their brains as well. Some of the (expatriate) supervisors were very good, but most were of average caliber.

2.2 Locations in East Asia

2.2.1 Facility: Liquefied Natural Gas Plant

This Liquefied Natural Gas (LNG) Plant was located on the coast. There were three production modules, where the natural gas was compressed and cooled, thereby liquefying it at -260°F. Steam turbines, each 9 MW in size, powered the nine refrigeration compressors. The plant generated its own electricity, using gas turbine and steam turbine driven alternators. Steam, at 60 bar gauge (barg.), was raised in 9 boilers. There were two liquid nitrogen generators to produce the nitrogen required for blanketing and as purging medium. The LNG was stored in double-walled cryogenic (extra low temperature) tanks. Dedicated LNG Tankers carried the cargo to the customers, from a company-owned deep-water jetty. The natural gas vapors, formed by evaporation from the storage tanks and by displacement from the tankers, were collected and compressed for use in the boilers and gas turbines.

Cryogenic plants require special materials of construction because low-carbon or low-alloy steels are prone to brittle fracture at low temperatures. The main materials used include aluminum and austenitic stainless steels. Aluminum is a difficult metal to weld and needs specially qualified welders and welding processes.

Most of the local people employed in the plant were middle-school or high-school graduates. They were young and enthusiastic, but with little exposure to heavy industrial or high hazard plants. Although expatriates held most of the senior technical positions, local engineering graduates were placed in supporting roles so that they could take over senior positions quickly.

2.2.2 Facility: Large Complex Oil Refinery

The refinery intake was about 14 million tons per year, received mainly by ships tethered to a single buoy mooring. The main units of this large refinery were: three crude distillation units, two reformers, a lubricating-oil (lube-oil) complex, a thermal gas-oil unit, a hydro-cracker, a long residue catalytic cracker, an isopropyl alcohol plant, and a large generation and utilities complex. There were a number of other smaller processing units and a large oil-movements area consisting of tanks, blending, and a pipeline operation. Operations were controlled from a number of control rooms. Waterfront operations moved thirty million tons per year of product over 11 wharves with 3,500 shipping movements. The facilities described above varied in age from the geriatric to brand new and had a replacement value of about US $4 billion.

About 600 people lived on the refinery site and another 2,000 came to the refinery each day.

The organization was as traditional as you could get with Operations, Technology, Finance, Personnel, and Engineering Managers. Under the Engineering Manager were discipline heads for Mechanical, Electrical, Projects, etc. Expatriates held a number of key positions, but most positions were filled by very competent local staff.

The refinery was well run and profitable but significantly overstaffed. Benchmarking studies showed there were areas of superior performance with excellent practices that others could copy to their benefit. However, they also showed that this refinery was, at best, an average performer and hadn't moved with the times. Reliability, manning levels, and operating costs needed attention.

2.2.3 Facility: New Medium-Sized Complex Oil Refinery

This was a brand new joint venture; an 8-million-tons-per-year refinery situated on the coast and designed to process a mix of Middle East crude oils (by ship, over a single buoy mooring loading facility) as well as indigenous crude oils over jetties. The main plants included atmospheric and vacuum distillation units, hydro-cracker, visbreaker, hydro-desulfurizing unit, hydro-treater, and a reformer (platformer). The electrical power generation capacity met internal requirements fully. Similarly, there was capacity to produce other utilities. Thus, the refinery could operate effectively on a stand-alone basis, although it was connected to the local electrical grid.

The refinery was designed to perform at world class standards in e.g., process efficiency, plant availability, utilization, organization style, manning levels, safety, environmental impact, and overall costs.

The management structure was traditional with Operations, Engineering, Finance, and Personnel functions. Maintenance engineers and technicians provided the core expertise at the working level. The philosophy adopted in recruiting staff was, however, definitely non-traditional. Plant operators were recruited from a craft background and then further trained in a specific craft skill, e.g., mechanical, instruments, or electrical. These operators then spent two thirds of their time in operations and one third in maintenance. During the latter period, they did the bulk of the maintenance work while specialized technicians provided the high-level competencies. A competence framework helped manage the concept, rewarding acquisition of needed skills.

2.2.4 Facility: A Medium-Sized Simple Petroleum Refinery

Commissioned in the late 1960s, this was a medium-sized, simple (hydro-skimming) refinery. Together with primary distillation capability, it had a platinum reforming unit to obtain high octane naphtha and hydrogen for its hydro-treating processes. A short period before the events described in this book, a refrigerated liquefied petroleum gas storage and export facility had been constructed and added to the refinery assets. As the country has very high literacy, the refinery had a well-educated work force. Traditionally, the population is highly skilled in crafts and people are proud of their handiwork. The value system also includes respect for elders, obedience to authority, and help and support to one another within the community.

2.2.5 Facility: A Regional Oil Company Operating Refineries and Downstream Operations

The company operated four refineries located in various parts of the country as well as a national marketing network. All refineries processed imported crude oil. The refineries ranged from medium–sized, semi-complex to large complex units. Between them, they had a full range of petroleum refining process plants, with sophisticated, state-of-the-art process control systems. Although the region was technologically very advanced, they did not apply sophisticated computerized methods to refinery maintenance. The refineries functioned with very hierarchical organizations in which communication was strictly via the official chain of command.

2.3 Locations in Europe

2.3.1 Facility: A Large Petroleum Refinery

This very large refinery was located on the coast. It was a 'swing' refinery, processing several types of crude oil. Because it had many process plants, there were a number of plant shutdowns every year. In this facility, they did a number of things very well, and others learned from them. Staff were disciplined and generally performed competently. Their technical knowledge base was excellent, but typical large facility silos had developed, leading to indifferent business performance. In benchmarking studies, they came out about average.

The refinery was located in an industrial belt, along with a number of other large chemical plants, refineries, and manufacturing facilities. All the companies used contractors extensively for shutdown work. Most of the shutdowns were scheduled in the April-October period, avoiding the cold weather as far as possible. Contractor manpower requirements peaked significantly during these periods. Inter-company agreements were in place to minimize bunching, which could result in dilution of skills.

2.3.2 Facility: Large Oil and Gas Production Company

The company carried out Exploration and Production (E&P) of hydrocarbon oil and gas. It was one of many companies in a large multi-national oil and gas group of companies.

The facility had several offshore oil and gas reservoirs. Some of these were exploited from fixed Platforms, others from Floating Production Storage and Offloading facilities (FPSOs). There were also a number of sub-sea installations feeding Platforms or FPSOs.

The company was innovative and used leading edge technology. It provided skilled staff on loan to other companies in the group.

It suffered from some of the large company problems. These included working in silos, optimizing things to benefit the department rather than the overall business, and slow speed of decision making.

2.3.3 Facility: Corporate Technical Headquarters

This facility was the technical headquarters of a very large multi-national oil and gas group of companies. From these offices, the corporate staff provided technical support to a large number of exploration and production facilities, refineries, gas plants, and chemical plants located around the world. A small maintenance and reliability team provided a benchmarking and consultancy service to the refineries and gas plants. The team identified maintenance best practices for sharing within the group to promote increased profitability and plant availability. They used written guidelines, newsletters, training courses, workshops, and conference events to transfer knowledge between locations. The three authors were founding members of a reliability improvement team, which worked with sites to promote performance improvements. This process proved to be very successful.

2.3.4 Facility: A Small Complex Oil Refinery

This was a small, rather aging asset onto which was being grafted some modern world-scale plants for product upgrading. It had the usual refinery plants of atmospheric and vacuum crude distillation and lubes along with thermal gas oil, reformer, hydro-cracker, catalytic cracker, bitumen, and a large electrical generation and utilities operation. There was a large tank farm, dated both in age and technology, with a fairly new blending facility and jetty area handling significant shipping movements.

Poor industrial relations with resistance to change sapped managerial energies. Outdated attitudes and work practices typified this location.

2.4 Location in Australasia

2.4.1 Facility: Medium-Sized Semi-Complex Petroleum Refinery

At the time of these events, this medium-sized, semi-complex refinery was owned as a joint venture by five partners who used it to process crude oil owned by them—either from their own fields or bought in the spot market.

The refinery was located in a breathtakingly beautiful natural environment. People there were very proud of this, and the quality of life it offered. Right from the time it was conceived, there was a strong lobby against its very existence due to the inherent potential threat to the pristine environment. In spite of the fact that it was always operated to the highest environmental standards in the world at that time, it was an eye sore in the perception of the local people.

2.5 Locations in Central and South America

2.5.1 Facility: Small Petroleum Refinery

This was a small and simple (hydro-skimming) refinery, containing plants for primary distillation, platinum reforming, and hydro-treating of naphtha, kerosene, and gas-oil. The refinery was wholly owned by a major multi-national oil company. It was one of the few technologically advanced industries in the country. For this reason and because it was also among the top quartile payers, it was one of the most sought after places of work. It attracted the best of the local people as employees who proved to be very loyal and were always willing to give their best to the company.

They had a very progressive management team, always on the look-out for improvements and trying to bring the best out of their individual staff. The staff responded enthusiastically to all the challenges put forth by their management.

LEADERSHIP

PART 2: LEADERSHIP

Chapter 3

Creating the Vision

...and seeing it through to fruition

Effective people are not problem-minded, they are opportunity-minded.

Peter Drucker, Management Guru.

Author: V. Narayan

Location[1]: 2.1.2 Automobile Parts Manufacturer

3.1 Background

On my first day at work in this company, I met my boss, the General Manager of Production (GM). My position had been vacant for a year, during which time the head of the production planning department had been managing it. During this interim period, a number of issues had arisen, which the GM listed for my action. When he finished, I requested a three-week vacation, and he nearly fell off his chair! I explained that I would come to work, but wished to be free of executive responsibility in order to evaluate the current situation for myself. This review would help me identify the expectations of all the stakeholders, including the people on the shop floor—but I did not share this thought with him.

The review would give me a first-hand impression of the current status. From these inputs, I would produce a master plan. Each item in the master plan would be a separate project, with its goals, cost, time, and resource estimates. When he heard this explanation, he accepted my request. He still negotiated the review time period downward to two weeks.

3.2 Review Process

I arranged meetings with about 70 production supervisors and their managers, in groups of 10–12 people. In these sessions, I listened to them and recorded their requests and complaints. There were additional meetings with the main service department staff as well, including those in the stores and main canteen. The canteen staff had several requests, some of which ap-

[1] For this chapter and all subsequent chapters, see Chapter 2 for additional information about location.

peared quite important for staff welfare. During factory rounds, I spoke to production and maintenance workers and union representatives. My own discipline engineers, supervisors, and contractors also provided their inputs. It appeared that many of the issues were related to stresses on the infrastructure. The company had seen rapid growth over the initial 15 years of its existence, but the infrastructure had not kept pace with the growth in production volume.

Analysis of the feedback highlighted some common problems. These included complaints about the utilities: provision of electricity, water, and air. Factory ventilation, dust levels in the ceramics department, and fume extraction in the plating department were also significant issues.

The main canteen provided food to more than 4000 people in the daytime, about 2500 in the second shift, and about 1500 people at night. Food was served in batches, as the seating was limited to 1500 people. Electrical heating was used for cooking, for which they needed a secure electricity supply.

These issues did not appear in the GM's list, which generally covered current projects, some staff issues, and a list of complaints from production managers.

I applied the first two project selection hurdles. Were these expectations related to production or welfare or safety, and were they feasible? This process narrowed the list down to about 10 items that could be handled as stand-alone projects. The next step was to evaluate them for importance. We had to find the money for items that affected health, safety, and environment (HSE) and staff welfare. Items that were critical to production were clearly important. We will not go into the details of how funds were obtained; that is a long story in itself. Suffice to say it needed lateral thinking and agile maneuvering.

The lead time for completing some of the items was two–three years. This allowed us to phase the work within a three-year budget window. The company had an annual budgeting system. This imposed additional challenges of phasing, accruals, and other familiar accounting handcuffs with which most engineers will be familiar.

3.3 Selected Projects

We selected the following projects based on the criteria discussed earlier. A brief description of the work is given along with its justification and timing.

1. Factory Ventilation (HSE)

With conventional north light roof trusses, the temperature inside the large factory buildings reached 85–90°F in summer. There were a few large column-mounted air circulators to provide artificial ventilation. We planned to install 40 additional air circulators to alleviate the problem. The lead time was 6 weeks and we could get 10–12 units per month. Summer was approaching, so this became a high-profile HSE issue. The costs involved were relatively low and people on the shop floor would see action being taken. The workers representatives helped decide the sequence in which the new units would be installed, giving them a role in decision making. The sequence was something I preferred they decided themselves, as it would minimize arguments. The project was justified as an HSE item.

2. Electricity Supply

The public electricity supply system was unreliable, due to a serious mismatch between supply and demand. There were frequent power cuts; to overcome this difficulty, the company had installed four 350 kW diesel generators, with a fifth on order. These worked as stand-alone units, supplying isolated sections of the factory. This limited our flexibility to provide power where it was needed to suit the (variable) production demand. With stand-alone units, we could never load the machines fully. To overcome these limitations, we planned to synchronize the generators and connect them to the distribution network. The latter was currently not a ring main; this was another shortcoming needing correction.

This project required major investment in new transformers, circuit breakers, and feeders. Due to the lead time required for procuring the hardware, this project was phased over three years. The cost of the project was high, but so were the expected returns. We expected to reduce the value of lost production due to electrical supply problems by 50–60%, giving a benefit-to-cost ratio of 5:1.

A different issue related to the cost of electricity purchased from the public supply system. The electricity supplier applied a three-part tariff, with charges for the connected load (kW), energy consumed (kWhrs), and a surcharge for power factor below 0.96 (kVA charge). In addition to the thousands of induction motors in service, there were large induction furnaces in the factory. Without correction, the power factor could drop as low as 0.91. We already had a number of power factor correction capacitor banks, which brought it up to 0.94–0.95. We planned a separate project to increase the power factor to a maximum of 0.98. This upper limit was set by the possibility of a large induction furnace trip when we could end up with a leading power factor. The new capacitor banks would be brought into service or disconnected so that the power factor never exceeded 0.98 or went below 0.96. The project was phased over two years, based on hardware availability. The costs were relatively low and the expected benefit-to-cost ratio was 5:1.

3. Air Supply

There were two problems, one relating to pressure fluctuations and the other to entrained water. The latter issue had been so serious in the past that the main air supply lines in the factory buildings were sloped in a saw-tooth fashion, with manual drains at the low points (see Figure 3.1).

Pressure fluctuations were due to peak demands exceeding installed capacity and because of pressure drops in the pipelines. The entrained water came from the humid air. The water should have condensed in the after coolers of the air compressors, but a simple calculation showed that the cooling water temperature was far too high to be effective. In turn, this was due to an overload on the closed circuit cooling system. The original cooling pond was suitable for two diesel engines and three air compressors. The equipment num-

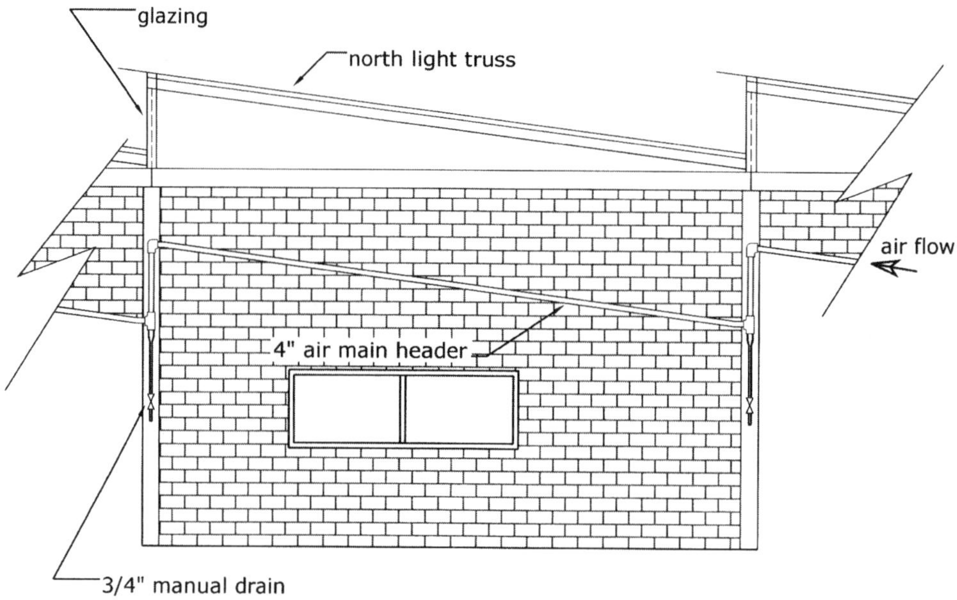

Figure 3.1 Original design of 4" air mains.

bers had grown to four diesel engines and four compressors. One more generator and two compressors were on order.

The air compression capacity was marginal and the projected demand increase was 30 percent. We decided that a third one would be needed to provide buffer capacity. In order to reduce the pressure drop in the pipeline distribution network, we planned to add four new air receivers located close to the main consumers. Peak demands could then be met from these receivers. They would also act as additional knock-out vessels to trap entrained water.

We planned to install industrial cooling towers to absorb heat from the cooling water used in the engine and compressor cooling jackets and after-coolers. This would eliminate the bulk of the entrained water at source.

These two projects were planned for completion in 18 months. The cost of the third compressor, air receivers, and cooling towers was in the medium-range. We expected to reduce the value of lost production due to air supply problems by 90%, giving a benefit-to-cost ratio of 15: 1.

4. Water Supply

The city municipal water supply system provided about 70% of the factory's requirements. The company had installed many bore wells to draw groundwater to meet the remaining requirements. The city accepted our justification for requesting additional water supply, but were not willing to invest in a new pipeline from an existing reservoir about four miles away. We offered to underwrite the capital costs while the ownership remained with the municipality. I convinced the finance manager that we should pay a grant towards the capital cost of a city asset that would benefit the company.

We also decided to accelerate investment in additional bore wells in plots of land owned by the company in the vicinity of the existing factory site.

These projects were also in the medium-range of costs. Most of the additional water requirements were for welfare facilities. Without these projects, production levels would eventually have to be drastically curtailed, but we justified the project on staff welfare and HSE grounds.

5. Dust and Fume Pollution

The dust pollution in the ceramics department and the fume problem in the plating department were potentially serious health issues. The existing extraction systems were clearly not effective, but the solutions were not obvious. At this stage, the project scope was to study the problem carefully, understand the causes, and identify solutions. We employed a specialist consultant to assist us, and the work took several months to complete. The problem was traced to the particle size of the ceramic dust. These were so small that much higher velocities were required at the extraction unit inlets. The project scope included the installation of cyclone separators and powerful extractor fans.

At the plating department, we found that the fume extraction issue was more complex. The extraction hoods had to be redesigned and repositioned. Extraction velocities had to be increased, so new fans were required.

The costs of these two projects were in the medium range, and the lead time of the equipment required meant that the project had to be scheduled in the third year. We justified it as an HSE project, but the results showed that there were other benefits as well.

6. Security of Energy Supply to the Canteen

The scale of the problems that the canteen faced on a daily basis was staggering. The local culture required that freshly cooked and piping hot food be served. The main staple was cooked rice, of which we needed on average, 10 oz. per employee. About 1500 meals were served in each batch.

The rice was cooked in large electrically-heated cookers mounted on trunnions. Each batch had to be cooked in 20 minutes, and the vessel cleaned and ready for the next batch in 5–10 minutes. The water temperature had to be raised from the ambient 60–70°F to 212°F, and this could take 10–12 minutes. The canteen manager was visibly under stress. If there was any glitch, food could not be served—to at least 1500 and possibly up to 4500 waiting people!

The electrical cooking system was excellent, but consumed significant amounts of energy. Because sunshine was available in plenty, we planned to install solar water heater panels on the concrete roof of the canteen. Each panel would be about 120 square feet in area. With four of them in series, even on a cloudy day we could get the water to 150–160°F in about 10 minutes. We decided to install two banks of four panels each along with an insulated hot water storage tank. This allowed us to supply hot water rapidly, and stored enough water for the second and third shifts as well. A structural design check of the roof confirmed that it was suitable for the additional roof loads.

The project costs were in the medium range. Delivery of the solar panels would take 6–8 months, so we phased the project into the second year. The primary purpose was to get rapid supplies of fairly hot water to the cooking vessels, so that cycle time could be reduced. This would give recovery time to the canteen staff in the event of a power supply glitch. The bonus was that electrical energy savings made it economical as well. The project was justified as a welfare item.

3.4 Results

We completed all the selected projects within three years. When computing benefit-to-cost ratios, we measured or estimated the benefits over a 3-year period (thereafter, they would be influenced by other initiatives as well). The results are described below.

1. Factory Ventilation (HSE)

The air circulation fans were installed more or less on schedule. Some installations were late, caused by delivery delays from the vendor, but all the fans were in place within four months. Our departmental credibility went up a notch in the eyes of the workers.

2. Electricity Supply

There were budget overruns, as the transformers and circuit breakers cost nearly 30% over the estimate. This had to be offset by savings elsewhere. On the plus side, the value of production lost due to electricity supply problems went down by nearly 80%. The benefit-to-cost ratio was 5.5:1.

The power factor capacitor banks and their control systems were very effective. The reduction in electricity bills was better than estimated, and the benefit-to-cost ratio was 6:1.

3. Air Supply

We installed pressure recorders at key points in the three factory buildings. The charts showed that after installing the air receivers, the pressure fluctuations were minimal and well within acceptable limits.

Once the new cooling towers were connected, more than 95% of entrained water was trapped at the supply end. A small quantity was drained from the air receivers, but there was no water to be drained from the low point drains on the air mains any longer. The saw-tooth pipeline design described earlier was abandoned whenever new air lines were laid.

Production loss due to air supply or quality problems all but disappeared once all the new facilities were installed. Computing the benefit-to-cost ratio proved difficult, as there were questions about the number of compressors to be included in the cost figure. The range was 11:1 to 16:1, depending on the cost figure selected.

4. Water Supply

Laying the new water mains proved very time consuming, as the municipality had complex and slow tendering processes for procuring and laying the pipe. There were city streets to be crossed; this required coordination with other city departments and utility companies. Eventually it was completed after about 30 months.

We made better progress with the additional bore wells, about half of which turned out dry while the rest yielded varying amounts.

Meanwhile, the demand was rising continuously. These two projects helped us to meet the demand, but there was no doubt that the problems would worsen in future. We did not compute a benefit-to-cost ratio as it was a survival and welfare issue.

5. Dust and Fume Pollution

The ceramics departments used to be in a permanent dust haze before we installed the new cyclones and larger extractor fans. The haze cleared visibly and quickly, so the workers were happy. But there was an attractive spin-off as well. Most of the ceramic dust recovered from the cyclones could be reused, allowing a small production volume increase and cost savings. What started off as a welfare/health project gave a benefit-to-cost ratio of 2.5:1.

The new fume extraction hoods and fans in the plating department worked well from the beginning. The number of workers reporting sick dropped significantly, so we felt quite pleased with the results.

6. Security of Energy Supply to the Canteen

The solar water heater panel project produced dramatic results. The canteen people were relieved from the tension that prevailed earlier. They could go about their work calmly and with less anxiety. The savings in electrical energy paid for the project within eight months, which was a bonus.

3.5 Lessons

When management gurus talk about vision, mission, and objectives, we may find our eyes glazing over. However, this experience taught me that the gurus are quite right. A systematic approach allows us to objectively evaluate what needs to be done and why.

As engineers, we do not always think in commercial terms; technical excellence is what most of us find appealing. Without an effort to do a cost-benefit analysis, I suspect these projects would have been shot down. When the benefit is 250% of cost (in some cases it was over 1000% of cost), it is easy to convince management. Funds suddenly become available to maintainers and engineers, instead of the much-favored Production and IT departments.

We found that shop floor workers can be quite realistic in their expectations. When it comes to recognizing infrastructure weaknesses, their inputs are often quite useful. Visible feedback that they can see through our actions helps build trust and confidence. Shop-floor staff helped identify the main weaknesses during the two-week review period, not outside consultants. The items they highlighted proved valuable, as all of them had excellent economic or HSE benefits.

That the boss is an important customer is not in question; not recognizing this can be career limiting! However, we should pay heed to the other customers as well, and include their ideas in our plans.

Expectations should be vetted to ensure that they add value and are manageable within existing cost constraints. Only those projects that pass the hurdles should be used to formulate the plan.

3.6 Principles

1. Deciding a line of action pro-actively is distinctly superior to playing catch-up. The vision and the current status give us the means to do a gap analysis and set our objectives.
2. Knowing the customer's expectations is important, whether these are from management or the shop floor. Asking them directly is better than making assumptions.

Chapter **4**

Setting Objectives

... why customer expectations matter

Energy is the essence of life. Every day you decide how you're going to use it by knowing what you want and what it takes to reach that goal, and by maintaining focus.

Oprah Winfrey, Talk Show Host

Author: V. Narayan

Location: 2.1.2 Automobile Parts Manufacturer

4.1 Background

The company designed and built many of the special purpose machine tools (SPMs) they needed for manufacturing their product range. This work was done by a separate division that had a design office, a large machine shop, and an assembly department. The design group was in close contact with the production and process planning departments. Castings and forgings required for these SPMs were made by third-party vendors to the company's specifications and rigorous quality standards. The 500 odd staff in this division occupied one building, approximately 60,000 square feet in area.

The company had a principal in Europe and affiliates around the world, making a similar range of products. The company's European principal decided that SPMs made in this plant were of comparable quality to those made in their European factory. They made a policy decision to increase SPM production in this plant with orders from affiliates being executed here, once additional capacity was established.

The SPM manufacturing, assembly, and testing areas had to be increased significantly. Additional machine tools were required along with overhead cranes, packing and dispatch bays, and a small increase in office space for a larger design group.

Demand for the company's main product range of fuel injection pumps and spark plugs was also rising rapidly. As a result, the company's own production process needed additional factory area. The company decided to relocate the SPM division to a new factory to be built on a green-field site. This would release an additional 60,000 square feet in the existing plot to cater to the growth in primary product demand.

They owned a plot adjacent to the existing factory, with a public road dividing them. The plan was to build a new factory building 100,000 square feet in area with its own infrastructure services such as electricity supply, water, and air. The SPM division would thereafter operate as a profit center.

For many years, the company had used a respectable and reliable firm of architects for their civil engineering work. At the time of these events, they were supervising the extension of an existing factory building (described briefly in Chapter 22). While observing this work, I noticed that our own civil engineers and the architects were operating well within their comfort zones.

The architect's designs looked very sound, but it was not clear whether more economic designs were feasible. We could resolve this question by opening the architectural and structural design work to outside bids. After obtaining approval to conduct a conceptual design competition, we invited other qualified architects. The successful submission would meet specified criteria: customer expectations, cost, and adherence to schedule.

4.2 Customer Expectations

In Chapter 3, we discussed the importance of getting the inputs of shop-floor staff when planning for the future. We applied this principle in planning this project. The starting point was 'market research' with our main customers, the workers in the SPM division.

We prepared a questionnaire to evaluate the requirements and expectations of all the people who would be working in the new factory building. The questions tried to identify their preferences with regard to working conditions and services. Specific requirements of specialized groups could be recorded in free text. We selected about 50 machinists and assembly technicians randomly, then interviewed them individually, using the questionnaire as a prompt. We interviewed trade union representatives, designers, and managers as well. The results were compiled and collated so that we had a good idea of the expectations of a cross-section of customers.

As in the earlier exercise described in Chapter 3, we were quite surprised at the number of common factors in their responses. The majority of those interviewed wanted the following:

1. Natural ventilation; in the existing factory, large column-mounted air cir culators were used to cool the work area; they did not want these in the new building.

2. Natural lighting.

3. Large spans between columns; some people specified 60–70 feet as the desired span in both directions.

4. Overhead gantry cranes to cover the entire assembly and dispatch bays.

5. At least 20 cubic feet of storage space per machine-tool, for tools, jigs, and fixtures.

6. Dry air supply for machines; in the existing factory buildings, condensed water in the air pipelines had been a major problem for some years.

7. Assured supply of power and water.

There were a few other requirements, but these were relatively minor and could be carried out at low cost during the detailed design.

4.3 Technical Criteria

In the earlier designs of the factory buildings, north light trusses were used. At the time this project was being planned, structural and reinforcement steels were very expensive and in short supply in the country. As a result, the roof structure costs were over 30% of the total whereas the foundation costs were relatively low, because the site was on a solid granite formation. In the most recent design, the weight of the roof structure was about 6.8 lbs/square foot of roof area; in earlier designs, it was nearly 7.5 lbs/sq ft. We decided to inform the participants in the competition that we would expect to see a significant improvement in the structural design over the current performance.

We told them of our desire for large spans, cranes, and other items highlighted in the survey results.

4.4 Commercial Terms

We paid a nominal fee to the competitors to cover part of the costs of preparing their proposals. Under the terms of the competition, they had to assign the ownership of their designs to the company. The company could ask the winning competitor to incorporate features from other designs if that was considered useful.

The competitors were to advise us of their fee structure, which we would incorporate into the final contract to the winning competitor. We included a preliminary project schedule in the invitation to compete, which we asked them to accept on a best-effort basis.

4.5 Selection Criteria

We informed the competitors in advance of the criteria which would be used in making the final selection. They had to meet our technical criteria or, if not, demonstrate why their design was superior technically and commercially. Their design had to be aesthetically pleasing; this of course was a subjective issue. Their fee structure should be comparable to those prevailing in the market, but this was negotiable, if other conditions were met. They had to demonstrate that our project schedule would be met.

4.6 Competition Outcome

All eight short-listed firms submitted their proposals. We opened these in the presence of the two executive directors. A three-person evaluation team selected the two best proposals, and listed their merits and shortcomings. The evaluation team presented their results and recommendations to the directors, who made the final selection.

The selected firm of architects had offered some innovative design ideas in their conceptual design. The roof structure design was even better than we expected, weighing about 6 lbs/sq ft. This would lower total costs by nearly 4%.

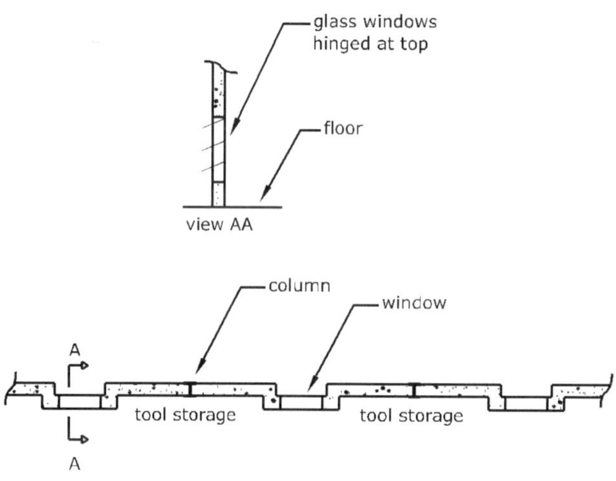

Figure 4.1 Folded Plate Design of Outer Walls

The outer walls were designed as a folded plate (see Figure 4.1). The folded plate design strengthened the relatively slim stone wall (about 18″ thick) considerably. It was strong enough to withstand the bending and buckling stresses caused by the wind and roof loads. Folded plate designs are normally used for concrete roofs, but using them for the walls was an interesting concept.

The wall section on the inner part of the folded plate had large outward facing top-hinged window panels. The section of the wall forming the outer part of the folded plate walls had a large recess, which could be used for storage of tools, jigs, and fixtures (see Figure 4.1). A thinner outer wall section meant that fewer materials were required, so the walls would be cheaper.

The main columns were spaced at 70 feet x 70 feet. With this spacing, we anticipated some problems with rainwater disposal because the down-take pipes could at best be spaced every 70 feet. This problem had to be solved during the detailed design, and was a situation about which we were aware. The provision for natural lighting was excellent, with large window areas in the outer walls and the north light roof structure design. Aesthetically, their design was pleasing.

4.7 Results

The factory building construction progressed quite well, in spite of many problems, some outside the company's control. Within a year of ground breaking, most of the building work was completed. For various personal reasons, I decided to take up a new assignment with another company, so I did not see the last stages of this project through to completion.

I visited the factory five or six years later, and was quite impressed with the design. Several machinists and technicians recognized me as I walked through the factory. They offered greetings and expressed their satisfaction with the building. They were proud in the knowledge that their ideas and contributions had helped make their work environment pleasant. The best part was that the overall cost was much lower than if we had persisted in 'doing business as usual.'

4.8 Lessons

1. Incorporating customer expectations in the design specifications helps optimize plant design. Objectives can be clearly set out at inception.

2. Specifying success criteria at the outset removes subjectivity in decision-making.

3. Paying competing architectural firms a small fee can help get better designs by releasing their creative juices. It also enables company 'ownership' of all the designs.

4. Better design features do not necessarily cost more. This example illustrates how they could save money.

4.9 Principles

1. People like to operate within their comfort zones. It is the leader's job to recognize the symptoms and shake them out of this situation.

2. Consultants (including architects) must pull their weight—they can add value and help make large savings. To do so, they must be given

freedom to exercise their creativity.

3. In order to establish trust in any partnership, we need clarity from the outset. If the objectives are clearly stated up front, people usually rise to the challenge.

Chapter **5**

Changing Paradigms

...with leadership and expertise

It is paramount for leaders to align the organization so that all are working together to achieve the same objectives.

Peter Wickens, Author.

Author: Mahen Das

Location: 2.4.1 Medium-Sized Semi-Complex Petroleum Refinery

5.1 Background

The refinery received its compensation on a cost-plus basis, i.e., it received all its operating cost, plus a fixed percentage of that cost. Each partner paid a sum in proportion to the amount of crude processed for it. In such an arrangement, there was no motivation for the refinery to function cost effectively. In fact, the higher the cost, the higher was the fixed percentage of compensation, i.e., the profit.

Over the preceding decade, the refinery had undergone a major expansion. As a result, it had focused on new construction and start-up activities, not maintenance of assets as a business process. Although the people were generally well educated and competent, expertise and leadership for maintenance of assets had been lacking.

5.2 Prevailing Culture

The working culture was quite similar to that in most other places at that time. Departments were securely compartmentalized. The Operations department called most of the shots. The Maintenance department was at their beck and call. Process technologists and advisory engineers had little to do with the overall efficiency of operations. The Materials department was within the Finance function and, with the mental make-up of typical bean counters, had little appreciation of consequential loss due to low quality of maintenance materials or their delayed delivery.

The Inspection section (within the engineering function) was very conservative, basing inspection intervals on a fixed-time schedule. Although the regulatory authority allowed considerable flexibility, they preferred to play safe. As a result, all process plants were subjected to annual inspection shutdowns during which almost all equipment was opened for inspection. At the time of these events, risk-based techniques such as RCM and RBI had not been introduced in this refinery, as in the process industry in most of the world. Conservative inspection and maintenance engineers only had past practice for guidance (see also Chapter 10 for some more detailed insights into a similar situation).

5.3 Infrastructure

Computerization of maintenance, inspection, and materials business processes was in its infancy. Computers were used largely as work list repositories. Work planning was fairly advanced. For major projects and plant shutdowns, Critical Path Planning with resource leveling was carried out using commercially available software, CASP®™. However, once the project execution began, there was little or no progress toward monitoring and updating the plan. The critical path charts remained as decorations on the wall.

5.4 Shutdown Work

Preparation for a shutdown mainly meant pulling out last year's work list, adding the current wishes of the operating and inspection departments, and having it estimated and converted to a critical path plan with CASP®™. The operators added tasks such as shutting down and gas-freeing at the front end, and starting up the plants at the back end separately to this plan. Technologists gave their requirements to the operators for adding to the plan. The project engineers made their own separate mini-plans and appended them parallel to the main plan. There was little coordination of the preparation activities between these departments. In the absence of a milestone chart, these preparations were never completed in time for proper award of work contracts—and contract work was required. This meant that there was never enough time for proper competitive bidding, so prices were higher than necessary. Local contractors maintained a skeleton work force of skilled craftsmen. During big projects, such as a shutdown, they hired temporary workers. Often, they hired whoever was willing to work, without regard to skills or experience. Contractors and their personnel were viewed with suspicion by the refinery and always kept at arm's length.

During execution of shutdowns, the maintenance engineer was supposed to be the coordinator. Other participating departments did not recognize his role because top management never announced it formally. As a result, the execution was as if there were many separate football games instead of one well-orchestrated team.

In the past, management did not formally spell out the objectives of the shutdown. The duration was fixed by the refinery schedulers on the basis of past history. Authorization of overtime work, extra work, additional contract work, etc., was done in an ad hoc fashion, without an overall guiding principle or premise. In short, there were no clear answers to the following questions, before and during the course of the shutdown:

- Why are we carrying out this shutdown?
- On what basis is maintenance and inspection work selected?
- Is the shutdown to be realized in the shortest possible time?
- Is the shutdown to be realized at the lowest possible cost?
- What are the economic consequences to the refinery and its partners if the shutdown is realized a day earlier or a day later than planned?
- Who is the person overall in charge of the shutdown activities?
- Is the work going as per plan?
- Are the costs on track?

Top management rarely, if ever, visited the shutdown site. Cost and time over-runs were accepted as inevitable.

The management team, including the Maintenance and Engineering Manager, considered maintenance as a necessary evil. In their perception, maintenance was primarily the act of fixing things when they broke down. As long as there was a credible story to explain to the outside world why a plant did not deliver its planned production, they were happy. There was no impact on the compensation to the refinery!

5.5 Economic Imperatives

The high cost however, was reflected in the selling price of the products of the refinery. By the time I arrived on the scene, the operating cost had reached such a level that the products coming from this refinery were barely competitive with imports. The refinery ceased to make economic sense and had thus lost its raison d'être!

Inevitably, the threat of a close-down of the refinery loomed, unless it took steps to become competitive with imports. For the first time in its history, the refinery was under pressure to improve its cost performance.

5.6 New Brooms—To Sweep Clean

A new refinery General Manager (GM) arrived at this time. As maintenance cost (always a major portion of the total operating cost of a refinery) was running at a much higher level than benchmark levels, he sensed the urgent need for expertise and leadership for the Maintenance and Engineering function. The person the new GM selected for this job had to have hands-on experience of several years in process industry at various levels of hierarchy in several locations around the world. I was fortunate to catch his attention, and so got the job.

When I arrived to take over the position of Maintenance and Engineering Manager, a planned major plant shutdown was about six months away. It took me about two weeks to sum up the situation. The next task was to change the prevailing mind-sets and behaviors.

5.7 Improvement Process

I decided to use a top-down and bottom-up approach simultaneously. I would explain a principle to the management team, convince them of its benefit, get their commitment to it, and then do the same in my own line. I went through this process applying five principles. Throughout this campaign, I used my daily walk-about in the process units and workshops talking to the people I met. I conducted several shop-floor meetings to explain the five principles.

Using an open door policy facilitated one-on-one debates on relevant topics with anyone who chose to come. This established credibility with staff and helped build a strong case for change. In a period of ten weeks, I explained these principles and obtained the support of relevant personnel and the management team. Cynics were silenced by peer pressure.

The refinery was now ready to try out the new paradigm on day-to-day maintenance, as well as the approaching shutdown, now about three months away.

5.8 The Five Principles

Principle 1—Define a Maintenance Philosophy

Through this step, we defined the purpose of maintenance and the manner in which it would be carried-out, and stated it as follows:

"The purpose of maintenance is to keep the technical integrity of assets secure, and to ensure that their operational reliability is at all times at the level which our refinery business needed. Maintenance activities should be carried out in a safe and environmentally responsible manner. This should be achieved with the maximum possible efficiency so that the overall cost, i.e., the sum of direct and consequential costs, is minimized."

It took some time and effort to make most people really understand the meaning of this definition; but once that happened, there was visible enthusiasm especially among the plant operators and maintenance workers. For the first time ever, refining economics reached people at the shop-floor level. Indeed these were the people who made things happen! Everyone could clearly see the direction in which the maintenance effort needed to be pushed.

Principle 2—Maintenance is a Business Process, Not Just a Department

This principle attempted to break down departmental barriers and promote team spirit. The general message was that Maintenance, as defined in Principle #1, was a business process which transcended departmental boundaries. Unless all participants in the process—namely, operators, technologists, inspectors, and maintainers—worked as a team, the process would not perform optimally.

Explained in this manner, the principle met little resistance and was readily accepted. The best measure of this was that representatives of all the disciplines started attending morning meetings in the main control room. In these meetings we reviewed the events in the past 24 hours and decided actions needed.

Principle 3—Contractors are Essential Partners in the Enterprise

It was not optimal to carry all the manpower required to do maintenance work using our own payroll. This was because of two main reasons. First, the work load varied a lot. Second, not all skills were required all the time.

Therefore, this refinery, like most of process industry, carried a base-load manpower on its payroll and did peak-shaving using contractors. In our case, the peak manpower, annualized by averaging over the shutdown cycle, was about 30% more than the base manpower. Individual peaks were many times more. Thus, contractors were major participants in the maintenance process; unless they felt part of the team, their performance could not be optimal.

This principle also found ready acceptance except from the Finance function. They needed further convincing that there were sufficient checks in place to prevent malpractice when contractors' personnel worked as a team with refinery personnel.

Principle 4—Define the Day-to-Day Maintenance Process

We defined the day-to-day maintenance process using the diagram shown in Figure 5.1

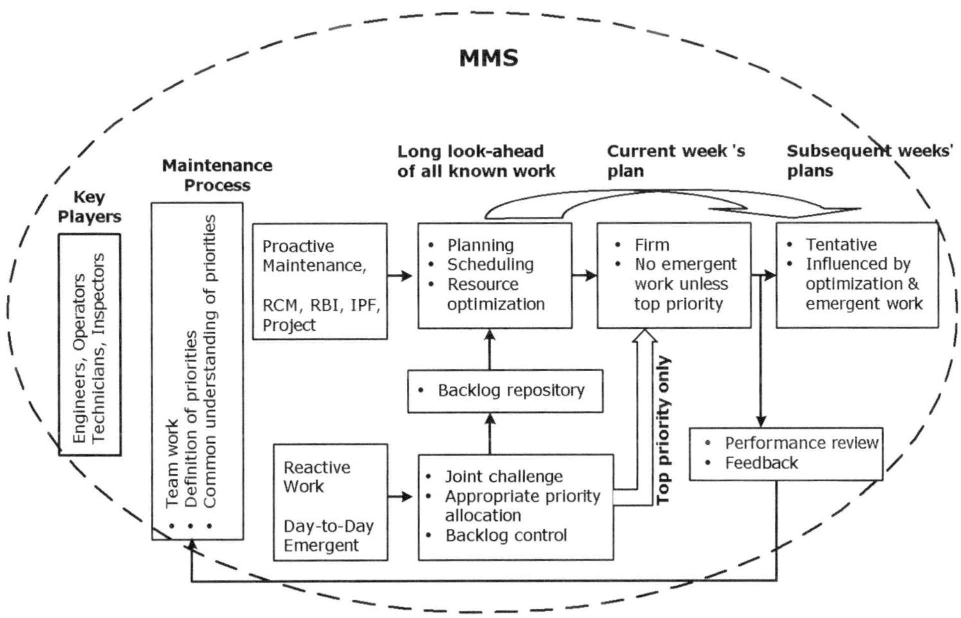

Figure 5.1 Day to Day Maintenance Process

The main features of this process are:

- The key participants in the Maintenance process work as a team; priorities are clearly defined and understood, and all work is screened by this team.
- Proactive work is determined with the help of risk-based methodologies; it is planned and scheduled for a long period. This is the long look-ahead plan of known work.
- Emergent work is subjected to daily scrutiny and appropriately prioritized.
- Backlog is used as a repository of work. It is managed within defined parameters, e.g., ceilings on total volume and residence time for each item.
- The current week's work plan is firm. It consists of proactive work and all high-priority emergent work which was known before issuing the plan the week before.
- Work on this plan will be displaced by new emergent work only if the team decides that it has high enough priority. Otherwise it will be put in the repository.
- The following weeks' work plans consist of proactive work and appropriately scheduled emergent work from the repository.
- After execution, every week's plan is reviewed; learning is extracted and applied in the future.

Principle 5—Define the Shutdown Maintenance Process

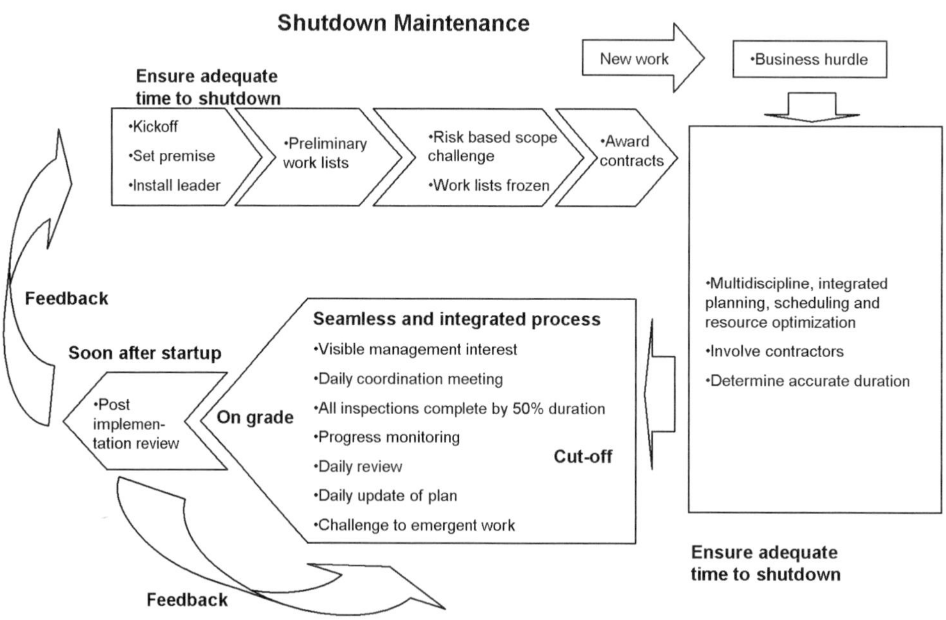

Figure 5.2 Shutdown Maintenance Process

We defined the shutdown maintenance process using the diagram illustrated in Figure 5.2.

The main features of this process are:

- Well ahead of time, management installs a team leader and identifies future team members of all disciplines. Their roles are clearly defined. The premise of the shutdown is clearly established, from which the objectives may be derived.
- Timely compilation of the work list, including a review of process is sues, e.g., catalyst regeneration.
- The team challenges all items in the work list using a risk-based approach. They freeze the revised work list and, thus, the scope.
- Any new work proposed after the freeze has to surmount a tough business hurdle.
- The team identifies contractors at this stage.
- The planner uses a multidiscipline-integrated planning, scheduling, and resource optimization of all work in the scope (people, equipment, cost, etc.). This results in one plan for all disciplines, optimized for all resources. Contractors are fully involved in this work.
- Use of brain-storming exercises to identify alternative solutions for expensive items of work at this stage, e.g., scaffolding rationalization.
- Actual shutdown execution is a seamless and integrated process from the time the feed is cut off until the time finished products start to flow to storage. During this entire period, the team leader is solely in charge and manages daily coordination meetings, daily safety supervisors meeting, completion of inspection before the halfway point, and daily update of plan. The team leader applies a tough business challenge for emergent work.
- Top management including the GM frequently visit the site, show visible support, and get a first-hand "feel".
- The team leader carries out a post-implementation review and feed back (improvement cycle), soon after completion of shutdown.

I published the new maintenance philosophy document, Principle 1 described above, within three weeks of arrival. The remaining four principles, which were based on this principle, followed in the next few weeks.

5.9 Results

Within a year, the new ways of working were firmly in place. The mind-set and expectations of staff were radically different from those seen just a year earlier. After three years, using an international four-quartile benchmarking scale, the maintenance performance of the refinery moved up two quartiles, and thus became a leader.

5.10 Learning Points

1. Importance and Power of Benchmarking

With the help of benchmarking, the new GM quickly came to realize that one area with poor performance was maintenance. The benchmarks he used were developed by an international petroleum company, also one of the partners in this refinery (see Chapter 8 for additional details). These were used to compare the performance of its numerous child companies all over the world. That maintenance performance needed improvement will be clear from the following popular benchmarking factors:

- ***Annualized Total Maintenance Cost (TMC) as % of Total Operating Cost, excluding fuel.***

In a petroleum refinery, most cost elements are independent of the activity level, i.e., the throughput. These cost elements add together to account for the fixed cost. Of the elements which are dependent on activity level—accounting for the variable cost—the cost of fuel is the most significant. Others, e.g., process chemicals, have a negligible effect in the context of this benchmark. Therefore, if the cost of fuel is removed from the total, the proportion that various cost groups such as production, maintenance, technology, and administration form of the total is nearly constant from year to year. The proportion of the annualized TMC should be about 30%. The GM noticed that in this refinery it was about 45%.

- ***Annualized TMC as % of Asset Replacement Value.***

As the replacement value of an asset varies with inflation and other market forces, so does the cost of maintaining that asset. This ratio, therefore, is quite a good indicator of maintenance effectiveness. When the new GM arrived, this ratio for the refinery was 2.5% as against 1.4% for an average performer and 0.9% for the best performers.

2. Maintaining Focus

Keeping an enterprise or an initiative in focus is a major factor for its success and good performance.

It has been experienced over and over again that an initiative or enterprise will fail unless it is kept in focus by people responsible for it. This focus is often expressed as 'keeping your eye on the ball.' Focus is a top-down thing. Unless the top management sends clear signals of interest, the organization below will not respond. In this refinery, the glamorous thing was to build new plants and then commission them, thus being in the limelight. The mundane task of maintaining the existing and the newly-acquired assets was out of focus—and rightly so because there was no reward for good performance in maintenance.

The lack of focus on maintenance was not hard to recognize. I asked the refinery economist to relate the direct and indirect effect of maintenance on the refinery bottom line. When I revealed these numbers to the shop floor level, maintenance suddenly acquired a new glamorous high profile. This also led to maintainers talking to operators on equal terms.

3. Providing Leadership and Expertise

It is not enough to have a group of competent people in an enterprise. Their efforts will be wasted unless there is a leader with relevant expertise who can give direction to their individual efforts. High-visibility direct contact with the rank and file, easy accessibility to them, and leading from the front speed up the rate of progress towards the goal. Daily walks through the plant, shop-floor meetings, an open-door policy, and one-on-one debates with the rank and file help re-establish the focus.

5.11 Principles

Leaders need to understand the true state of affairs and, when necessary, have the courage and energy to take corrective actions. Lack of focus is a fairly common problem and sometimes happens over time due to the plethora of emerging ideas, projects, or external pressures.

The first step is to take stock and unambiguously define the purpose of the enterprise and the philosophy guiding its conduct. Good communication will ensure that every one concerned with the enterprise understands the issues.

Chapter 6

Applying Business Best Practices

...focus, alignment, and speedy implementation help to reap benefits

There are costs and risks to a program of action, but they are far less than the long-range risks and costs of comfortable inaction.
John F. Kennedy.

Author: Jim Wardhaugh

Location: 2.2.2 Large Complex Refinery in the Far East

6.1 Background

The organization had a very traditional functional structure. This structure is shown in a simplified way in Figure 6.1. Many of the senior managers were expatriates, but local people were very competent and were rapidly taking up senior positions. The refinery was making lots of money and, at the time, could sell all the products it could make. The focus was very much on throughput.

The company's attitude was certainly not one of complacency, but neither was there a real thrust to be maximizing profitability. The entire operation was waiting for a spur that would goad it into action. Then it came. A review by an American consultancy company, specializing in process plant benchmarking, showed that the refinery was a relatively poor performer in many important areas. In school report terms, it could do a lot better.

6.2 Reaction

The results of the benchmarking exercise were embarrassing. Nowhere was this more so than in the maintenance area, which was depicted as a very overstaffed and high-cost operation with low equipment reliability (although delivering respectable levels of plant availability). The first response to the benchmarking was one of denial. Many of the hard-working occupants of positions

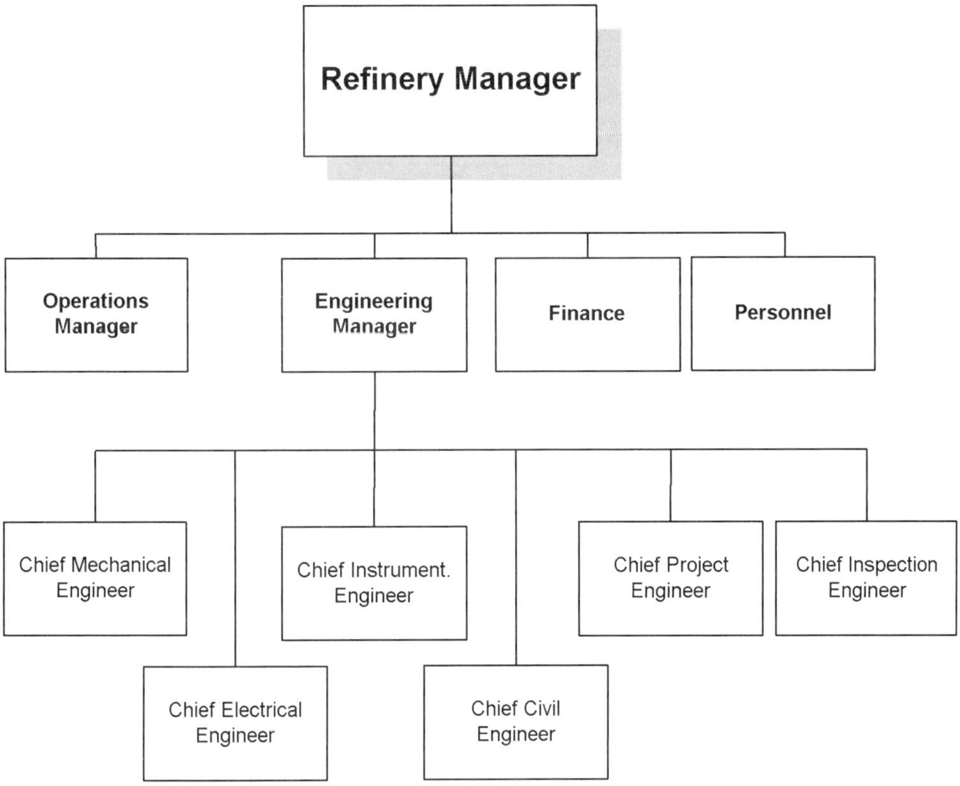

Figure 6.1 Simplified Organization Chart

in the maintenance department saw this as an attack on their personal competence and commitment to the company's performance. The results could not possibly be true. Their second reaction was fury. It was totally absurd that hard-working, committed, and competent people could be shown as poor performers. This just did not make sense. Their third reaction was to seek explanations and excuses. There must be input errors or errors in the analysis and comparison processes.

However, the results could not just be ignored. Interestingly, action was called for by personnel at all levels, from the top to the bottom of the organization. All had different motivations, but none could live with this slur; all demanded action.

What I didn't realize at the time was that the responses demonstrated by the workforce were following the classic Bereavement Curve (See Figure 6.2). This curve originated as a result of research by bereavement counselors and is usually attributed to Elizabeth Kubler-Ross. Change managers soon realized that this curve also fitted the classic reactions to many traumatic events in business; it has been used extensively by consultants to track responses to significant change.

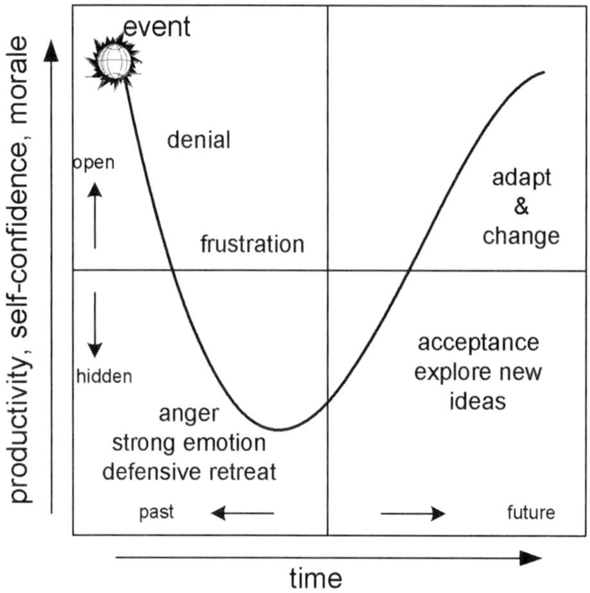

Figure 6.2
Bereavement Curve

6.3 Review Process

I was given the task of unraveling this mystery, and identifying the issues and charting a way forward. Because this was my first brush with benchmarking, my first step was to understand the benchmarking process as a totality. I needed to understand how it worked, and what both the terminology and the definitions meant. The second step was to scrutinize the input process which we had used.

The benchmarking input document took the form of an extremely detailed and structured questionnaire which was sent to a number of refineries in the area. Each completed the questionnaire for their own facility. There was some degree of validation built into these questionnaires, but anyone completing them properly needed a good understanding of the benchmarking firm's terminology and definitions. Teasing out the required information from many sources in a refinery was not easy. It was made more difficult because the in-house information was presented in many different formats and with many inconsistent definitions often slanted to the needs of particular users of this information. For example, we found about four different definitions of overtime, with different variants focusing on hours worked and hours paid and added complexities being introduced if the work was done by shift workers on national holidays.

Handling this data gathering process effectively was not easy; it required a good understanding not only of a number of underlying management and financial measurement concepts, but also the concepts and assumptions behind the benchmarking firm's definitions. In retrospect, and knowing the importance of correct inputs, a person of high competence with a good overview of the business should have been allocated to the job. However, completion of this sort of questionnaire is not the most glamorous of jobs and it certainly wasn't seen as career enhancing. So, unsurprisingly, we found that the task had been given to a fairly inexperienced individual.

Once an awareness of all of the above had been gained (understanding would only come later), the next step was to confirm that the input data was accurate, or at least reasonably so. This is where things started to become even more difficult. Although the location was reasonably sophisticated in the use of computer systems, the data sought did not seem to be retrievable in any sort of straightforward way. There were no consistent definitions between any of the computer systems or the various manual systems used in the location. Indeed, definitions were often totally absent; many individuals had concocted definitions as required in an ad-hoc way.

Certainly it was not easy to get the input information required. The sort of information being sought included details of plant utilization, availability, reliability, reasons for downtime or failure, overtime, and costs. It became clear that inspired (and some not-very-inspired) guesses had been made to feed the questionnaire. Indeed, there had been a large number of errors in answering the questionnaire; the data input contained significant inaccuracies. But it was impossible without a lot more effort to make more than guesses as to whether the inputs painted a black or a charitable picture of our performance.

6.4 Characteristics of Refining Industry Top Performers

Information from our benchmarking company and scrutiny of top performers showed that some of the excuses we were toying with as partial justifications for poor performance were invalid. It became clear that the following aspects of a refinery had little impact on performance:

- Age
- Size
- Geographical location
- Feedstock
- Extent of use of contractors
- Organization (functional or business unit)
- Unionization

There were top performers (and poor performers) of all sorts of shapes, ages, sizes, etc. However, what was apparent was that a move away from the traditional command and control regimes of the past would be beneficial.

The "Characteristics of Refining Industry Top Performers" were identified as:

- Clear organizational goals
- Flat organization with increased span of supervision
- Data-based self-management systems
- Good management systems with small management staff
- Emphasis on improved operational reliability

- Intelligent risk taking
- More collaboration and teamwork
- Emphasis upon value-added aspect of each position or policy

6.5 Results of the Review

We had many manual information systems, a Computerized Maintenance Management System (CMMS), and a lot of other computer systems. But these made up islands of isolated data with incompatible definitions. It was common practice to use different terminology to discuss activities around the refinery. We could not easily access such factual data as who was doing what and why. Thus, we had a high technology refinery run by well-educated and competent staff, but groups in the refinery were each speaking their own language. There were some common business objectives defined by senior management, but by the time they had gone through the translation filters of the disparate groups, they were no longer common.

We did not have the ability to define and measure performance in anything other than the crudest terms. We had bought a CMMS and many other computer systems but we hadn't bought increased visibility. We had almost no idea who was doing what, or why. We didn't know what the end results were. We certainly weren't measuring performance and, the more we looked at things, the more convinced we became that we definitely weren't managing performance. Indeed one manager, when faced with this dilemma, said that we were on autopilot. This seemed doubtful as autopilots do have one target destination. We had many different ones.

We still didn't have concrete answers or a good understanding of the picture that the benchmarking firm had painted of our performance. However, we found that there was enough factual evidence to identify numerous significant problems:

- Although the refinery had both an overall vision and targets, these had little impact on the efforts made by refinery management and staff. Departments in the refinery were optimizing their efforts based on their own aims rather than on the overall refinery business aims.
- Plant availability at 96% or so was reasonably good for the time. But this level came about largely by providing excessive redundancy of equipment.
- Our reliability effort was unfocused and ineffective. For example, Mean Time Between Failures of pumps was about one year rather than the four years attained by respectable performers.
- Overall costs of doing business were high. For instance, maintenance costs as a percentage of the replacement value of process plant and equipment was over 2% rather than the 1% or less of top performers.
- Maintenance and operations were significantly overstaffed, with too many layers of supervision, and most of the hands-on work was done by over-supervised contractors. Indeed, many of our own fit-

ters had stopped carrying tools as if it was beneath them. This inherent overstaffing was exacerbated by high levels of overtime.

- A huge number of contractor employees arrived at the refinery each day. It was unclear who they were working for, what jobs they were going to work on, and how competent they were. What was certain was that they were having too many accidents.
- There was significant over-management at all levels. Unnecessary authorization hurdles were found; these were causing delays in carrying out fairly mundane activities.
- Productivity was poor with a lot of apparently unnecessary work being done and much of the work being done by a few people. Hands-on-tools time was estimated to be about 30% of the possible time,with many delays.
- A low-risk culture permeated the refinery.
- The Inspection department was consciously acting as a police group separate from the refinery. They looked like employees of the regulatory authorities. There was little apparent business benefit coming from them.
- The Safety department had become emasculated, had no authority, could get little done, and was staffed by inexperienced personnel. The accident rate was increasing.

We wanted to publicize the problems in an easy-to-understand form, so we summarized the findings of the benchmark study and the further internal scrutiny, as in Table 6.1 below.

Issue	Performance
Departmental alignment on business	☹
Plant availability	☺
Reliability effort	😐
Cost of doing business	😐
Manning levels	☹
Contractor management	☹
Supervision / delegation	☹
Productivity	😐
Inspection effectiveness	☹
Safety / accidents	😐

Table 6.1 Findings of Benchmarking and Internal Scrutiny

6.6 Strategic Decisions on Effecting Change

Contact with the benchmarking firm had jolted us out of our complacency. What was frightening was how easily we had become outdated in our thinking and management styles. The findings of the benchmarking firm and the results of the in-house scrutiny were put to refinery management together with a set of proposals. This prompted a watershed in our thinking about how we were going to operate in the future. Four key decisions were taken which would change things forever:

1. The modern management styles advocated by the benchmarking firm would become our target organizational style (see Figure 6.3, Organizational Characteristics of Top Performers).

2. We would migrate to these as quickly as possible, so that effective change management would be essential (see Section 6.7 below).

3. A set of new computerized information systems would act as enablers (see Section 6.8).

4. The engineering group would be the engine to drive these changes refinery-wide.

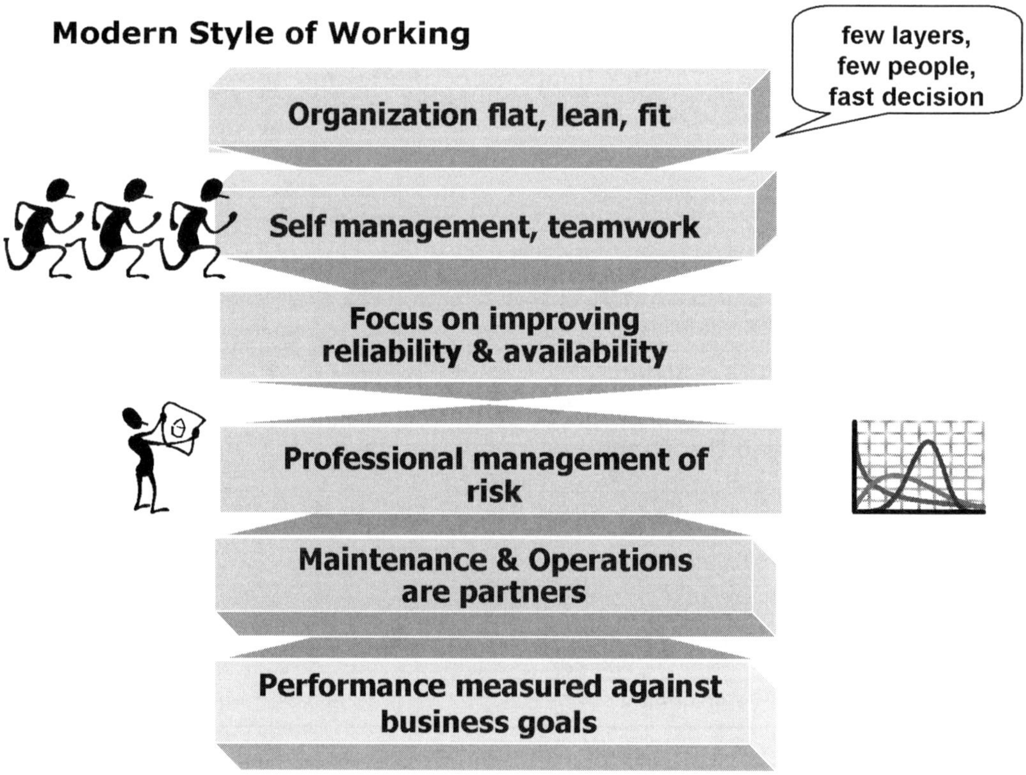

Figure 6.3 Organizational Characteristics of Top Performers

6.7 Managing the Change Process

We knew that we were heading for a big upheaval. We were going to change the way the business was run and this would have a significant impact on people at all levels. So we needed a consistent framework in which to manage the changes. The Kotter[i] approach was chosen.

John Kotter[i] studied over 100 companies going through change processes and identified the most common errors made as:

- Too much complacency
- Failing to get enough allies
- Underestimating the need for a clear vision
- Failing to clearly communicate the vision
- Allowing roadblocks against the vision
- Not planning and focusing on getting short-term wins
- Declaring victory too soon
- Not anchoring changes in the corporate culture
- Too much management and too little leadership

He made a clear distinction between:

- Management, which is a set of processes to keep complex systems running smoothly and
- Leadership, which defines the future, aligns people, and inspires them to pursue the vision

However, these were the recipes for failure. We were interested in success. For this, Kotter presented an eight-point recipe which we adopted:

1. Establishing a sense of urgency
   - Identifying actual and potential major risks or opportunities
2. Creating a guiding coalition
   - Assembling a group prepared to act as a team and with enough power to lead the change
3. Developing a vision and a strategy
   - Creating a vision to help direct the change effort and developing the strategies for achieving that vision
4. Communicating the vision
   - Making effective use of all opportunities to communicate the new vision and strategies, and teaching new behaviors by the example of the guiding coalition
5. Empowering a broad-based action
   - Getting rid of obstacles to change; changing systems that seriously undermine the vision; encouraging risk-taking and non-traditional ideas, activities, and actions

6. Generating short-term gains
   - Planning for visible performance improvements; creating these improvements; recognizing and rewarding employees involved in these improvements
7. Consolidating gains to produce more change
   - Using the credibility achieved from the short-term gains to change systems, structures, and policies that don't fit the vision; continuous re-invigoration of the transformation process.
8. Anchoring new approaches in the culture
   - Getting all parties to recognize the connections between the new behaviors and corporate success; ensuring that the commitment to change was embedded within the leadership succession process

6.8 Effective Computerization

A key part of our vision was either to buy off-the-shelf, or to build and implement quickly in-house, a number of information systems which would act as enablers of new and more effective ways of working. We had already had some success in using computer systems. Although we were fairly low on the learning curve, we were confident that we knew what ingredients were needed in the systems:

- Clear business objectives translated into department and individual performance targets
- Focus on value-adding work
- Best practice business model and workflow processes
- Good organization and execution of NECESSARY activities
- Visible performance measurement to:
  - Show what is important
  - Show where problems are
  - Drive the improvement process

The benefits we were seeking, and were confident of getting, are shown in a simplified way in Figure 6.4.

Earlier we had investigated several computer implementations in a number of refining sites, including our own, in a search for the recipes for success. We found that most systems developed in the traditional way by IT departments had been relatively unsuccessful. Indeed that approach seemed to be a recipe for failure and produced systems which:

- Were large and over-specified
- Formalized traditional work methods
- Became substantially unchangeable when the designers left the site
- Had little positive impact on site culture

More successful projects were run by users, with the assistance of the IT group. They had considerable visible managerial support as well as senior

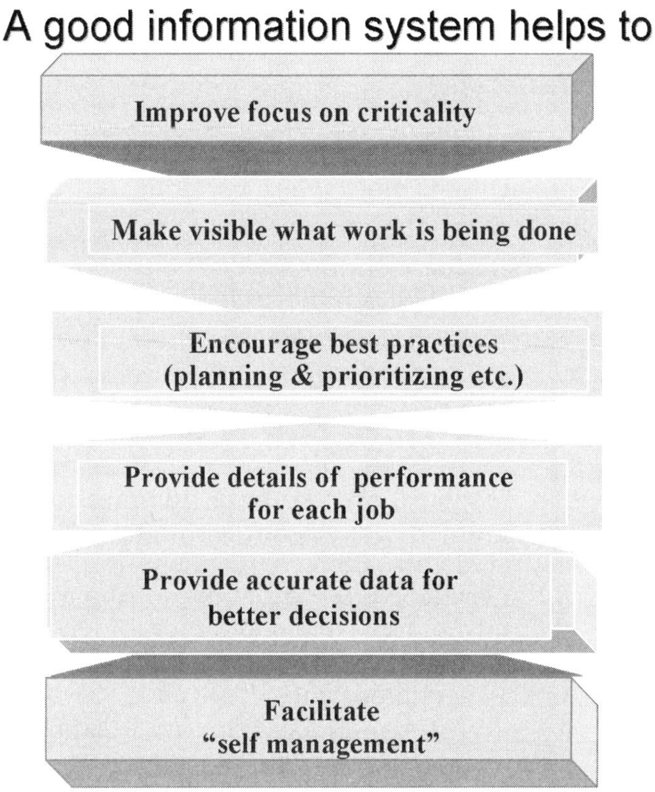

Figure 6.1 Benefits from a Computer System

user involvement and commitment. This approach became our model for all system implementations.

We bought systems off-the-shelf whenever possible. When we had to build our own, we used the unconventional approach of prototyping. A modern programming language enabled a manager and system developer to sit together and quickly make a working system, albeit at the expense of computer efficiency. This "draft system" could be modified quickly, as many times as necessary to produce the required result. The use of prototyping produced effective systems quickly. It also brought some scathing remarks from the IT professionals in other locations. They called our approach "kitchen computing." Further details are given in Appendix 6-A.

Success was brought about by focusing on the following key aspects of the systems:

- Small simple solutions to problems
- Focus on the key players who use the system most
- System consciously designed to effect an agreed transformation (e.g., to make planned work easy and unplanned work difficult)

- Presentation (screens) exert psychological impact on users
- Real time data, where necessary
- Fast implementation

The results of these efforts were:

- Cheap, simple systems which were dynamic, living, relevant.
- All significant actions of site personnel became visible.
- Poor performers in the workforce (whether engineers, supervisors, or fitters) were identified, embarrassed, and isolated.
- Psychological impact on site tradition, culture, attitudes, norms, etc.

6.9 Initiatives to Improve Performance

The four decisions in Section 6.6 acted as a framework for action. Large numbers of issues were identified. Equally large numbers of corrective actions were initiated and integrated to make step changes in performance. There were too many to cover in this book, but a number of the most significant issues and the related initiatives are explained in some detail in the chapters detailed below:

- Chapter 8 Benchmarking
- Chapter 10 Integrating Inspection & Degradation Strategies
- Chapter 15 Managing Surplus Staff
- Chapter 27 Workflow Management
- Chapter 32 Overtime Control
- Chapter 33 Managing Contractors
- Chapter 44 Pump Reliability

6.10 Lessons

1. Making lots of money does not necessarily imply that you are a good performer (but it can hide the truth and dull your desire to improve).

2. Improvements in performance need to be managed by defining vision and strategies.

3. Misalignment, however well intentioned, must not be allowed.

4. A sense of urgency needs to be established; otherwise, nothing happens.

6.11 Principles

Significant events and changes trigger responses, which follow the bereavement curve. This is true for people at all levels.

Facts demonstrate reality and drive alignment in a way that opinions never can.

Reference

Kotter, J. P. 1996, *Leading Change.* Boston: Harvard Business School Press. ISBN-10: 0875847471.

Appendix 6-A

Rapid Creation and Use of Simple Cost-Effective Computer Systems.

6-A.1 Business Aims

The aim of each system should be to bring increased business benefits. Any other reason is probably invalid. If we need to improve the profitability of our plants, we need to improve reliability and availability, and to optimize the capital and revenue costs of our operations.

6-A.2 Effectiveness Of Information Systems

We studied a number of information system implementations in various locations in the company. Few opportunities had been grasped to use computer systems as enablers of new ways of doing business. We became convinced that many of these implementations had actually made things worse. Relatively user-friendly paper systems had been replaced by unfriendly computer systems. However, in a few locations, we saw how modern computer systems could be effective vehicles for significant improvements in performance and cost.

Because we had seen so many poor implementations, we sought the recipes for failure and success. These are shown in Sections 6-A.3 and 6-A.4 below.

6-A.3 Features of Poor Implementations

- Implementation run by IT group
- Benefits intangible
- Replicates old paper systems and past work procedures
- System is avoided by workers and largely ignored by management
- Almost unchangeable

6-A.4 Features of Most Successful Systems

- Project run by users with assistance of IT group
- Considerable visible managerial and senior user involvement and commitment
- Shared ownership between Operations and Engineering, etc.; the systems are seen as a site wide repository of data and a facility which can be used by all
- Clear focus on the benefits expected, how these will be achieved, and by whom
- Systems designed for easy use
- They give benefits larger than the input effort to all levels of user and have support of all levels in the organization
- They are a good cultural and organizational fit
- One-stop shop for all required data
- Critical to day-to-day activities, to ensure use; this prevents them being bypassed and valuable data and history lost
- The systems contain "used" indicators of performance
- All significant activities, events, and performance are made visible

6-A.5 Creation and Implementation of the Computer System

- Project management: A key factor in bringing success was that users ran the project in partnership with the IT group. This notion, which these days is called "client-led," is very different from "client-centered" where users (the clients) are consulted rather than direct ing and managing. The modern term "client-led" is chosen to em phasize that clients are in control of the total process. System analysts and other specialists provide the clients with methodologies, tools, and techniques necessary to manage and control the process.
- We felt instinctively that this was the right approach; modern system development is aligning on this style. Today's arguments for this approach include:
- An organization's information system needs are difficult to define (especially by an outsider)
- IT analysts tend to drive for technological solutions
- Modern information systems must take account of the intertwined mix of hard needs, soft issues, and individual agendas
- Introducing new methods brings feelings of insecurity that need to be managed effectively
- Data and System Development Roadmap: It had been found that data was in a whole series of unconnected data islands with a vari-

ety of data definitions, preventing effective transfer and correlation of data. A project to rationalize data definitions and produce a consistent refinery data model was set up to run in parallel with the creation of small business systems. Prime focus was put on data that would be used in performance indicators to drive business improvements. A simplified overview is shown as Figure 6-A.1

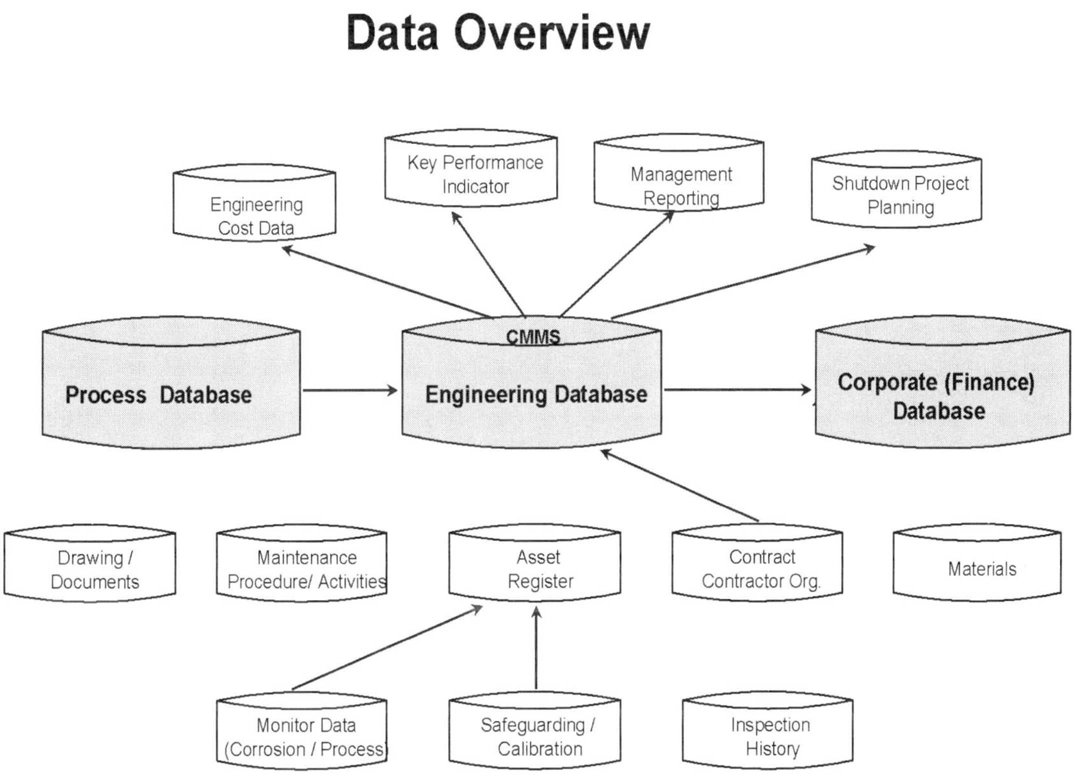

Figure 6-A.1 Data Overview

It was necessary to have an overview of the refinery needs. This overview is shown in Figure 6-A.2

We prepared a road-map of the systems to give a visual picture of the end results (see Figure 6-A.3).

6-A.5.1 Use of Prototyping Techniques

There are some myths, which, if not recognized, lead inevitably to problems in developing systems:

- Users know exactly what they want.
- All users have identical needs.

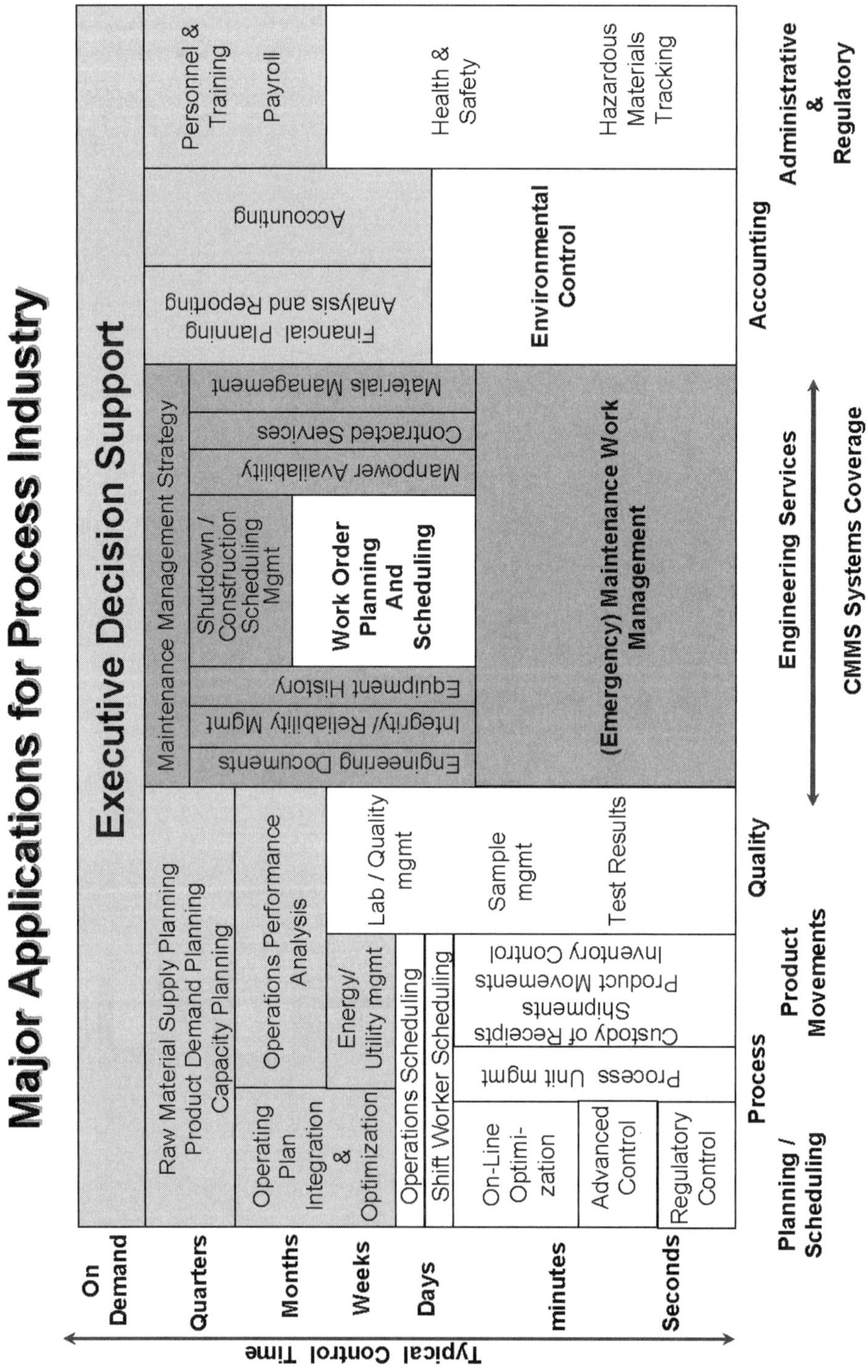

Figure 6-A.2 Major Applications for Process Industry

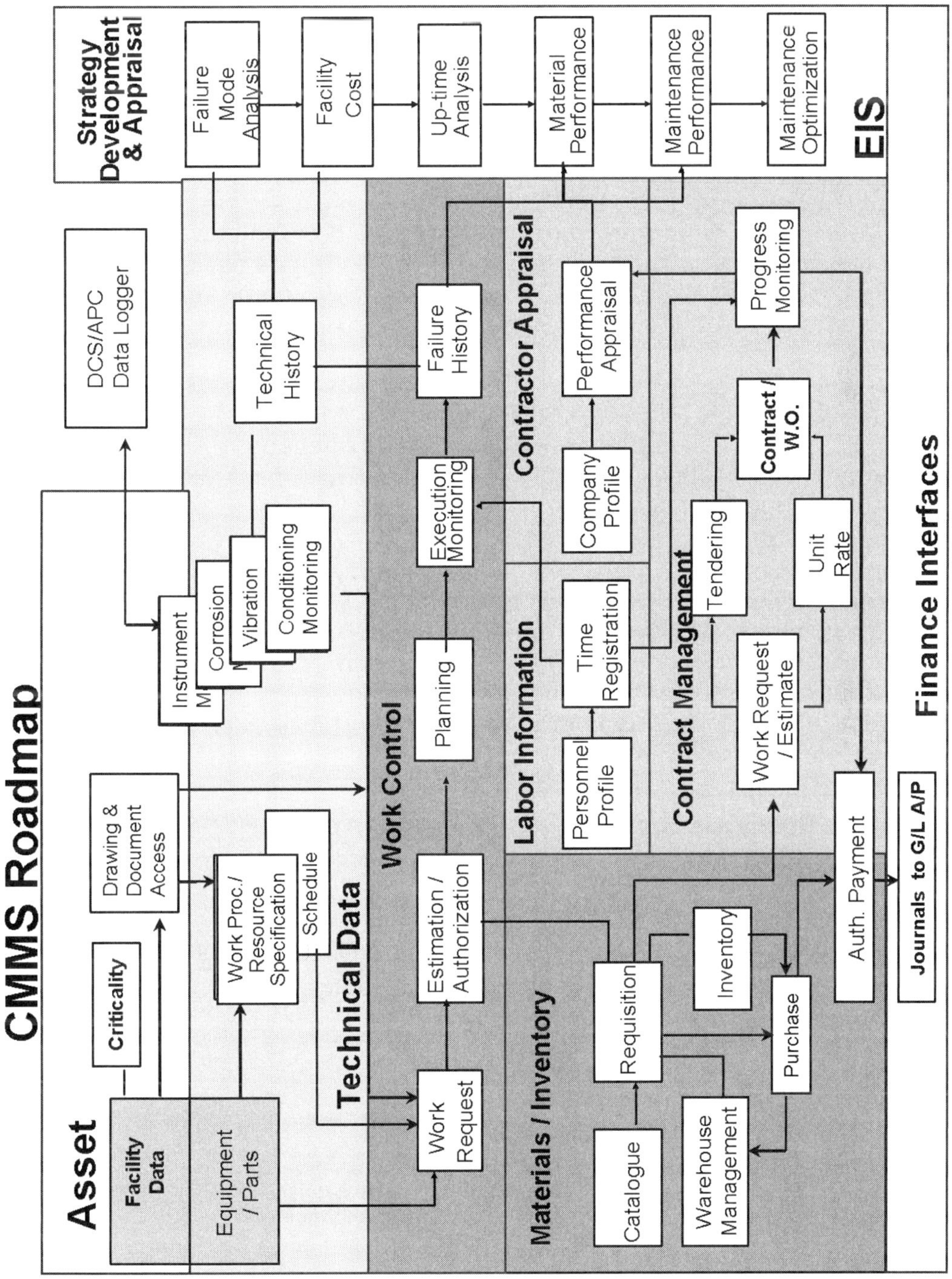

Figure 6-A.3 System Roadmap

- Users can effectively communicate their needs to computer people.
- Users needs never change.

Few users can design a new information system in an abstract atmosphere. But this is what the standard "efficient" way of developing an information system asks for. It asks you to identify and freeze requirements. If you can't visualize it but are forced to guess anyway, it is not surprising that end results are unsatisfactory.

A tangible demonstration to the users of what the system will do, and how, is essential to build confidence. Also essential is the ability to quickly change things to achieve a better optimization, either because the world changes or because your idea of what you want the system to do, and how, changes.

If you are buying an off-the-shelf system, things can be easier; we bought these where possible. We seemed to be ahead of the game in a number of cases so we had to create a number of our own systems, in areas such as

- Overtime.
- Scaffolding
- Permits
- Maintenance workflow and productivity
- Contractor management
- Equipment data
- Risk-based inspection

The prototyping approach to system development can provide a flexible way of creating systems. It assumes that change is inevitable and uses software which produces working systems quickly, but with an inefficient use of computer resources. It does this by its approach (see Figure 6-A.4) and through the use of a 4th or 5th generation language.

The steps in the approach and time scale for developing a typical small system are shown in Figure 6-A.5

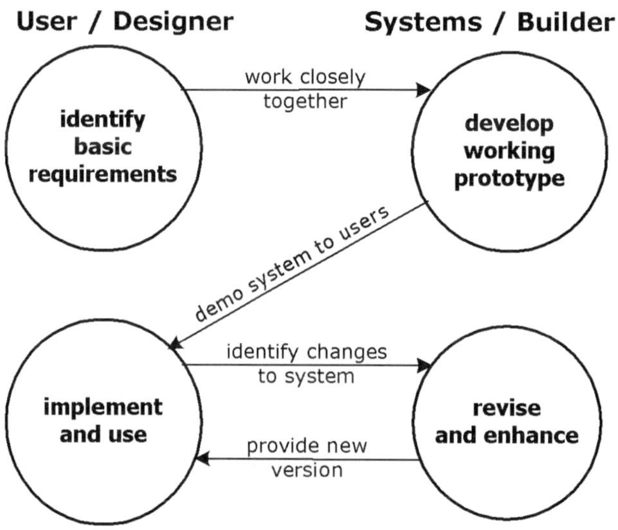

Figure 6-A.4 Prototyping Approach to Application Development

	Weeks
Senior manager(s) focus on problem *Discuss, decide what is core problem *Conceptualize solution / change required	2
Discuss requirements with systems development Produce draft screens and specs	2
Defend proposals against doubters Firm up screens, specs and justify systems	3
Write programs in high level language	6
Implement pilot de-bug and modify Spread to all disciplines, areas, zones	3
Total	16

Fine tune, enhance, add sophistication, professional gloss

*Consultation with selected knowledgeable leaders down the line

Figure 6-A.5 Prototyping Development Time Scale

6-A.6 Training

People need familiarity and confidence in the work flow, new business processes, procedures, and the new computer systems which are supporting them. We found that it is not effective to leave all this to the vendor.

A coordinated approach was found to be a winner, where the vendor acted as the technical expert training IT people; and site personnel were taught by the site's focal points.

Notes:

1. Site focal points were chosen because of commitment to the cause and interest in success. They were usually people of stature and informal leaders in their groups. There would be a focal point in each geographical and discipline area. We did not choose those who were computer geeks.
2. IT personnel received extensive training on hardware, software, back ups, software, and language from the vendor. This could take a month depending on prior knowledge.

3. The system administrator had a week of intensive training at the vendor's office on configuration and optimization.

4. User focal points had a week of on-site training by the vendor.

5. Managers and supervision had a day learning how to use the system and a day on how to extract benefits.

6. Technicians, operators. etc., were trained by focal points. The amount of training for an individual varied with the number of functions she or he used, and could vary from two hours to a week. We found we needed to allow one hour of training for each piece of functionality used and then allow for another hour of practice.

It is important to focus training on the specific needs of the group. It is not cost effective to try to train everyone to do everything. If the function is not practiced within a few weeks (or, in all likelihood, days!), the learning is lost. Several training packages, each focused on particular user types, should be made up from combinations of basic modules. For example, for a CMMS, you might select

- Work request creation, scheduling, executing
- Updating history
- Equipment register
- Getting material
- Getting permits
- Queries and reports

A mix of classroom training (maximum 6 participants) plus guided self-learning

- Training (rather than practice) should not be programmed for more than three hours a day
- Put a training system (simulator) filled with relevant data on site for users to practice on
- Training sessions to be "just in time" and no more than three weeks before hands-on opportunity
- Concentrate on core users at first
- Don't just explain how to use the computer system. Explain in simple language the cultural and work practice changes to be expected.

6-A.7 Lessons

1. The traditional approach to systems development seems to be a recipe for failure and produces systems bringing little business benefit.

2. The unconventional use of prototyping produced effective systems quickly.

3. The result was cheap, simple, quickly implemented systems which were dynamic, living, and relevant.

Chapter 7

Evaluate Contractors' Unit Rates

A cynic is a man who knows the price of everything and the value of nothing.

Oscar Wilde

Author: Mahen Das

Location:2.2.1 Liquified Natural Gas Plant

7.1 Background

In my capacity as an internal maintenance and reliability consultant, I visited this LNG plant to review their performance. Contracting efficiency and value for money obtained was one of the items reviewed. The company was a fairly mature operation and had a number of contract companies for maintenance work. These contractors had been established during the construction of the facility and had grown with it.

The company had set up norms for the effort required to carry out various types of maintenance work. These included man-hours required for or cost of:

- Manual excavation of 1 m^3 of earth
- Thermal insulation of 1 m of 4″ pipe at ground level
- Building tubular scaffolding from ground level, per m^3
- Inserting a 4″ 150# spade
- Grit blasting per m^2 of steel surface at ground level
- Painting per m^2 of steel surface at ground level

There was a tiered quantity-discount scheme in place for all types of work. They also had agreed rates per man-hour for different trades, including

- Pipe-fitters
- Welders
- Scaffolders
- Grit-blasters/Painters/Insulators

The unit-work rates had been established some years earlier. These had never been reviewed. The man-hour rates had also been established some years ago and regularly increased, based on inflation. For the past two years, however, the contractors had voluntarily foregone inflation correction, claiming that inflation would be neutralized by improved productivity of their workers. The management was pleased with this position.

7.2 Evaluating the System

Using call-off contracts, supervisors could easily farm out most of the day-to-day maintenance work with selected contractors. On completion, they could measure the executed work in the specified units. The contractor would submit an invoice based on the approved rates.

Together with an engineer from the company, I followed a maintenance job from initiation to completion. The job was to pull a spade from a 4" 150# line containing product after it had been prepared and made safe for maintenance work. The job was executed by a contractor. The work permit was obtained, the necessary precautions were taken, and the job was completed efficiently by the two contractor's fitters assigned, without any incident or hold-up.

From the moment the contractor's fitters were involved up to the time they went away, it took a little less than one hour. At this time, I was not familiar with the agreed rates, but on the basis of my observation, I expected that the contractor would invoice the company for 2 man-hours of work. When the actual invoice arrived, prepared strictly in accordance with the agreed norms, it was for 8 man-hours. The schedule of rates indeed specified an effort of 8 man-hours for removing a 4" 150# spade from a line at ground level, and for remaking the joint. The company's engineer who accompanied me was more embarrassed than shocked. His embarrassment was caused by the fact that such gross discrepancies had not been discovered earlier. They had simply been accepting the norms which had been agreed between them and the contractors.

7.3 Reviewing the Existing Norms

After this observation, the maintenance and engineering manager of the company agreed to carry out a review of the existing norms immediately. He then realized that there was no one in his organization who was sufficiently confident to make time estimates of maintenance activities. This explains why no one had thought of reviewing the norms until now. I suggested a two-man team be formed to work under my guidance. One would be an experienced supervisor and the other the engineer who accompanied me earlier. They soon realized how simple estimating was if one used real-life experience and common sense. I guided them for the first few items, after which the two of them carried on, on their own.

The review revealed that all items of work were grossly over-estimated; some, such as the de-spading work we had observed, were over by a factor of 4! No wonder that the contractors had "voluntarily" given up the inflation correction for the past two years.

7.4 Corrective Actions

When the contractors were confronted with this, it was not difficult to get them to accept that the existing norms were indeed grossly over-estimated and should be reviewed. They agreed to reduce the existing norms by 25% across the board with immediate effect, while the review got under way.

A joint company/contractor team was set up to formally review and agree revised norms on an urgent basis.

7.5 Benchmarking and Results

On return to my base, I initiated an intra-group benchmarking exercise. The purpose was to compare the norms for unit maintenance activities which were agreed between other associate companies and their contractors. A number of companies welcomed this and agreed to participate. Data gathering and processing took some time and effort. Once accomplished, however, this proved to be very useful. The product was regularly used during subsequent performance reviews. Many companies realized for the first time how far their norms deviated from that of their peers. Although some deviations could be explained by special local conditions, these benchmarks provided a basis for constructive discussion between contractor and company.

Some of the results, together with the question which generated that unit rate, are illustrated in Figures 7.1 to 7.7. Locations are marked AAA, BBB, etc., to protect their identity.

Excavation

Carry out excavation activities to expose an underground pipeline—to in-

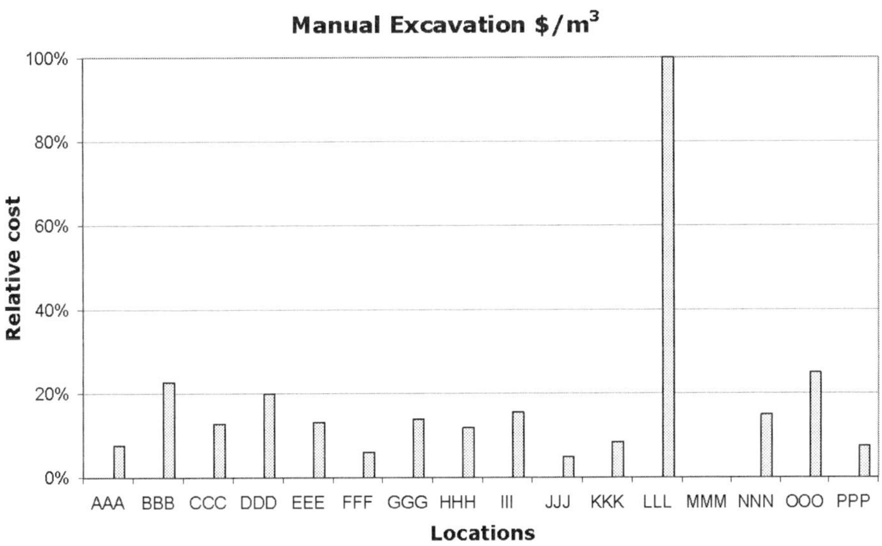

Figure 7.1 Excavation & Refilling—Relative Costs

spect the protective coating system, check for corrosion, and take wall thickness measurements. The soil surface is not covered by pavement or any other cover; the excavated soil can be put along the trench (the soil is not contaminated so the excavated soil in total can be put back). The total amount of soil to be excavated and backfilled is approximately 30 cubic meters. See results in Figure 7.1.

Insulation

Removal of cladding (galvanized iron or aluminum sheeting) and rock wool insulation over a length of 30 meters of a 6″ and a 12″ pipe, lying next to each other, in a pipe bridge of approximately 6 meter height. The lagging and rock wool insulation are in good condition and can be put back after inspection of the pipe. Scaffolding and grit blasting or power brushing are excluded from the contract. See results in Figure 7.2.

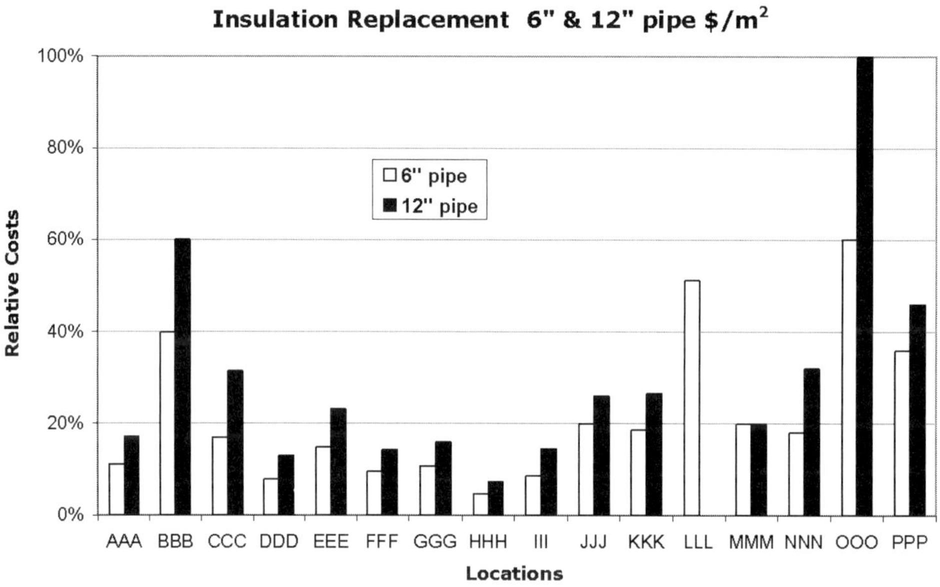

Figure 7.2 Insulation Removal and Replacement—Relative Costs

Scaffolding

Erect and, after use, remove tubular scaffolding for the above-mentioned example (insulation work on a pipe bridge) to the local safety requirements. See results in Figure 7.3.

Grit Blasting

The 6-inch pipe mentioned in the example for insulation needs to be grit blasted to SA 2.5. Estimate man hours. See the results in Figure 7.4.

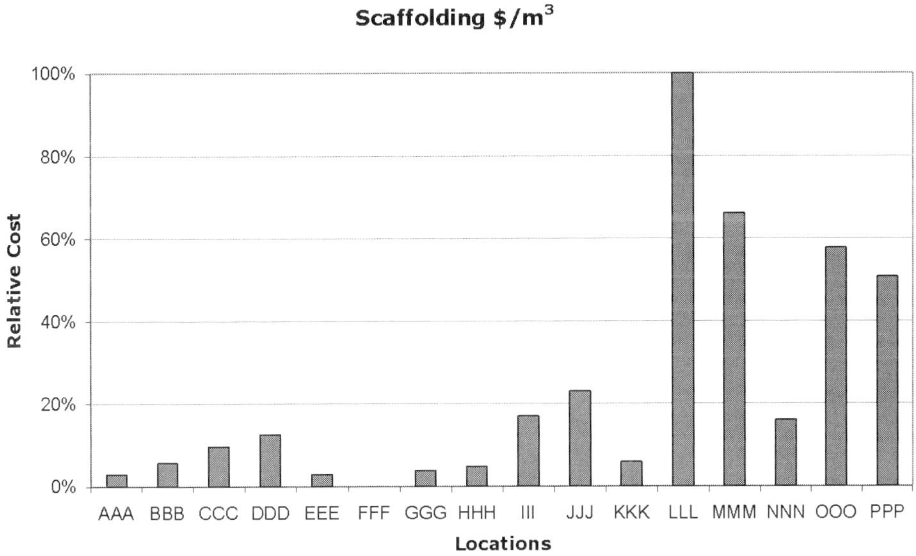

Figure 7.3 Scaffolding—Relative Costs

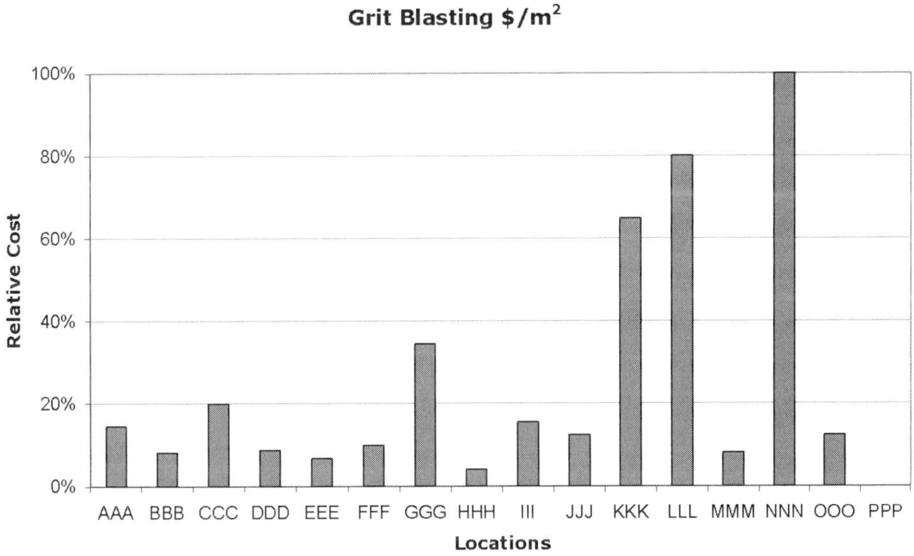

Figure 7.4 Grit Blasting—Relative Costs

Spading/Despading

As part of a job, spades have to be placed to isolate a vessel. For this purpose, 4 nos. 8" 300#, 4 nos. 6" 300#, and 6 nos. 2" 150# spades have to be installed in the existing line work. Estimate man hours required per spade of each size, including cleaning the flange faces, placing new gaskets, placing new stud bolts, and de-spading after the job is completed. See results in Figure 7.5.

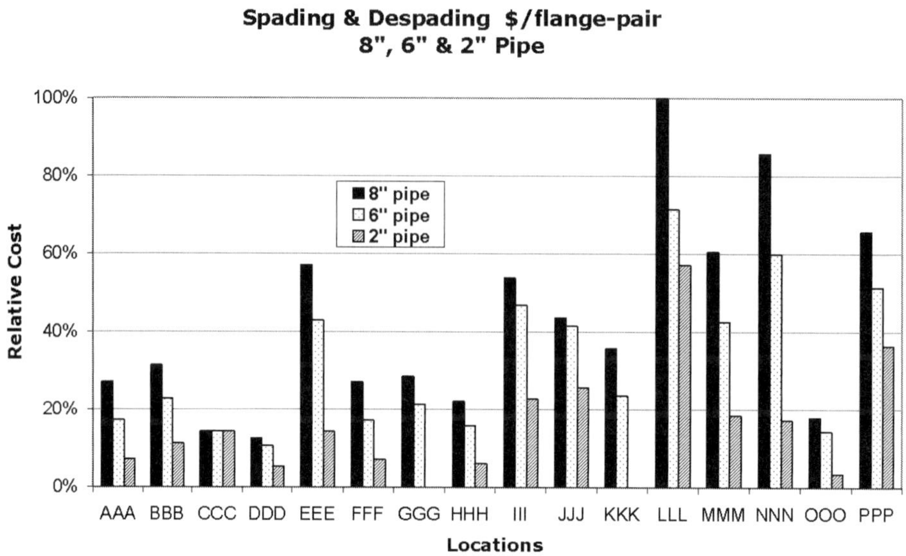

Figure 7.5 Spading/Despading Pipes—Relative Costs

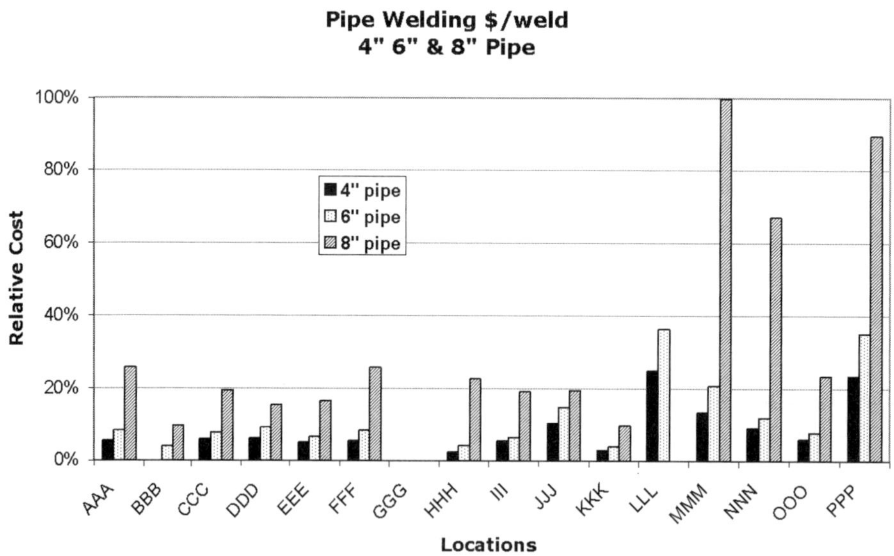

Figure 7.6 Welding 4", 6," and 10" Pipes—Relative Costs

Welding

A few lines in the pipe bridge mentioned in the examples above need to be renewed; each has a length of approximately 30 meters. These pipes are 4",

6", and 10" in size; all are schedule 80 carbon steel. Safe-to-work preparations, scaffolding, and insulation work are done by others. Please estimate man-hours per completed weld of each size, including joint preparation, grinding, alignment, and welding. See results in Figure 7.6.

Valve Gland Packing Renewal

During a shutdown, various types of gate valves need to be repacked (all old packing rings to be removed from the stuffing box and renewed). The total number of valves to be repacked is approximately 40 pcs of sizes 4, 6, and 8 inch. Estimate the man-hours required per piece of each size. See results in Figure 7.7.

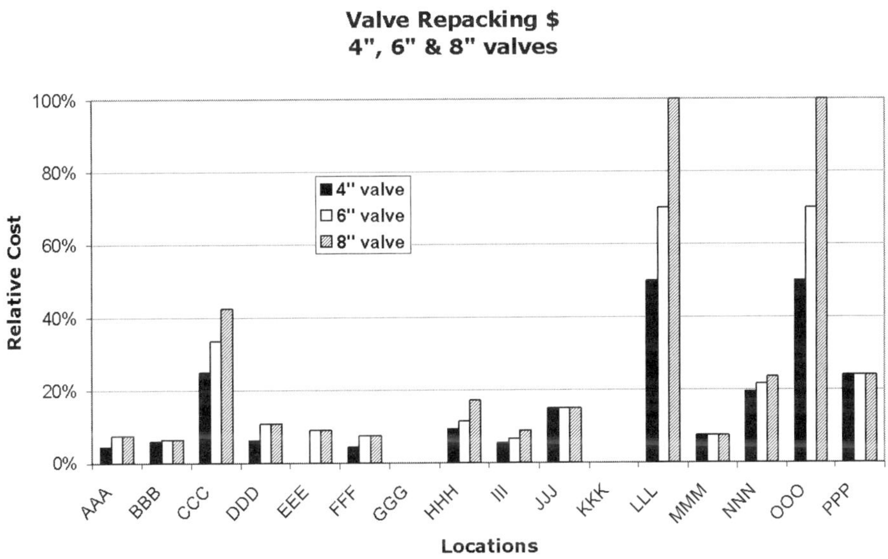

Figure 7.7 Valve Gland Packing—Relative Costs

This kind of benchmarking proved quite simple to carry out. It proved useful in checking contract prices and in preparing estimates prior to inviting competitive bids.

7.6 Lessons Learned

1. Although competitive bidding is a safeguard against overpricing, it fails when contractors form alliances.
2. All norms should be reviewed regularly and updated if necessary.
3. An outside pair of eyes can reveal weaknesses in your systems, which you yourself are too close to observe.
4. Benchmarking is a powerful tool for assessing comparative performance.

7.7 Principles

Without in-house capability for making realistic estimates, there is no way of knowing whether you get value for money from your contractors.

Externally-enforced maintenance cost reductions can hurt the long-term viability of the company, cutting away some flesh and bone along with the fat. Internal audits of current practices can help identify out-dated procedures that add costs without adding value. Some of these practices may have started off as well-intentioned streamlining exercises, to improve efficiency of repetitive work. Periodic audits will demonstrate that controls are constantly reviewed, and thus minimize external pressures.

Chapter 8

Benchmarking

Benchmarking is about being humble enough to admit that someone else is better at something than you; and wise enough to try to learn how to match and even surpass them at it.

American Productivity and Quality Center.

Author: Jim Wardhaugh

Location: 2.3.3 Corporate Technical Headquarters

8.1 Background

Our little group was providing a benchmarking and consultancy service to our own facilities and to a few others with whom we had technical support agreements. These sites were scattered around the world. They operated in different geographical areas, under different government regulatory regimes. They were of different ages and sizes; they used different feed-stocks to make different portfolios of products. Our task was to scrutinize data from these locations, identify those whose performance could be improved, and arrange to help those who needed it.

8.2 Company Performance Analysis Methodology

We had a systematic methodology for capturing performance data from the sites. There were structured questionnaires asking for relevant data. These were backed up by copious notes explaining in detail the methodology, terminology, and definitions. Some returns were required every quarter while the rest were required annually. Each client location would then send the requested data, which was checked rigorously for any apparent errors. The data was used by a number of different groups in the head office, each looking at different aspects of performance. Our group looked at aspects of maintenance performance.

We did not want to ask a site for data that it was already sending to the head office in any report. So we took great pains to extract data from a variety of sources. In this way, the input effort by the sites was minimized and little additional information was needed from them.

When satisfied that all the data looked sensible we massaged the data to identify the performance of each site (or a facility on that site) in a number of ways. The main performance features published for each site were:

For each of the major plants on site [e.g., Crude Distillation Unit (CDU), Catalytic Cracker (CCU), Hydro-cracker (HCU), Reformer (PFU), Thermal Cracker (TCU/VBU)]:

- Downtime averaged over the turnaround cycle (whether 3, 4, or 5 years). This smoothed out the effect of major turnarounds (also called shutdowns)

For the whole site:

- Maintenance cost, averaged over the turnaround cycle, as a percentage of replacement value
- Maintenance cost, averaged over the turnaround cycle, in US$/bbl.
- Maintenance man-hours per unit of complexity.

This information was published annually and provided in a number of forms, but the two most common provided comparisons with their peers and were:

- A straight-forward bar chart showing a ranking from best to worst (see an example in Figure 8.1).
- A radar diagram which sites found useful because it could show at a glance a number of aspects (see idealized version in Figure 8.2). Comparisons could then be made against the performance of the best (see Figure 8.3).

On each spoke of the diagram, the length of the spoke represents the actual value for each facility. The shaded polygon shows the data points for the best performers; these are the values of the item in the ranked order, one-third of the way from the best to the worst performer.

Comparisons were made against two yardsticks:

- The average performance of the group of plants or refineries
- The performance of the plant or refinery one-third of the way down the ranking order.

Because the facilities were of different sizes and complexities, we had to normalize the data. We used a number of normalizing factors to achieve this. For example, when measuring maintenance costs, we used factors such as asset replacement value and intake barrels of feedstock as the divisors.

These divisors gave different answers and thus somewhat different rank-

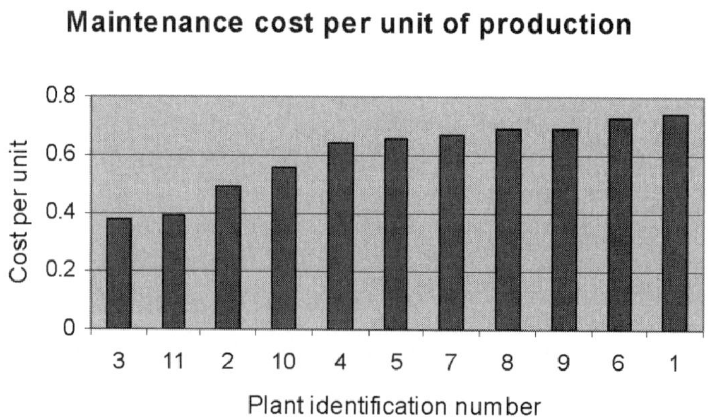

Figure 8.1 Example ranking on a bar chart

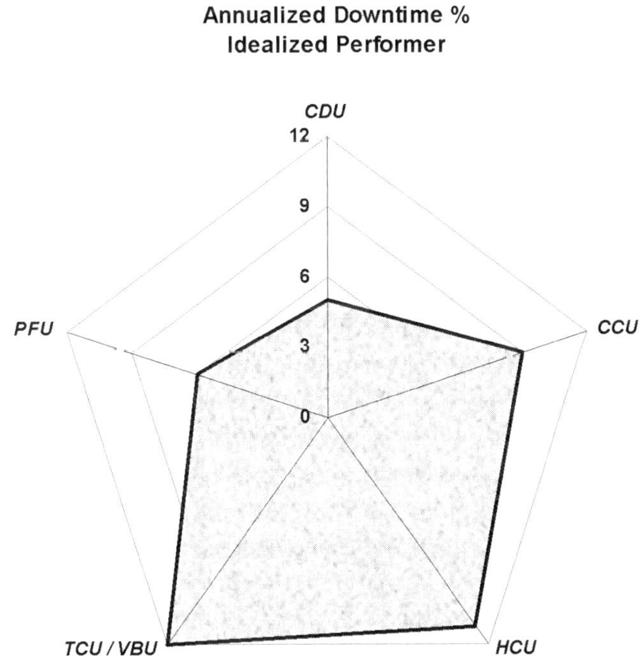

Figure 8.2 Idealized radar chart

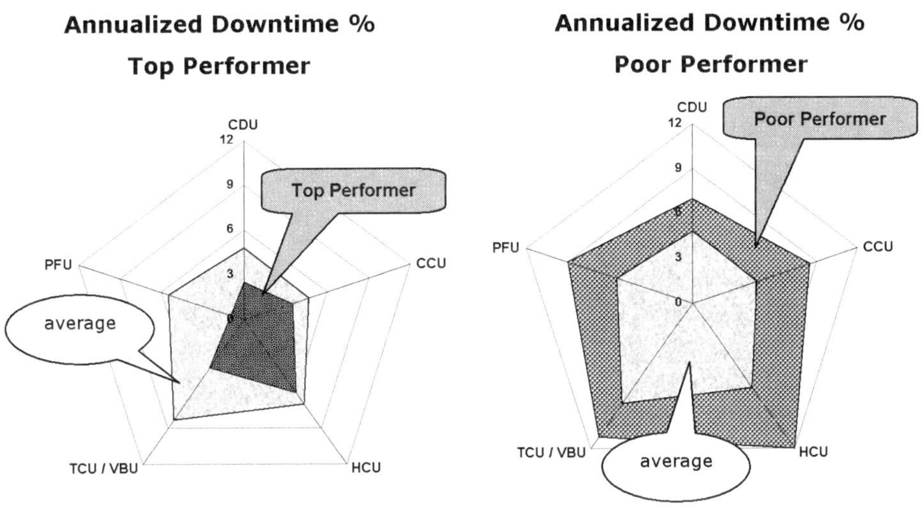

Figure 8.3 Realistic radar chart showing plant performance

ings. Not surprisingly, those deemed to be top performers, liked the divisor we used. Those deemed poor were highly vexed. Although there were exceptions, whatever the divisor used, those in the top-performing bunch stayed at the top, those in the bottom bunch stayed at the bottom. Only minor changes in position or performance were identified. Those in the middle of the performance band could show significant movement, however. Normalizing methods are discussed in Appendix 8-B.

8.3 Benchmarking Consultant's Methodology

As noted above, we had minimized the input effort for the in-house methodology. The benchmarking consultant, however, scrutinized in detail a much wider area of refinery performance. Each site had to make a significant input effort. This effort was made even greater because the consultant used terminologies and definitions that were different from those used in the regular company reports.

This benchmarking exercise was carried out every two years. Although we invited all refineries to participate, not all did. As explained, this was because of the cost and effort involved. However, enough did participate to enable us to rank company performance with those of peer competitors.

8.4 Recipe for Top Performance

By using data available from in-house returns and from benchmarking studies, it is possible to make comparisons/rankings of individual facility performances in a number of specific areas.

However, this number crunching can only take you so far. It does not tell how good performance is achieved. What do the top performers do that makes them different and more successful than their poorly-performing peers? Figure 8.4 shows where top performers differ from poor performers.

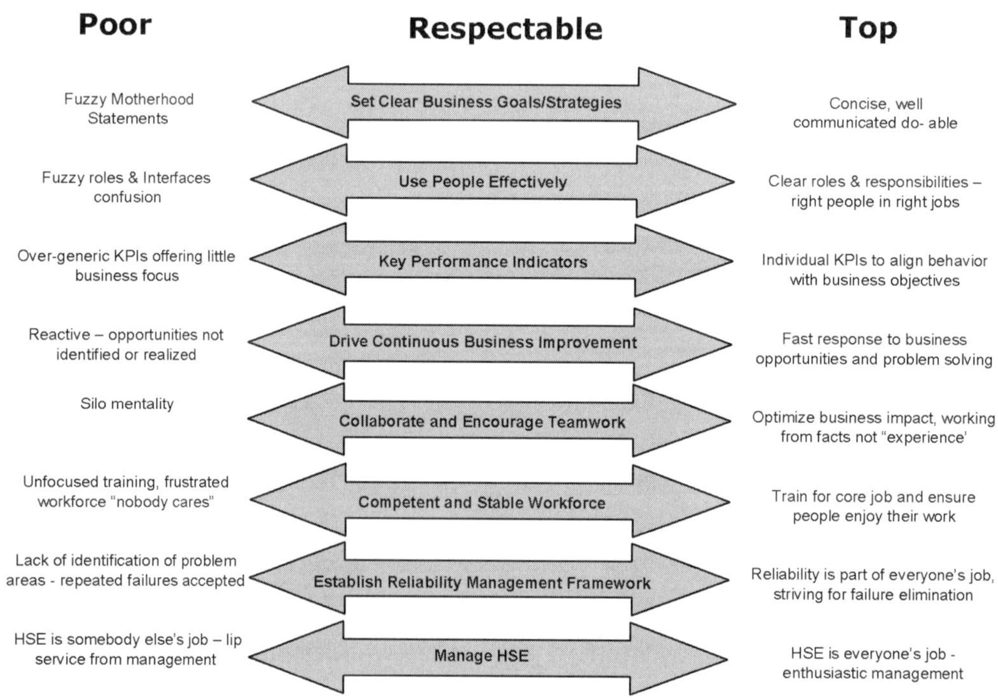

Figure 8.4

8.5 Driving Improvements in Individual Locations

Benchmarking is about improving business performance, so there are more aspects to consider than simply measuring some readily available numbers. Essentially the steps needed are as follows:

- Identify the key business processes that you need to do well to bring success.
- Understand your business processes thoroughly.
- Measure your performance.
- Measure the performance of good-performing peers (making sure terminology and definitions are reasonably consistent).
- Understand the business processes that bring this good performance.
- Consider whether these practices will work in your own company.
- If so, manage a change process to make it happen.

A simplified overview of the benchmarking process is given in Figure 8.5.

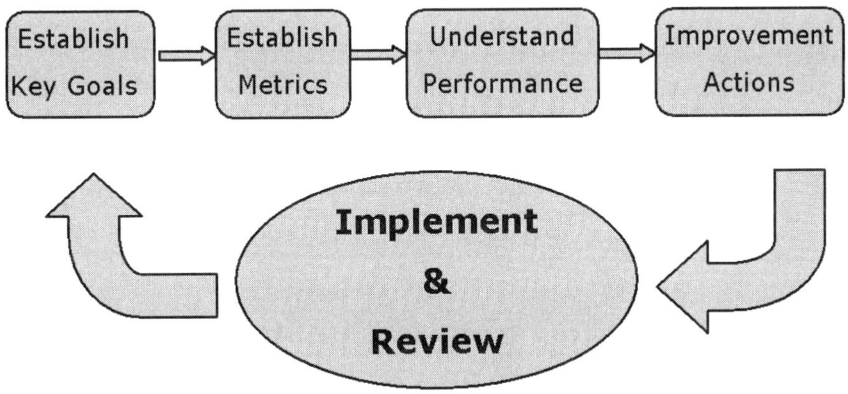

Figure 8.5 Benchmarking Overview

8.6 Partnering Process

Conceptually we thought we knew how to bring top performance to a business. We wanted to start delivering this know-how to the refineries and start them off on an improvement track. What we didn't have was the essential detailed information carried by staff in each location. Obviously walking into a location with a "We know it all" attitude would not work. Some partnering arrangement was vital to complete the picture and provide synergy. Conceptually this is shown in Figure 8.6.

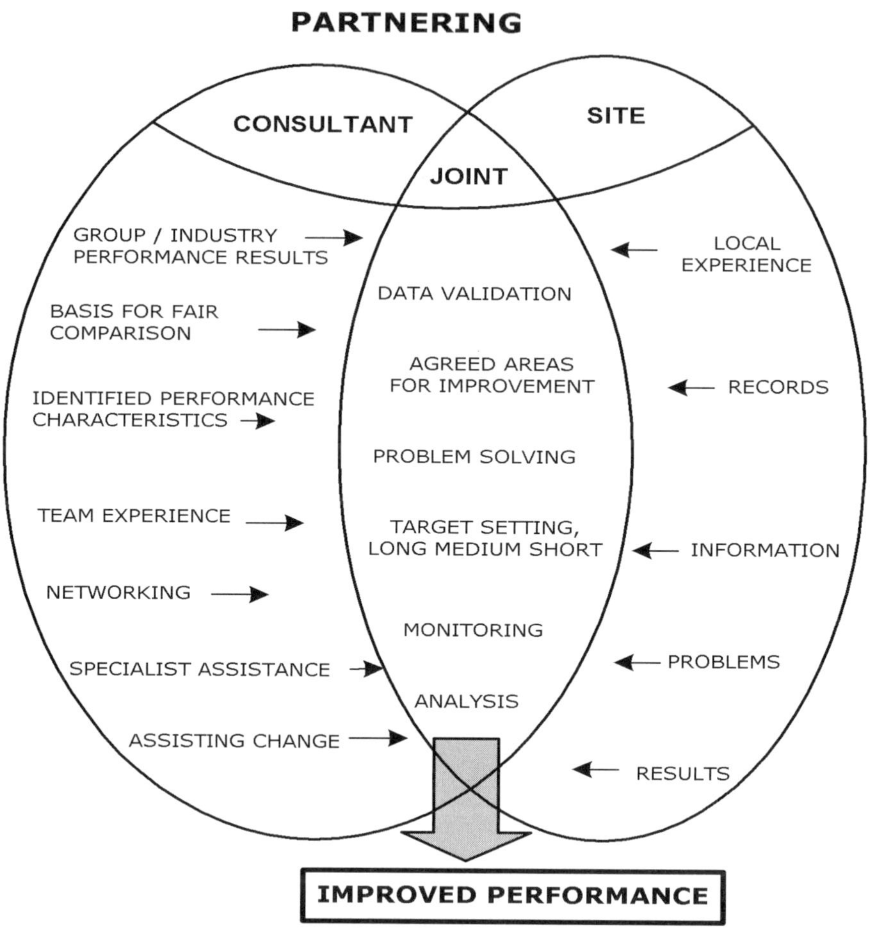

Figure 8.6 Partnering Overview

8.7 Delivering to the Sites

Workshops and Publications

We broadcast the message in workshops, papers, and company publications. People became used to the terminology and the general idea. We offered consultancy visits to assist them in the improvement process. Not surprisingly, most of those who replied were top performers. However, we managed to get a mix of locations so that our visit did not label a location as "failing" in any way.

Preparing for a Site Visit

Arrange a scouting visit to the site to smooth the path for a full visit. This is best done by one person or two as a maximum. They must be prepared for in-depth discussions with the site management. There should be no hidden agendas so it is important for both parties to be open and honest about the aims of the visit.

Each party must table information from all relevant sources to highlight perceived problems. The head office had only a limited amount of information about the site so it was important to get detailed site information to analyze before the formal visit. This may not be immediately available so a standard list of required information is useful.

There may be a complete understanding between the team from the head office and the site management but that doesn't always exist for site supervision and the workforce. Briefings or a simple mail-shot to advise the site what is going on, are essential. Also it is necessary to agree what information can be released to site personnel.

Agree on the team composition. This should reflect the focus of efforts, but areas scrutinized would always include Operations, Maintenance, Inspection, and Instrument/Electrical staff.

Physical facilities. Arrange for a room big enough for the visiting team and possibly a clerk. Additional space needs to be available for discussions.

Get a "gopher." These are people who can go for this or that and do it effectively. They can help you identify and arrange access to sources of information. Effective gophers will tell you also how the organization really works and who are the movers and shakers. They should arrange one-to-one interviews for team members so that they can hit the ground running.

Team Visit

Start with introductions to as many people as possible so your faces become familiar as soon as possible. Then you are into interviews to collect data. Initially go for a neutral data collection in the identified key areas. Use non-threatening but competent questioners. Understand and use site vocabulary and definitions to make the site comfortable, but ensure you can correlate these with your own.

There are five golden rules:

- Always interview on interviewees' home territory.
- Don't make inter-site comparisons while collecting data (otherwise you get into competitive and defensive modes).
- Never ask someone for information that you wouldn't give yourself.
- Cross check all information from a number of sources.
- Show draft conclusions to "partners" in the target location.

It is easy to see what you want to see. Therefore, analyze the data thoroughly; don't jump to conclusions. Make sure you have captured the real issues, the real performance, and how it is achieved. People too often tell you what they'd like to believe themselves.

Roadmap and Follow Ups

Before leaving, identify action items, action parties, timetable, and follow up methods (possibly a role for the gopher). It is good to quantify the benefits, however roughly, as this does add impetus.

8.8 Results

This proved to be a very successful process. In the initial stages, there was some resistance, especially from the poor performers. Within two years, results from those locations who took part in the process demonstrated the value of the process to the rest. We had refined the process itself continuously during this period, so new entrants experienced a mature approach. We had a high demand for this service, with a waiting list of about 18–24 months. The head office expanded these services to third parties on a commercial basis. Eventually the unit became a major global service provider in this area.

8.9 Lessons

1. Busy people at busy facilities do not welcome demands from the head office for information, especially if that information might be used to show them in a bad light. So when collecting data, try as far as possible to use information in existing reports and returns. The fewer times that demands are made on a site, the more likely it is that they will report accurately and on time.
2. Perfection and absolute accuracy are the Holy Grail, but not worth the benefit. The aim should be to produce a good enough result with as little effort as possible.
3. Top performers run their businesses in a similar way, focusing on a number of key aspects, which bring business benefit. These are well known, as are the factors, which don't matter significantly.
4. Poor performers put their efforts into excuses and the wrong things.
5. When giving advice to sites, it is important to be welcomed. You can't force people to take advice.
6. Measurements need a consistent set of definitions measured in a consistent way.

8.10 Principles

1. Knowing where you are is the first step in an improvement effort. Knowing what to do about it is not always obvious. We can seek recipes from top performance—and use them.
2. In-house and inter-site comparisons will take you a long way, but it can become incestuous and self congratulatory. The market is the best leveler; it is necessary to benchmark against others.

Appendix 8-A

Vocabulary and Terminology

8-A.1 In-House Terminology

Plant availability, reliability, and utilization are shown diagrammatically in Figure 8-A.1.

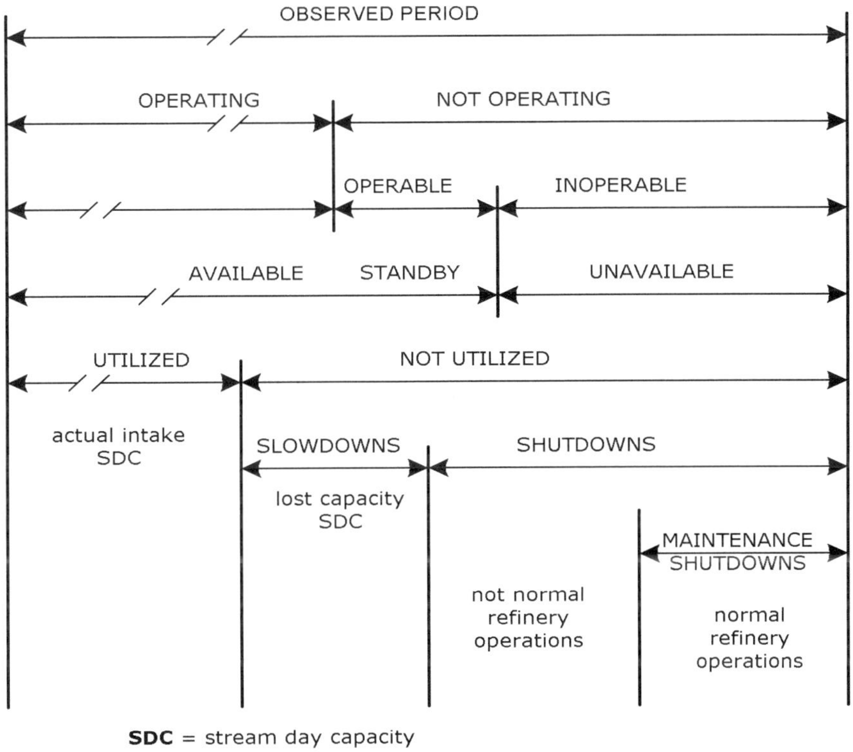

Figure 8-A.1 Availability, Reliability, and Utilization

8-A.2 Benchmarking Consultant Terminology

One leading consultancy dealing with refinery benchmarking uses a term called Equivalent Distillation Capacity (EDC) as a measure of size and complexity. The following definitions are based on their web site, presentations, and publications.

Annualized Turnaround Cost. Total turnaround costs divided by the turnaround cycle in years.

Availability. 100% minus (annualized turnaround downtime plus a two-year average for routine maintenance downtime)

Average Run Length. Mean on-line time of a process unit between stops.

Complexity. Refinery equivalent capacity divided by crude intake capacity.

Equivalent Distillation Capacity. EDC of a unit is capacity multiplied by complexity factor. Total EDC of a refinery is the sum of the EDCs of individual units. This refinery EDC is used as a divisor to normalize aspects such as costs, personnel numbers, etc.

Equivalent Maintenance Personnel. Total number of company's own man-hours + number of contractors man-hours + annualized turnaround man-hours including all overtime divided by 2080 (52 weeks x 40 hours).

Maintenance Costs. Total maintenance costs including capital replacement items, averaged over two years for routine maintenance and over a complete cycle for turnaround maintenance.

Maintenance Index. Maintenance costs divided by EDC.

On Stream Factor. 100% minus percentage of all downtimes (maintenance and others).

Replacement Value. Investment needed to replace refinery in its same location.

Routine Maintenance Index. Routine maintenance costs averaged over two years divided by EDC.

Turnaround Maintenance Index. Annualized turnaround costs divided by EDC.

Utilization%. 100 x total annual intake in bbl divided by (365 x annual design capacity in bbl/day).

Appendix: 8-B

Discussion on Normalizing Different Facilities

8-B.1 Possible Normalizing Factors

The following lists a number of normalizing factors usually used to try to cope with the differences found in facilities:

- Intake barrels or tons
- Replacement value (RV)

- Mechanical complexity of the plant (MC)
- Equivalent Distillation Capacity (EDC)

People often suggest (especially the poorer performers) that as a basis of inter-refinery comparison, such lists can be misleading, unfair, and certainly limited in their use as league tables of performance.

8-B.2 Comparison of Methods

Conventional financial reporting is extremely difficult to use for performance comparisons between refineries for the following reasons:

- **Profitability**. Return on investment or cash flows are not generally reported for individual facilities separate from their marketing and ancillary functions. They are too far removed from the supply sources and the market place to use actual trading values; therefore, transfer values have to be devised. These, however, are often driven more by taxation and where it would pay to take profits, than reality. Also, because of the volatility of market prices, results achieved do not necessarily reflect operational efforts.

- **Asset values.** Such book values are distorted by financial practices as well as by varying capitalization and depreciation methods. Yet they do reflect the age of assets to a degree. Although useful in the business world, asset values are not a particularly useful basis for intersite comparison worldwide.

- **Volumetric divisors.** Divisors such as intake quantities may reflect the size of a facility, but they do not make allowance for downstream costs of operating added-value processes, i.e., complexity/conversion factors. One leading consultant uses an Equivalent Distillation Capacity (EDC) which attempts to relate size and complexity. Many people express reservations about this methodology. It is rather artificial and difficult to sell the EDC concept to middle management and supervision as a motivating tool. People can see barrels, but not EDC. However, it has become widely used and accepted in management circles

- **Complexity indices.** These indices, which use the amount of equipment, are liked by engineers as they can easily visualize these as maintenance workload. By giving suitable weightings, these can be made to correlate with other normalization factors. These tend not to have credibility in fields outside engineering

- **Replacement Value (RV).** This is a popular divisor. Values show wide differences between similar plants located in different regions. UK and Japan show significantly higher replacement values than Continental Europe and Australia.

- In Australia, the RVs are deliberately kept as low as is believable because the local authorities use them to determine local taxes.
- When derived for insurance purposes, there is an equally powerful drive for minimization, as long as it is believable.
- One consultant calculates an estimated replacement value(RV) for each process facility considered in their studies. Although the methodology is proprietary, their RVs are almost directly proportional to their EDC values. We found at the time that they were approximately 60% of our company's quoted RVs. Possibly this is because of their simplification of ignoring facilities such as their own utilities and generation, and assuming optimized size of tank farms, jetties, pipelines, and other peripheral activities. The valuations seem to be based on modern designs and technology; this penalizes the older plants with their piece meal modifications over the years.

8-B.3 Concluding Thoughts

Each divisor has some advantages (easy availability) and disadvantages (varying degrees of inaccuracy).

As a leading benchmarking firm once remarked, "A dog is a dog however you measure it!"

Reality brings us to three divisors:

1. Intake barrels per day as it is immediately available and free; but the answers should be viewed with caution.

2. Equivalent Distillation Capacity (EDC) as it is becoming widely adopted in the refining world.

3. Replacement Values, but see the caveats below:
    - They should not be too influenced by insurance, financial, or rating considerations.
    - The traditional method of escalating original as-built costs, using published international construction indices, should be used with caution as the final values can be unevenly distorted and defeat their purpose.
    - The calculation should take reasonable account of standard and special-process plants, utilities, off-sites including major pipelines and jetties, and offices.

PART 3: PEOPLE

Chapter 9

Staffing Levels

Only those who will risk going too far will ever know how far they can go.

T.S. Eliot, Author.

Author: Jim Wardhaugh

Location: 2.3.3 Corporate Technical Headquarters

9.1 Background

When I first started work, I had no computer and I shared a telephone with six other engineers. I had a real-time information service in the form of an engineering clerk; I also had over a hundred employees with a bevy of foremen. As I progressed through my career, I have acquired a computer and a telephone of my own. I lost my real-time engineering clerk and the number of people reporting to me has shrunk every year. Indeed for most of my career, how many people I could get rid of each year and still get the job done seemed to be the most important factor in setting my salary increase.

What I have learned over the years about staffing levels is that:

- Doing unnecessary work unproductively requires a horde of people, but
- Doing only the necessary work efficiently requires amazingly few people.

In addition, running a lean, mean, empowered type of operation focused on the things that matter brings:

- High morale (people feel like winners)
- Easy communication
- Reduced support and logistics effort
- Enjoyment to the job

It also makes good business sense.

I would always advocate having as few people as you can get away with. But you need to retain core competencies and have enough physical bodies to do the job. It does not matter whether these people are your own staff or contractors, as long as they stay with you. Over the years I've had contractors working for me who have been just as loyal as my own people.

Taking this view, how do you work out how many people you need—whether you are setting up a new operation or trying to slim down an existing one? I suggest that there is no algorithm which will bring exactitude, but I am going to suggest a few approaches which will bring you into the right ballpark. You will have to work from there to find a happy minimum; I suggest that you only know when this point is after you have gone just that little bit too far.

Let me tell you a story to illustrate the point. I worked a number of years ago in a facility which was in a rather remote location. Because of this, a rather conservative view had been taken of the size of fire-fighting infrastructure and force we needed. As a result, we had about eight fire appliances and, as the Irish would say, more firemen than you could shake a stick at. As managers, we knew little about fire fighting, but common sense told us that we had too many people. Over the next few years we gradually reduced the force and put out many fires. Then one night we had a somewhat bigger fire and we couldn't put it out. We now had found our irreducible minimum so had to add in another two firemen.

While in the head office, we were often asked to review staffing levels in existing facilities and proposals for staffing new operations. Almost always, these were too high so we put together a number of ways of looking at this problem in a pragmatic and believable way.

9.2 Organizational Levels

In any modern business, you will find an arrangement of employees as shown in Figure 9.1, which illustrates in a simple way the necessary roles that people play.

When translating this model into a real organization for a facility, the first rule is to have no more than five hierarchical levels. This is shown in a simplified form in Figure 9.2.

Performance is relatively independent of the type of organization. However, top performers do tend to be of the traditional organizational style or at least centrally directed in key business aspects. Because this style is the simplest to understand, I am using it in my examples.

Throughout the world and between poor and top performers, there is a significant variation in maintenance staffing levels. Table 9.1 compares these variations.

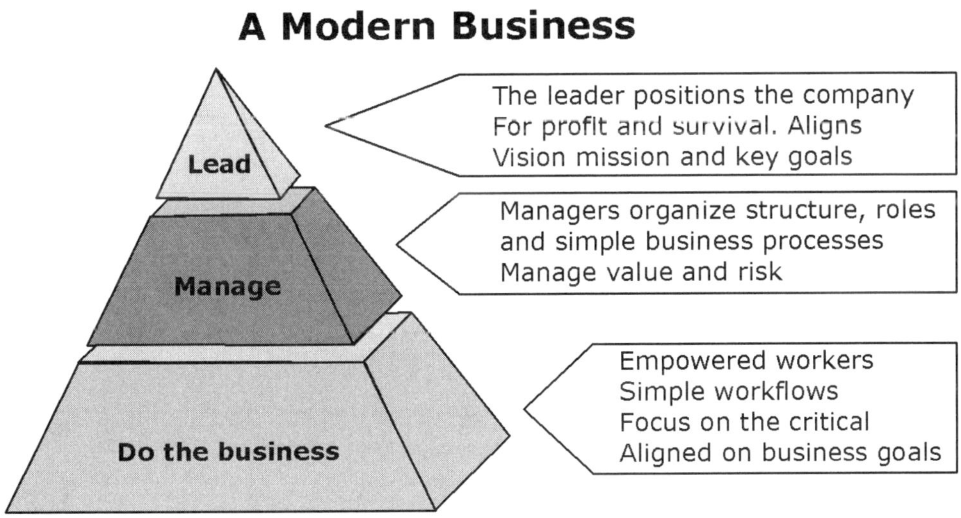

Figure 9.1 Modern Business Model

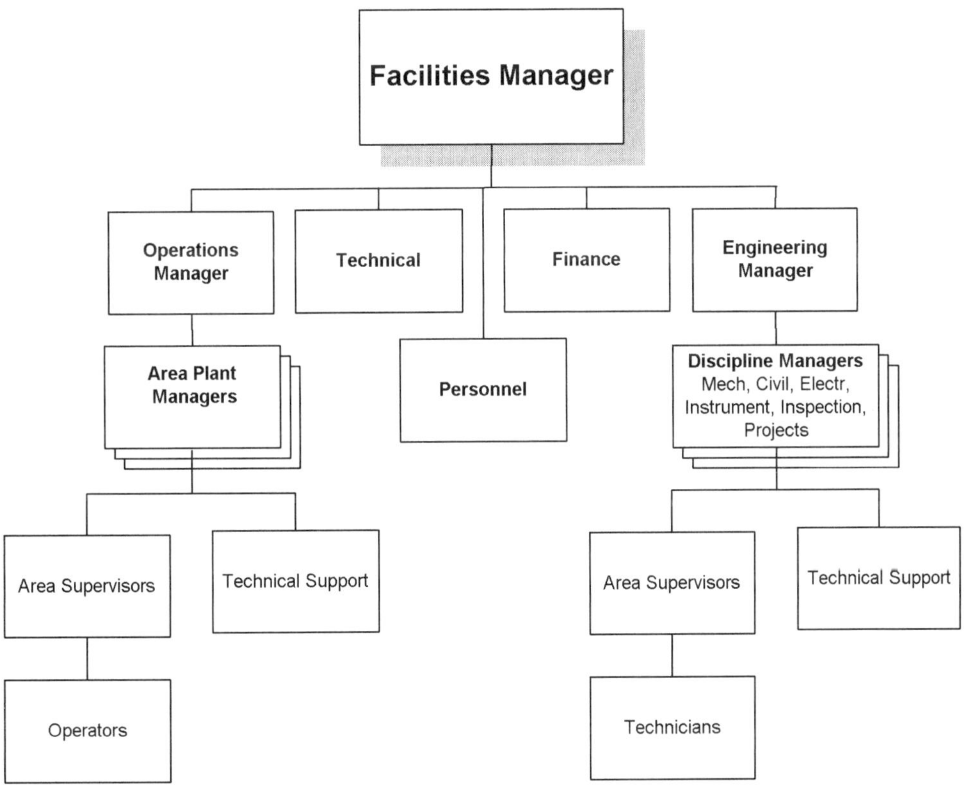

Figure 9.2 Simplified 5-Level Traditional Hierarchy

	Asia	Europe	USA
Top Performers	160	130	100
Average	300	160	150

Table 9.1 Staffing Level Variations—Comparison with US Top Performers

9.3 Numbers of Managers

Almost inevitably there will be proposed (or actual) separate managers for:

- Facility (overall management of the facility)
- Operations (managing the production processes)
- Engineering (managing Maintenance, Projects, Inspection, etc.)
- Technical Support, Personnel, Finance, etc. (separate managers for each of these)

It is possible to have shared roles depending on the skills of the individuals; at smaller sites, this arrangement should be considered seriously. However, the most significant problem areas are usually the supervisory, technical support, and workforce numbers, so let us move on to those.

9.4 Supervision Ratios

A flat, lean, empowered organization is the preferred one. The Department of Trade and Industry in the United Kingdom provides a benchmarking service for British industry. This has a data base of literally thousands of companies, mostly manufacturing of some type. This data base shows that poor performers tend to have a supervisor to worker ratio of 1:4 while the top performers have a ratio of 1:12. Where you decide to pitch your ratio is a matter of judgment, but 1:8 or 1:10 would seem reasonable. All the evidence shows that over-supervision does not improve performance. We have a case study covering some elements of this issue in Chapter 27.

9.5 Approaches to Staffing Numbers

Companies and consultants use a number of methodologies and algorithms to calculate numbers. The main approaches are:

- Activity/work volume based where each element of supposedly necessary work is identified, the time to do it is calculated, and that time is multiplied by the frequency to give an annual man-hour requirement. The method is hugely time consuming. It also tends to inflate work requirements and the man-hours because conservatism adds in lots of low-value activities and because of a tendency to always round up fractions of hours or workers. It does sell well to the workforce though, as they can understand how the answer has been arrived at. This is almost a work-study process.

- At a slightly higher level is a synthesis of the work hours needed to operate and maintain specific building blocks of a plant. This data is industry specific.
- At an even higher level, we can work out an equivalent "plant com plexity" Using a proprietary multiplier, we can derive the staff re sources needed for the whole operation. More complex or bigger plants need more people than smaller, less complex plants. You will see numbers per complexity unit or per unit of production.
- Specific, easily-counted elements of the business are a way to esti mate the resources using a set of standardized ratios. For example, for an Instrument group, we can use the number of control valves; for electricians the number of motors, etc.

The validity of all of these methods is always open to challenge and hangs to a large extent on the quantity and quality of relevant data available. A consultant's detailed knowledge of large groups of similar plants is much more convincing than snapshots of a few different plants across a number of industries. As discussed in Chapter 8, arguments will rage about the divisor used to normalize complexity and size of an operation. Does it really take that much longer to maintain a big pump than a small one?

What we are really doing is working out the resources needed to maintain your plant professionally. We need to count all the human resources used. You are fooling only yourself if you hide work and the effort to do it. It is important to include not only normal hours, but also overtime, and effort expended by contractors either on site or performing activities in their premises. For instance, if you send a pump out to a repair workshop, the hours spent on the repairs should be counted in these calculations.

To get the number of heads, you need now to work out the attendance hours of our workforce. By the time you have knocked off holidays, sickness, training, etc., of the working weeks, you may well be down to something like 1700 hours per year. Note: That does not mean hands-on-tools time (or wrench time as it is sometimes called) which can be as low as 30% of this number.

Production facilities may, or may not, carry out a wide range of secondary activities such as refurbishment, gardening, purchasing, and sales. These are usually excluded from the norms, but treated and calculated separately. A focus is put on the standard types of activity, i.e., production (operations), maintenance, inspection, and technical support.

9.6 Getting our Operator Numbers

In a production type of business we always have operators. So let us start with them. We discount shift patterns, spare operators, and such, and ask how many operators do we need for the minimum staffing? You can do this the hard way by building blocks and equipment counts. Usually the data base needed will be proprietary and you will need to hire a consultant to gain access to it.

You can, however, get a good feel for the operator numbers needed by getting together a few knowledgeable people. Putting some outsiders in the team is always a good idea. It may not provide a very elegant approach but it gives a reasonable answer and, at least, site involvement will bring some buy-in.

Another way to cross check and validate staffing numbers in the production area is from approaches such as the following. In a production process with local control panels, a panel operator would typically control about 50 instrument loops. When digital control systems and central control rooms hit the scene, this leaps to about 200. Now in a state-of-the-art control system, we can get one panel operator looking after 400 control loops. For a middle of the road DCS in a control room, you need one panel operator and one and a bit outside operators for every 200 control valves. The good news is that these valves are very easy to count. They don't hide and they don't breed. This of course only impacts the operators at the heart of the production process.

ACTIVITY	TRADE	% man power
Management & Supervision	Mechanical	4
	Civil	1
	Electrical	1
	Instrumentation	1
Hands-on Tools Labor	Mechanical	40
	Civil	13
	Electrical	6
	Instrumentation	10
Cleaning, waste disposal		10
Inspection (mechanical & civil)		4
Planning & Scheduling		3
Plant Changes		3
Office Engineering & Advisory (drawings, library, rotating, corrosion, DCS specialists)		4

Table 9.2 Typical Staffing Distribution in a Refinery

There are other operators involved in many other activities and we have to make provisions for their numbers.

Once you have your base case minimum number, you add in spares to cope with reality.

9.7 Maintenance Personnel Numbers

Let us assume that you have worked out that you need a minimum staffing of 20 operators per shift. There is a correlation between operator numbers and the size of the maintenance workforce for each specific industry. Hence it is possible to derive maintenance numbers reasonably accurately. As an example, for the petroleum refining industry, 20 operators per shift would correlate to a total maintenance workforce size of 150–200, depending on the efficiency of the operation. Table 9.2 shows how this total might typically be broken down for specific activities in a refinery.

9.8 Staffing of Instrument Group

There is a reasonably good correlation between the number of instrument control loops (or valves) and workload, hence, staffing levels. This can be used as a check on the numbers derived above.

9.9 Staffing of Electrical Group

In a similar way, the electrical workload correlates with the number of electric motors. This ratio again can provide a check.

9.10 Lessons

1. You can always manage with fewer people than you think, but few people will agree with you and they will have many excuses.

2. No plant start-up managers ever get fired for having too many people so they will have too many. Make sure this surplus does not become a problem for the permanent organization.

3. Many consultants produce high-level algorithms with which to derive staffing levels. No one at the work level will believe these so you need to derive and explain your numbers in a more practical way.

4. The resource numbers situation should be kept under continuous review. This should cover not only labor, but also supervisory and managerial grades. It is much easier to make gradual reductions rather than step changes.

5. Adding more incompetent people to cover competence shortages will not make your life easier. Nor will it bring business benefits.

9.11 Principles

1. Have only five layers from facility manager to hands-on tool workforce.

2. Work out what you will need for an efficient operation and set your staffing levels at 90% of this.

3. Keep reducing numbers at all levels until there is clear evidence that you have too few.

4. Make your algorithm saleable to the staff at site.

Chapter **10**

Integrating Inspection and Degradation Strategies

...A launch pad for a bigger reliability initiative

Insanity is doing the same old things in the same old way
and expecting different results.
Rita Mae Brown, Author and Social Activist.

Author: Jim Wardhaugh

Locations: 2.2.2 Large Complex Oil Refinery

10.1 Background

A benchmarking exercise had shown our performance to be inadequate; and an internal review had shown a lot of problems. One of the more significant problems that we found was the compartmentalization of work groups, with each doing their own thing. There was little synergy between departments. In the climate of the time, this would have been called interference in another department's affairs. We identified another problem, namely that the Inspection group was acting in a policeman role rather than contributing to the business. In adopting this role, they had consciously raised barriers with other groups. In this chapter we will look at a move toward a joint three-party approach to the management of degradation of the static mechanical equipment and piping in a plant. This approach developed as a launch pad for a wider reliability initiative.

10.2 Previous Approach to Degradation

Regular inspections have always been a major defense against degradation of plant integrity and potential accidents. The inspection group did regular inspections, handled the evaluation of the condition, and identified potential problems in static equipment like columns, vessels, tanks, and pipe-work. They reported to the Engineering Manager.

They had very little involvement with any of the other groups in the facility, even though there was a huge amount of knowledge and expertise residing in those groups. Even the operators and the process engineering group were involved only in a small way. Decisions were based on the inspection and maintenance findings from shutdowns and other experiences.

This compartmentalized approach was common across the facility, so it was not considered abnormal. The result was a significant non-alignment of responsibility and an inability to influence events. Once these facts were put on the table and the approach challenged, the absurdity of the situation became apparent. This was a large facility with different plants operating in different modes with constantly varying raw material feed-stocks. As the benchmarking firm had highlighted, we needed to chase reliability and availability much harder than we had been doing so far. What we didn't need were frequent failures, unplanned shutdowns, and the resultant downtime. We wanted to minimize degradation and make things predictable.

10.3 A New Way Forward

During the review of the location's problems, we had identified many issues. The corrosion management issue provided an opportunity to make inroads into three of these problem areas.

- Work groups were compartmentalized.
- Inspection group worked as policemen, rather than proactive business associates. We wanted to bring Inspection into the "business" (see Figures 10.1 and 10.2).
- It was difficult getting out a message that synergistic decisions were good for the business.

All parties agreed that our present approach was ineffective and led to an inadequate reliability performance. They also agreed that a much more integrated and proactive approach was needed if we were to manage degradation professionally (see Figure 10.3).

We initiated a pilot scheme in one area focusing on the management of a specific aspect of degradation, i.e., corrosion. If this was successful, we aimed to extend the corrosion management concept across the whole site. Once this

Monitoring compliance and symptoms with no business responsibility leads to many inspections and much downtime

Figure 10.1 Traditional Inspection Role

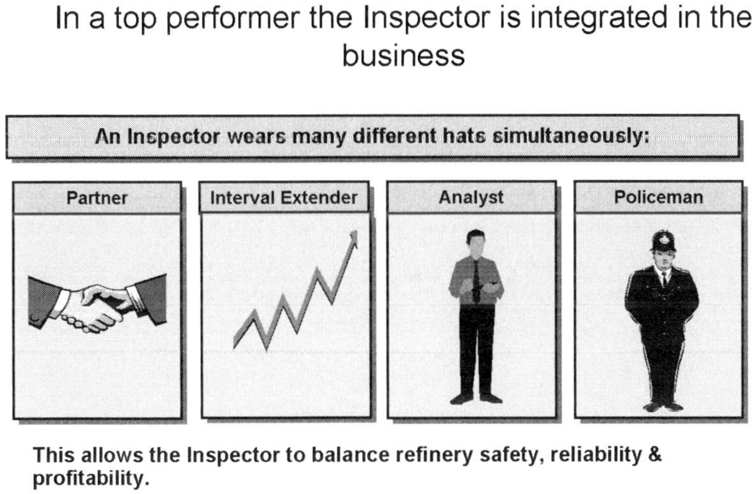

Figure 10.2 Widened Inspection Roles

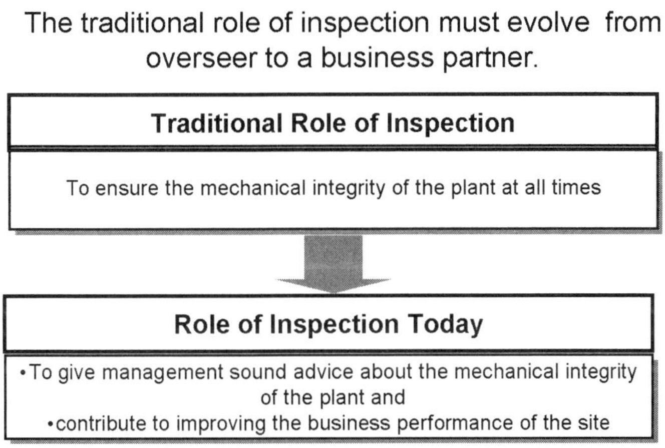

Figure 10.3

was bedded in, we aimed to translate these concepts and experiences from corrosion management into a more generalized management of degradation and reliability. This was our road-map.

10.4 The Pilot Scheme

The key ingredients we looked for in this corrosion management initiative were:

- A move away from a reactive approach to problem finding
- A move toward a preventive approach underpinned by a belief that failures were not necessarily inevitable
- To capture all relevant knowledge from whoever had it

- As a first step, to bring together in a series of briefings and meetings those parties with the most to contribute to the declared aims, including:
    - Inspectors for the area under review
    - Facility corrosion engineer
    - Technologist for the area
    - Plant maintenance supervisor
    - Senior Plant operator
    - Plant Manager (Chairman)

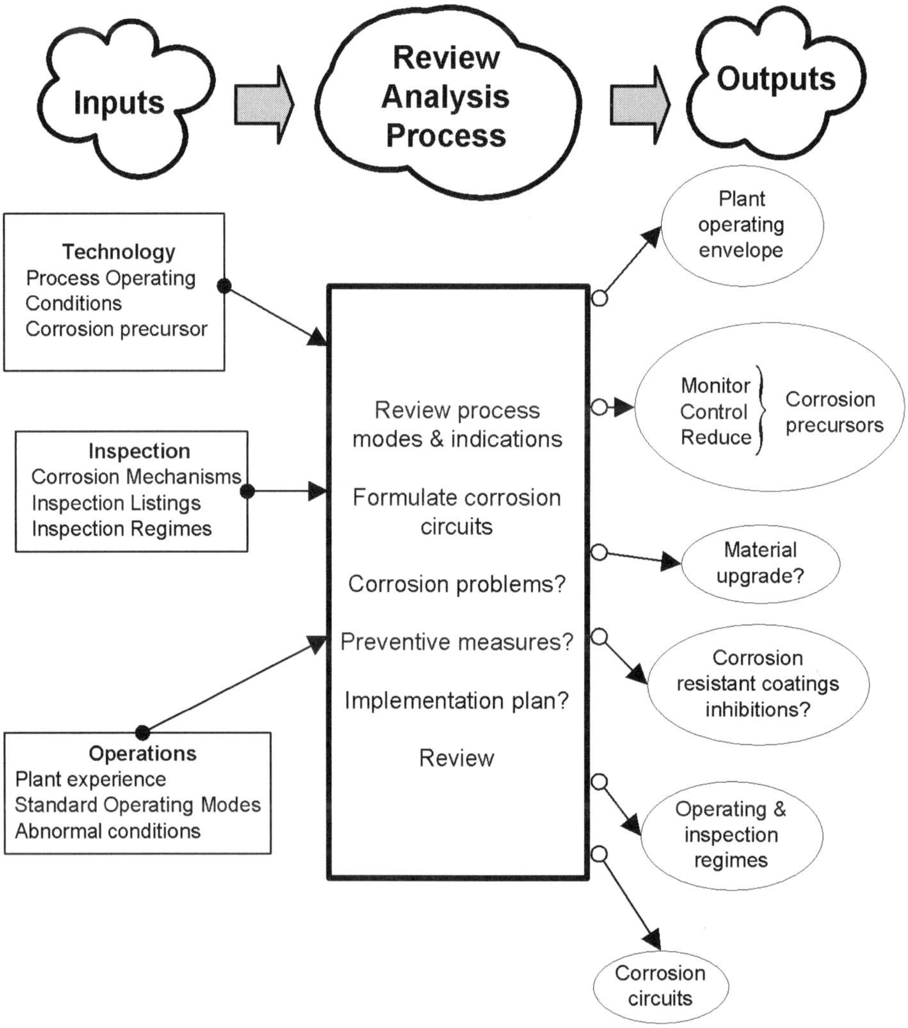

Figure 10.4 Corrosion Review and Analysis Process

Much debate took place before the meeting in management circles on the tactics that should be adopted which would bring the most chance of success. We decided that the plant manager had to be clearly identified as the owner of the equipment and made responsible for reliability end results. To that end, he had to manage a process of bringing together many disparate strands of activities (operators, inspectors, and maintainers) and integrating them into an effective whole. We tried to promote a move away from the compartmentalized past where Production operated and broke the equipment and Maintenance then mended it.

Our next step was to get the concepts agreed by all parties and the group acting as a team in concert. We chose the Crude Distillation area to pilot this effort. The people there were open-minded and positive, so our chance of success was better there. Not surprisingly, the ways of the past were quickly seen as absurd and enthusiasm grew for the new method.

The next step involved a structured approach to identify jointly potential corrosion problems in the plant. Inspection, operations, and maintenance contributed in an atmosphere of openness. This synergy brought a learning experience for all and an increased awareness and understanding of the corrosion issues in the plant. Counter-measures could then be developed jointly.

The review and analysis process is summarized below.

- Review all process modes and operating conditions, and understand implications
- Agree operating envelope for the plant[2]
- Establish corrosion circuits[3] for the plant
- Identify all potential corrosion problems
- Formulate preventive or remedial measures
- Make an implementation plan
- Review and follow up

Figure 10.4 summarizes the corrosion monitoring review and analysis process for the plant.

[2] The operating envelope of a plant consists of the set of operating conditions (temperature, pressure, flow, corrosivity etc.) which balances the plant degradation against the economics of production. Operating within this envelope ensures that the failure modes, mechanisms, and rates remain reasonably consistent. When this is done, the time to failure is predictable with a high degree of confidence. Lack of confidence is a major determinant of inspection (and hence shutdown) intervals. The lower the confidence level, the more frequent the inspections tend to become. Increased predictability of degradation brings increased confidence. so intervals between inspections can be increased, reducing the need for shutdowns and increasing uptime.

[3] A Corrosion Circuit covers that section of the plant that has a similar corrosive environment and is made of similar materials. The Corrosion Circuit concept helps to do five things:

- Provide clear corrosion footprints in the plant
- Show where the different corrosion problems start and stop
- Highlight the most vulnerable areas
- Assist monitoring of different types and severities of corrosion
- Assist use of most appropriate remedial measures

10.5 Results—The Good News

1. The involvement of Operations, Technology, and Inspection brought a new awareness of how poorly the corrosion area had been handled in the past and the potential for significant business benefits through a synergistic approach.

2. Teamwork and communication significantly improved.

3. The concept of the plant operating envelope and the important role of operations in preventing excursions outside this became clear. This brought a new understanding of potential corrosion problems and their implications.

4. Operating and inhibitor regimes could be optimized easily.

5. Analysis of the issues was done in a more structured way and corrosion was being managed.

6. There had been a move from reaction to pro-action and prevention.

7. There was a move away from the previous mindset, which believed that failures were inevitable, to one that saw failures as something to be managed (though not all could be economically prevented).

8. The process and the mindset above were then spread across the site with a significant impact on degradation-driven failures.

9. More significantly this proved an effective launch pad for a more generalized approach to reliability improvement.

10.6 Results—New Problems

Success had come, but a number of problems had been identified. We had used paper and manual systems. These worked at a certain level but we had significantly underestimated the effort involved. There was a need for something a lot less labor intensive and better structured. We needed:

- A comprehensive and reliable data base
- A good documentation system to provide an up-to-date set of in spection records, changes, and decisions, etc.
- A system to manage recommendations and requests to maintenance for action
- A system to trigger an alarm if the plant operation went outside the agreed envelope (from plant instrumentation and laboratory results)

We had some success using computer systems in this type of setting and had recently installed a very effective Maintenance Management System. So we decided to write an in-house program to enhance this system and provide these additional functions. A brief overview of the system functions is given as Appendix 10-A.

Data entry to the system was a big effort; it took us about 4 man-years.

10.7 Spreading the Degradation Management Message

The pilot project on corrosion management was successful. The time was right to build on this to get our message on degradation management across. We decided to spread a simple message around the site to prepare the ground for a big reliability initiative and change the mindset of the group at all levels. In its essence, the message was:

- Like accidents, failures are not inevitable...degradation can be managed.
- Like safety, degradation is a joint responsibility.

There is a simple recipe for managing degradation:

- Buy and use inherently-reliable equipment.
- Operate and maintain it well.
- Focus on critical equipment.
- Use risk-based inspection techniques to optimize efforts.
- Eliminate repeat failures by effective use of RCA.

10.8 Lessons

1. Compartmentalization of departments produces sub-optimal decisions and, all too often, conflict.

2. Inspection emphasis should be on problem solving rather than symptom monitoring.

3. When embarking on a significant change, start small and pilot the change in an area with a high probability of success. Advertise success and then spread the change quickly to other areas.

4. Gain stakeholder agreement of all new approaches. Have a roadmap and a change management plan.

5. Define plant operating envelopes and operate within them.

6. Highlight excursions and any significant implications.

7. Use the knowledge available in all sections of the facility and align on required end results.
8. Make reliability and degradation management (like safety) everyone's responsibility.

10.9 Principles

Large organizations have a tendency to work in silos. These create data islands. By working in isolation, organizational synergies can be lost. Therefore, breaking down these barriers is an important role for management. Computer systems can help or hinder data integration within the facility, so they have to be designed carefully.

Appendix 10-A

Functions of the Computerized Corrosion and Inspection System

The main functions were:

- Defining corrosion circuits
- Monitoring of corrosion circuits
- Highlighting excursions
- Inspection scheduling
- Printing reports

These are briefly described below:

- Equipment Register (shared with CMMS) giving fixed details of equipment, design, criticality, corrosion precursors, etc.
- Corrosion Circuit Register tied to tag numbers, also generic corrosion graphs
- Key Points, which are the monitored point of each tag so that the appropriate NACE, ASME, or ANSI formula is used to calculate minimum thickness
- Thickness Measurement Input/History, which forms the data base of all thickness measurements and is where inspection intervals, etc., are calculated
- Equipment History, which gives the details of all significant events and recommendations/actions
- Equipment Schedule, which provides a means of defining periodic inspection or maintenance activities; all scheduled jobs are generated automatically
- Links to plant and laboratory systems; comparison with pre-set values, alarms, excursions outside agreed plant operating envelope
- Standard and ad-hoc reports

Chapter **11**

Technician Training Challenge

The way managers attempt to help their people acquire knowledge and skills has absolutely nothing to do with the way people actually learn ...we learn by doing, failing and practicing.

Roger Schank, Author.

Author: V. Narayan
Location: 2.2.1 Liquefied Natural Gas Plant

11.1 Background

The plant was located in a heavily forested area in a relatively undeveloped part of the world. There were very few industries. Agriculture, fishing, and timber logging were the main sources of livelihood. The most advanced machinery items in the local area were chain-saws and bulldozers. Maintenance meant replacing complete engines or other large sub-assemblies. There were no repair workshops other than those for automobiles. Perhaps not an ideal location for a large LNG plant with all its complexities, but that was the way the cookie crumbled.

Workers had been recruited locally. Most of them were young men in their 20s, with middle-school or high-school education and no exposure to technical training or process equipment. We had to train them to work in a high-hazard industry with sophisticated equipment and materials with complex metallurgy. We needed skilled mechanics, machinists, welders, electricians, instrument technicians, insulators, riggers, and crane-operators, and this had to be done in three years. English was a foreign language and some of them struggled to cope. We had to start at the very beginning to improve literacy and numeracy standards before tackling the technical elements. On the plus side, the workers were enthusiastic, and very keen to learn.

11.2 The Challenge

LNG plants use advanced cryogenic technology. The low temperatures involved (up to -260°F) require stainless steels, 9% nickel alloy steels, and aluminum. Parts of the plant operate at high temperatures, which require the use of different alloy steels (e.g., 5%Cr 1/2% Mo and 1%Cr 1/4%Mo). All these construction materials require special welding processes and technology. The incoming feed gas was at 160 barg. A number of vessels and piping were operating at pressures ranging from 50 to 160 barg. Some high pressure vessels and columns had 2 1/2" thick shells.

The plant had several pieces of large rotating machinery, some operating at high speeds. These machines require skilled craft workers to maintain them.

We had to generate all the electrical energy and steam required for the process in-house; there was no public utility to provide these services. The total load on various steam turbines driving compressors and alternators was about 120MW. Steam for these drives and for process use was raised at 65 barg., using 9 boilers. There were desalination package units to provide potable drinking-water requirements and for process use. Large pumps in the cooling water pumping station supplied cold sea-water for process cooling and to the condensers of the steam turbines. Cooling water was transported in 3 over-ground steel pipes, each 8' diameter and about 3 miles long. Liquid nitrogen was required for blanketing and purging; this was also made in-house using package units.

As the plant was operating at temperature extremes, good insulation quality is vital, not just for thermal efficiency, but even for process viability. Equipment repair would invariably require removal and replacement of insulation, sometimes 8–12" thick. This work required skilled and motivated craft workers. Their ownership of the work could be critical to its success.

While training was in progress, we had a plant to operate and maintain. As part of their learning process the trainees had to work on the plant equipment to test their knowledge and skills. On occasion, we deliberately disturbed the alignment of standby equipment to let trainees gain practice with real equipment. In this situation we introduced the risk of losing production.

11.3 Training Facilities

The company had built a superb, well-staffed training center in anticipation of the scenario described earlier. It had mechanical, electrical, and instrument training workshops and classrooms. We had a very capable training manager and qualified and experienced expatriate trainers.

I had two experienced expatriate field trainers to help in this program. They acted as mentors to the trainees.

11.4 Training Strategy

The trainees knew almost nothing of the theory or the practical skills required of them. In one sense, this was an advantage. Starting with a clean slate, we could give them just the skills and theory they needed. We decided to use a structured approach to the training program. It would include a 'training needs' analysis, a suitable set of training packages to fill the gaps, tests to establish competence, and on-the-job training to provide confidence. Close alignment between theory classes and practical training in the plant were necessary to promote rapid learning. Competence testing, recording, and follow-up were essential.

We estimated that by the end of the first year of training, we would build up a fair level of expertise and skills. In order to bridge the demand during this period, we hired technicians from the main construction contractor. They would do the bulk of the maintenance work during the first year, while our trainees would assist as apprentices. At the end of the first year, we extended this contract to retain about 40% of the contractor's staff for one more year.

Training Process

We had 60 mechanical trainees and 12 civil engineering trainees. Every week, 12 of these 72 men would be in the training center. Here they would learn basic theory and do some exercises in the training workshop. For example, in one week, they might learn simple trigonometry in the classroom, and do some alignment exercises in the workshop. Next week, they would be back in the plant. The field trainer followed this up closely by giving them practical alignment exercises. We followed a similar approach with machinists, welders, and other craftsmen.

Electrical and instrument trainees pursued a similar process.

11.6 Progress Assessment

The field trainers assessed the progress every month. In doing this, they consulted the training center staff and the maintenance supervisors. The progress was recorded on a simple spreadsheet, an example of which is shown in Appendix 11-A. The symbols used to show each trainee's progress were as follows:

(/) designated a basic understanding and ability to assist under close supervision

(/\) meant an ability to work following instructions and limited supervision

(Δ) meant an ability to work independently by following instructions with little or no supervision

Each of the symbols was put in the cell formed at the intersection of the rows identifying the skill element and the column identifying the trainee. When all three symbols were in a cell, they formed a triangle. When trainees reached a level of competence where they could guide others, the triangle was filled in, so that it now showed up as a bold object.

Trainees who had all the skill elements marked with 'empty' triangles were deemed to have completed their basic training and were eligible to work independently. Those who reached the filled triangle status in all the skill elements were suitable for senior technician or supervisory roles. Appendix 11-A shows a typical status report for machinist training.

11.7 Results

The chart gave us a real time status of training progress. Those with only one or two sides of the triangle in any skill element received extra training and mentoring. Some learned much faster than others, achieving a 90–100% filled triangle status within 18 months. Others took three years or more to reach a 100% empty triangle status. Their career progression was closely linked to their successfully completing their training, which continued as long as needed by each person. About 90% of all the trainees reached the empty triangle status within three years.

In this period we had trained 2 welders, 6 machinists, and 18 fitters to very high skill levels. The remaining 34 mechanical craftsmen did not reach these levels, but were not very far behind. The two welders, for example, were certified to weld low alloy steels, and could use specialized welding processes (using inert gas shielding, termed TIG and MIG). They could weld aluminum, a very difficult metal to work with, and stainless steels. Similarly, two of our civil craftsmen produced close to perfect insulation repairs. Here it was the ability to inject polyurethane foam uniformly and with the right bubble size that mattered. Thermo-graphic images showed that their work was as good as or better than that of the expert insulators who did the project construction and defect repairs.

The reputation of four of our machinists was so high that a new Fertilizer Plant set up nearby two years later, borrowed them from time to time. Our welders were also used by associate companies in the country.

The training center itself eventually became a national venue, where all the associate companies sent their people for training.

11.8 Lessons

1. A proper 'training needs' analysis is the first step to planning a successful training program.
2. A well-structured plan helps impart training properly, especially when large numbers are involved.
3. The plan must include on-the-job experience to supplement formal training; this helps build confidence and verify competence. As Schanki says, people learn by doing, failing, and practicing.

11.9 Principles

The most important ingredient in a successful training program is the willingness and enthusiasm of those being trained. Suitable facilities and good trainers are obviously essential. A well-formulated training strategy and plan form the framework to guide the program. Measurement, gap analysis, and course correction for each trainee on a continuous basis is what makes the whole group progress rapidly.

Appendix 11-A

Craft - Machinist | **Status of progress in skill set** | **Month 03**

Number	Skill set \ Trainee	Takit Chatpuran	Andrew Wu	Mohd. Mustafa	Han Chi Fung	Rosli Ahmed	Thrang Suchatran	Ng Chan	Kim Jo Park	Vijay Ganesan	Ramon Mendoza	........etc.
1	**Safety**											
1.1	Handling tools	Λ	Λ	▲	Λ	Δ	Δ	Λ	Λ	Δ	Δ	
1.2	Clothing	Δ	Δ	Δ	Δ	Δ	Δ	Δ	Δ	Δ	Δ	
1.3	Personal protection	Δ	Δ	Δ	Δ	Δ	Δ	Δ	Δ	Δ	Δ	
1.4	Hazard awareness	/	Λ	Λ	Λ	/	/	Λ	Λ	/	Λ	
1.5	Emergency response	/	/	Λ	/	/	/	/	Λ	/	/	
2	**Measurement & Fits**											
2.1	Calipers, scales	Λ	Λ	Λ	Λ	Λ	Λ	Λ	Λ	Λ	Λ	
2.2	Micrometers,Vernier calipers	/	Λ	Λ	/	Λ	Λ	Λ	/	/	/	
2.3	Taper guages	/	/	Λ	/	Λ	Λ	/	Λ	Λ	/	
2.4	Go-No Go gauges	/	Λ		/	/	Λ	/	/	/	/	
2.5	Fits & Tolerances											
2.6	Allowances											
3	**Lathes & Turning**											
3.1	Plain turning	/	/	/	/	/	/	/	/	/	/	
3.2	Taper turning	/	/			/	/		/	/		
3.3	Sleeves and bushes											
3.4	Clearances/interferences											
3.5	Screw cutting											
3.6	Surface grinding											
3.7	Machining special shapes											
4	**Shaping**											
4.1	Surface levelling											
4.2	Slotting & keyslots											
5	**Milling**											
5.1	Surface milling											
5.2	Gear hobbing											
5.3	Use of indexing plates											
5.4	Racks and pinions											
6	**Boring Machines**											
6.1	Setting up large workpieces											
6.2	Surfacing											
6.3	Boring, honing											
6.4	Drilling deep holes											
7	**Cutting Tools**											
7.1	Tool geometry											
7.2	Use of Tool Grinders											
7.3	Lubrication											
7.4	Storage of tools											

Table 11-A.1 Periodic Assessment of Trainees

Chapter **12**

Competence Profiles

...Rice Farmer or World Expert?

...for a good leader... the greatest responsibility is to appoint good people.
Peter Wickens, Author.

Author: Jim Wardhaugh

Locations: 2.2.3 New Medium-Sized Oil Refinery
2.2.2 Large Complex Refinery
2.3.4 Small Complex Refinery

12.1 Background

Some years ago, I was on a panel interviewing school graduates for craft positions in one of our locations. The sort of jobs they would initially fill would be as fitters, welders, electricians, etc. Some would progress to be technicians and supervisors, while a few could attain engineer and managerial positions.

We were a big-name company and paid good salaries, so we were always flooded with applications—many more than we needed. The personnel department would filter these into a short list for me and others to interview. This filtering was largely on the basis of academic excellence; I ended up interviewing candidates with a potential to do quite well at university. Every year I would argue that this was nonsense as much of the work we wanted these people to do was rather mundane and there was only a small amount of really challenging work. The personnel department's approach meant that we were taking on people who were too good, would soon get bored, become troublesome, and leave.

I still believe this and any recruitment I have done since has been on the basis of producing a preferred profile of required talents and abilities. I believe this to be valid whether we are looking at directors, managers, engineers, or fitters.

Therefore, when taking on personnel for our grass-roots refinery, I tried to make sure we had a profile which roughly fitted the work profile. However, I modified it so that we took on a few people who could eventually be moved into more senior positions.

12.2 Case Study of a Technician Competence Imbalance

This particular location had an all-technician electrical work force with a grading structure, as shown in a very simplified form in Figure 12.1. There were six grades and detailed skill descriptions had been written for these. These descriptions acted as yardsticks to judge whether a person was fit to be promoted in the progression system. The intent of the system was to encourage the acquisition of new skills and the development of personnel to their maximum without artificial barriers such as qualifications, years of service etc.

T1 -	Low engineer activities
T2 -	Improvements, investigations of complex problems
T3 -	Troubleshooting all faults (except most complex) Contractor supervision
T4 -	Troubleshooting common faults
T5 -	Routine go/no-go testing Replacement of obviously faulty parts
T6 -	Basic craft skills

Increasing hands-on (T1 → T6)

Increasing supervision / organization, Troubleshooting (T6 → T1)

Figure 12.1 Technician Grades and Skill Descriptions

In the past, recruitment had consisted of taking on the best people that could be found. It resulted in a grading distribution for technicians, as shown in Figure 12.2. During annual staff appraisals, we made an estimate of the ultimate grade each technician was likely to achieve. Figure 12.3 shows the distribution that might be attained in an environment with unlimited opportunities for promotion. We were looking at a time horizon of less than 10 years and assumed that a technician would spend about 2 or 3 years in each grade.

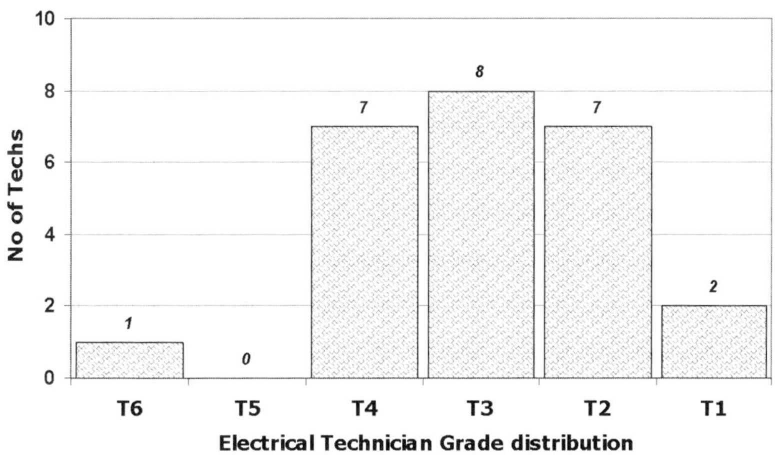

Figure 12.2 People's Skills Distribution

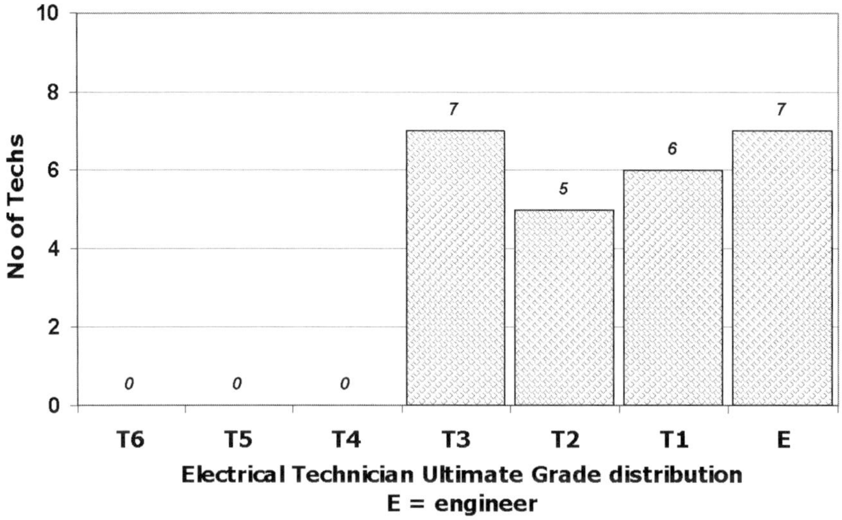

Figure 12.3 Technician Ultimate Grade Distribution

The recent introduction of a new computerized maintenance management system (CMMS) had made it possible for the first time to get detailed information on activities. Some three hundred typical maintenance jobs were scrutinized by a small group of engineers and supervisors. They then decided for each job the lowest technician grade that could do a job effectively and safely without direct supervision. This distribution is shown in Figure 12.4.

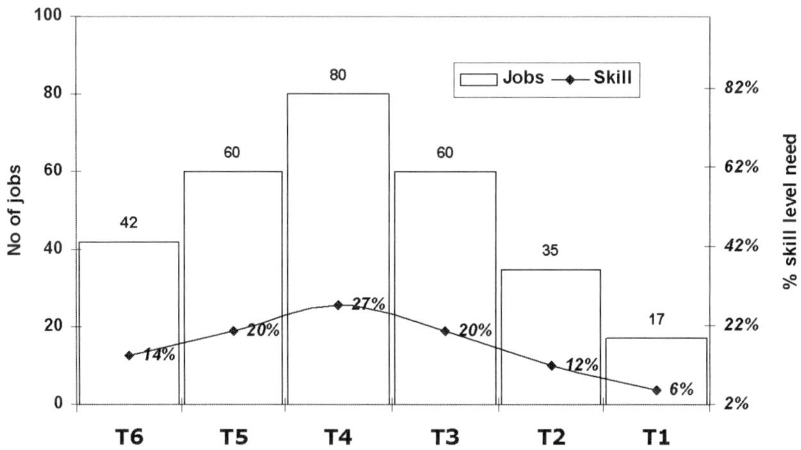

Figure 12.4 Profile of Needed Skills

Looking at these distributions, it is clear that people's capabilities were much in excess of job requirements. Options for action were limited. Many people's aspirations could not be met and the better people started to leave.

12.3 An Approach to Contractor Competence

In the Far East, it is not easy to ensure that the individuals you get have the competence they claim. This is a particular problem when you have a big job like a shutdown. Then you get the absurd situation of rice farmers claiming to be rotating equipment technicians.

There is an often-expressed view that, if you go for fixed price contracts, it really doesn't matter who the contractors use to do the job, how many people they use, and, indeed, whether you get different people every day. This approach with its ever-changing labor force causes a huge administration effort in issuing passes and in induction and safety training. It also brings a high need for toilets and canteen facilities, both of which can add significantly to facility overhead. The doubtful competence of this mob of workers causes a tremendous amount of rework and unacceptable accident rates.

Laws allocate responsibilities in different ways in different countries. Quite often, if you are managing a project (and a shutdown would come under this definition), you as the company (the client) will be held responsible for incidents. You cannot hide behind the contractor. The company then must take all necessary steps to ensure competence.

We used the gate access system with a swipe card to restrict access onto the site to only a reduced core of contractors with defined competencies. This made the numbers manageable. More on this is given in Chapter 33.

For this reduced group, we defined competencies that we considered critical, e.g., supervision, various grades of welding, alignment of rotating equipment, and making up flanges on pressurized systems. Workers were then tested and interviewed against set criteria. If they passed, they were defined as competent and authorized to do certain work. This competence level was captured in the gate access system and personnel records could be reviewed on line. Unannounced competence audits were carried out at regular intervals.

Many other companies have now adopted systems like the above. For example, a number of companies now give formal competence certification for flange work. This has resulted in leak-free start ups where shutdowns had once become almost a norm. Some have even extended this system by gaining inter-company agreement to accept each other's competence certificates. This eliminates the need for constant re-testing every time a contractor moves to work in a different company.

We found that starting the process in this low-key way with simple competence demands made for an easy acceptance. Once this basic system was in place, extending the process came easily. As with all changes, organizations can handle a series of small steps much more readily than one giant leap.

12.4 Competence Profiling—A Bit of Theory

Have a look at Figure 12.5; this shows in a few words the essential steps of competence profiling. It looks for the business requirement in terms of competencies, identifies individual competencies, and makes the gap visible. Then what we need is focused training to bridge the gap. This training does not necessarily imply a training course, which is often the knee-jerk reaction

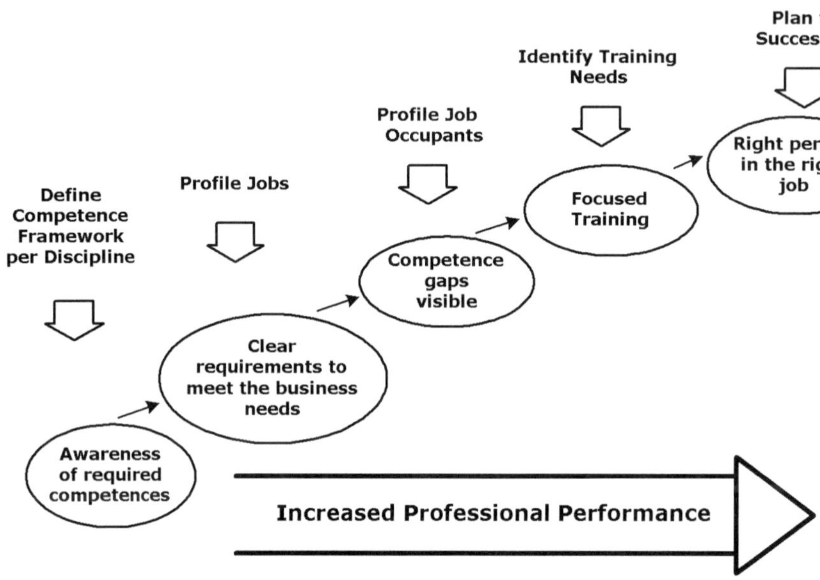

Figure 12.5 Job Competence Profiling

to identification of a training gap. There is much to be said for on-the-job training and mentoring.

What should not be contemplated is wall-to-wall training of all people in all aspects of the business. Yet paper after paper on the Internet purports to show a correlation between the amount of training and performance. In my studies, the converse is true. I have found that those who train the least have the highest performance. This might sound like heresy but let me explain.

My thesis on business success is quite simple. Top performers have identified what are the important factors which they have to get right to make their business a success. They focus their efforts on what is important. This carries over into training and they do small amounts of focused training for people who need it, when they need it, and only in areas that matter. Poor performers don't seem to know what is important, don't prioritize, and don't focus their efforts. The same happens in training. In brief:

- Top performers do small amounts of (cost) effective training
- Poor performers do lots of (cost) ineffective training.

12.6 Competence Grading Scale

Competence is the combination of knowledge, skills, and attitudes necessary to carry out a job to the required standard of performance. Figure 12.6 gives a vocabulary and shows a simple way of grading competence through a spectrum of increasing effectiveness in specific areas of competence.

We tried some mechanistic scoring systems, but quickly abandoned them. They were too complex and did not provide answers to which we could relate.

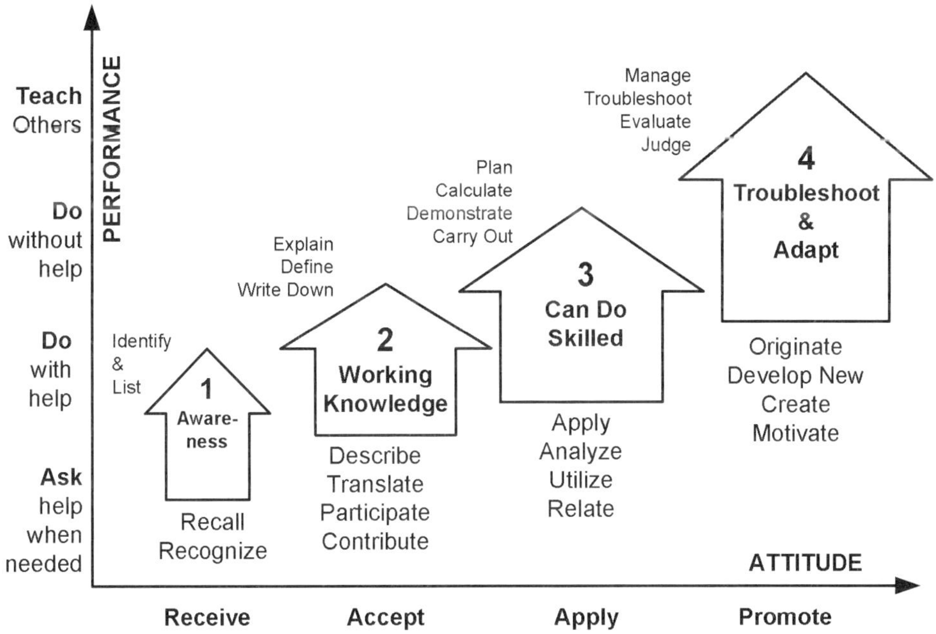

Figure 12.6 Competence Grading Scale

12.7 Competence Management

It was important to start up our new grassroots refinery with a minimum staffing level, but an adequate level of competence. Using the approach shown in Figures 12.5 and 12.6, we defined specific job competencies as the combinations of knowledge, skills, and attitudes necessary to carry out that job to the required standard of performance. Four competence levels were defined in simple everyday language. These are shown in Table 12.1 Competence Levels.

COMPETENCE LEVELS		DEFINITIONS
Awareness	1	Knows what is involved and when to ask for help
Working knowledge	2	Can explain the knowledge and use it with some help
Can do (skilled)	3	Competent able work with minimum help
Troubleshoot Adapt	4	Expert in this area and can diagnose & resolve significant & unusual problems

Table 12.1 Competence Levels

Scores: 0 = Ignorance, 1 = Awareness, 2 = Working Knowledge, 3 = Can Do, Skilled, 4 = Troubleshoot/Adapt
M = Manager, S = Senior, A = Assistant

COMPETENCE		INSPECTION			MECHANICAL			CI
		M	S	A	M	S	A	S
A	**DESIGN & ENINEERING**							
1	Dept's International Standards and Codes	3	2	1	3	2	2	3
2	Safety in Design	3	2	2	3	2	2	1
3	Calculations	2	2	2	2	2	2	2
4	Equipment Design	2	2	2	3	2	2	2
5	Drawings	3	3	2	3	3	3	3
6	Fabrication	3	2	1	3	3	3	3
7	Legal Requirements	3	3	1	2	2	2	3
B	**MATERIALS STANDARDS**							
1	International Standards	3	2	1	3	2	2	2
2	Material Properties	3	2	1	3	2	1	2
3	Quality Requirements and Certification	3	2	1	2	2	2	2
C	**PRODUCTION PROCESSES**							
1	Common Processes	2	2	1	2	1	2	1
2	Raw Material and Products	2	1	1	1	1	1	1
D	**WELDING & FABRICATION**							
1	Welding & Fabrication Processes	2	2	1	1	3	2	2
2	Code Requirements	3	2	1	1	1	1	1
3	Qualification Processes	2	2	1	1	1	1	1
4	Heat Treatment Processes	2	2	1	2	3	2	0
5	Repair and Maintenance	3	3	1	3	3	3	3
E	**MATERIAL & CORROSION**							
1	Material & Corrosion	3	2	1	2	2	1	1
2	Material Degradation Phenomena	3	2	1	3	2	1	1
3	Material Selection (Lubricants)	3	2	1	3	2	1	1
4	Coatings / Insulation / Linings	2	2	1	2	1	1	2
F	**PERSONAL & MANAGERIAL**							
1	Managing self	3	2	1	3	3	3	3
2	Managing others	2	1	0	2	2	2	2
3	Managing the Business	1	0	0	1	1	1	1
4	HSE	3	2	2	2	2	3	3
5	Interface with Bodies outside Company	3	2	1	3	3	3	3

Figure 12.7 Example: Job Competence Matrix

For the technician and operator grades, we devised a grading scheme which defined competence levels and rewarded the acquisition of skills. This is discussed in detail in Chapter 13.

All too often when looking at competence, the focus is on the technical aspects. We found that delivering a professional performance required a combination of personal and managerial competence as well as the discipline specific competence. The level of each and emphasis varied with the job. Small teams of knowledgeable individuals used this methodology to define the needed competency in each role whether as a mechanical technician, a mechanical maintenance engineer, a project engineer, or any other position. This set the target for each person so any gap and training needs could be defined.

To establish and record competence levels for the technical and managerial grades, we used matrices, as shown in Figure 12.7. This figure shows an example of a job competence matrix we used to map the required competence for Inspection, Mechanical Maintenance, and Civil positions. Three job levels are shown here; for ease of understanding, I have called them Manager (M), Senior (S), and Assistant (A). Scores in the matrix follow the definitions and theory given earlier with

Ignorance = 1
Awareness = 2
Working Knowledge = 2
Can Do and Skilled = 3
Troubleshoot/Adapt = 4

We found that some temporary facilitation was needed to attain a common understanding of these terms, but it was not a big hurdle. Only a sample of a total matrix is given here and it needs significant expansion to reflect complete reality. The competences themselves and levels needed are specific to the type of business. The matrix will need to reflect that.

Figure 12.8 shows a matrix which we used to compare the individual job holder's competence with the job requirements. This gave a structured way of capturing this information.

12.8 Lessons

1. The pool of competent contractors has always been quite small and is growing smaller. This pool can easily be used up. Contractors hungry for work are reluctant to highlight this problem.

2. The company is often legally liable for ensuring competence and liable for the actions of incompetent contractors.

3. Sometimes we felt we were writing down a statement of the obvious, but it soon became apparent that what was obvious to one was not necessarily obvious to others.

4. A common vocabulary and approach was a good way of achieving consensus on needed competences. It took us away from descriptions that called up super-heroes who could walk on water. It also took a lot of emotion out of the debate when assessing the ongoing performance of individuals.

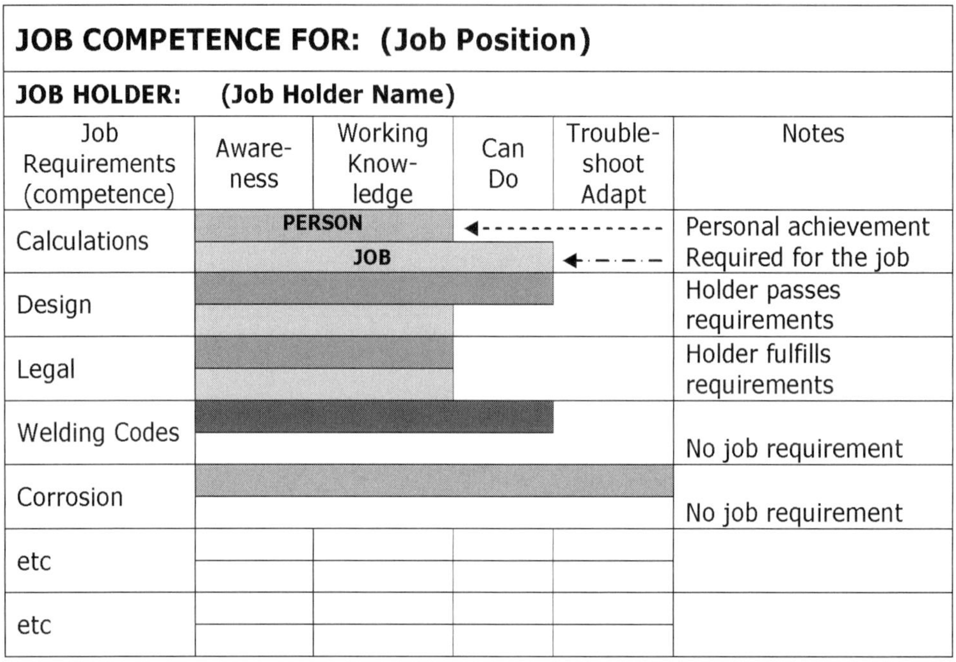

Figure 12.8 Comparison: Position Requirements and Competencies of Job Holder

5. Have systems to ensure sufficient competence in all significant job positions, whether filled by a contractor or own staff.

6. Have simple, easy-to-understand, definitions of the competence levels.

7. Avoid complex, mechanistic scoring systems to establish competence levels of job position or job holder.

12.9 Principles

Do not blindly take the best people available and be prepared to reject those who are clearly too good. Recruit people who can grow to just meet future job needs.

Contract out low grade and extremely specialist tasks wherever possible.

Chapter 13

Operators and Maintainers

It is not the strongest species that survive, nor the most intelligent, but the ones that are the most responsive to change.
Charles Darwin: The Origin of Species.

Author: Jim Wardhaugh
Location: 2.2.3 New Medium-Sized Complex Refinery

13.1 Background

All too often, roles at the working level in a plant are those traditionally derived from union demarcations. There is a certain level of comfort in operating within such limitations, but demarcations are often neither sensible nor efficient. We decided to break this mold when designing roles at the working level in our new grassroots refinery in the Far East.

Plants in Scandinavia had recruited plant operators from personnel trained in a craft background (often from ship's engineers looking for shore jobs) and then trained them in plant operation. This had been successful, so we decided on a similar path. In our case, we extended the role of the operator to that of operator-maintainer. In addition, there would be a few specialist maintainers forming a high-skill nucleus.

It was essential that our new local recruits acquired knowledge and skills quickly, so we had to make this process:

- Attractive by the use of a grading system which brought appropriate rewards
- Attainable by ready access to appropriate training

Credits and associated salary increases could be gained by attaining meaningful skills in depth, in breadth, or in a combination of the two. Our refinery was designed to attain top performance in all significant aspects. Through this structure, top performer's staffing levels would be attained.

13.2 Concept and Initial Training

We decided to make the operating shift lean and mean, but having sufficient staff with competencies to be largely self supporting. A 'day' work force would back up the shift, but we designed it to provide a level of support significantly lower than that traditionally found in the industry. The initial idea was that at any one time about 60% of the operators would be operating the plants on 12-hour shifts, 30% of the operators would be attached to maintenance doing necessary maintenance on days, and about 10% would be spare.

To achieve the above, we would train operator-maintainers to be competent in two disciplines:

- Operations (panel and/or outside operations)
- Maintenance (Mechanical, Instrument, Electrical)

We recruited local staff with five or six years of craft experience backed up by a vocational qualification such as a diploma. The profile of craft backgrounds was roughly 60% mechanical, 25% instruments, and 15% electrical. Only about 25% of the recruits had some operational experience.

Training of the operators consisted of spells at other operating plants, classroom training on specific aspects of the new refinery, simulator training, mentoring under the wing of experienced expatriate operators, and involvement in the start-up. A small group of trainers taught the basics of maintenance and we used on-the-job training to enhance their skills. At every opportunity available, they saw the internals of equipment and participated in the commissioning and overhaul of complex equipment.

A small number of specialist maintenance technicians supplemented this operator-maintainer group. The operator-maintainers received in-depth training in specific aspects of the refinery equipment. They became the source of high level maintenance expertise (along with the engineers).

13.3 Initial Experience and Findings

Presence of the operator-maintainers for twenty four hours a day brought immediate benefits in that they prevented a number of trips and shutdowns. Repair of equipment did not have to wait for a maintenance call–in, but could be started immediately, even if in some cases the expertise was not available to complete the job.

Operators took considerable ownership of the equipment. Their craft backgrounds brought sensitivity to equipment needs, which the more traditional separation between operator and maintenance roles tended to diminish. They were eager to participate in reliability initiatives which, although set at a fairly basic level, gave immediate dividends.

We found quite soon that we had overestimated the maintenance requirements. The number of dedicated maintainers could be significantly reduced. We set a target to ultimately reduce this number by half.

The cycling of unnecessarily large numbers of trainees through day maintenance had two unfortunate effects:

- Operations training, development, and competence attainment of local staff were slowed down. As a result, we could not release expensive expatriate operators as quickly as we would have liked.
- With maintenance specialist technicians doing the higher grade work, operator-maintainers were left with only low-skill tasks. This situation brought those with significant craft backgrounds little satisfaction. The result was a lowering of morale.

Table 13.1 Operator-Maintainer Progression Routes

OPERATOR / MAINTAINER PROGRESSION ROUTES		
A progression route is agreed for each person based on business needs and individual aspirations		
Operating Specialist	**Generalist**	**Specialist**
Only basic engineering skills required Possibility of becoming highly-skilled outside operator Option to develop as either panel operator or Team Leader	Route to be selected for those who prefer working as outside operator Opportunity to develop engineering skills at a high level	Route to become highly-skilled maintenance technician Future source of engineering expertise

13.4 Revised Scheme

We set up a team to review the scheme and produce modified proposals to get over the problems outlined above. These proposals were accepted and implemented.

The scheme consisted of the following:

- Three separate, but compatible and integrated progression routes—see Table 13.1. None of these routes were blind alleys. All the skills learned were useful, useable, and portable within the grading system. Maintenance people were incorporated in the scheme and a correspondence of competence values made between operating and maintenance tasks.
- The scheme allowed the development of both operational and maintenance skills and allowed flexibility between the ratios of skills acquired.

- We created an operations-only fast track route, aimed at developing panel operators and team leaders quickly. This encouraged rapid development of operating skills in the local workforce and enabled reduction in expatriate numbers. See Table 13.2.
- We developed a maintenance skills only route giving credit for in-depth skills gained, e.g., instrument technician, electrician, or fitter. See Table 13.3.
- A large number of competence building blocks allowing skill development by increases in breadth of knowledge, depth of knowledge, or by a mix of both. This allowed for more flexible progression routes. See Figure 13.1. This figure shows how operators could gain three Level 1 credits by being competent in three different plant areas. They could then go on to gain Level 2 and Level 3 credits for developing skills which were portable across the three areas. By doing a panel operator's job at Level 4, they could gain 5 credits. In this way, they could rise up the skill ladder and gain salary increases. Thus, there were many routes to the top.
- We based the scheme on people acquiring measurable competences without being restricted by artificial hurdles or time scales. Progression was limited only by the rate of skills attainment of individuals. In fact, the best were limited by our training resources.

OPERATING COMPETENCES	
LEVEL 1	Monitors and maintains steady state conditions
LEVEL 2	Starts up and shuts down equipment and makes preparations for maintenance
LEVEL 3	Able to respond to abnormal process conditions, can interpret DCS information and provide on-plant instruction
LEVEL 4	Panel operation or in-depth panel understanding and capable of diagnosing process faults
LEVEL 5	Supervisor (team leader) positions

Table 13.2 Outline of Operating Competences

- Supervisory positions, such as team leaders, maintenance supervisors, and junior engineers at Level 5 (see Tables 13.2 and 13.3) were, however, vacancy limited. This meant that promotion to that level would only become possible when the company declared a job vacancy and advertised a position.

Table 13.3 Outline of Maintenance Competences

MAINTENANCE COMPETENCES	
LEVEL 1	Basic craft skills
LEVEL 2	Routine testing of defective equipment and replacement of faulty parts
LEVEL 3	Trouble shooting and repair of most common faults
LEVEL 4	Trouble shooting and repair of all faults (except most complex) and supervision of contractor labour
LEVEL 5	Supervisor & junior engineer positions

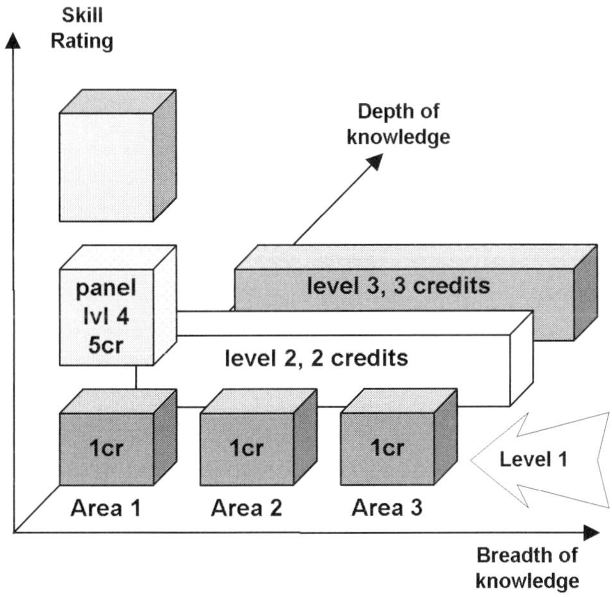

Figure 13.1 Competence Acquisition Routes

13.5 Credits for Competence, Grades, and Salary Bands

Space precludes a complete discussion of all aspects so only the salient points are given below. We wrote competence descriptions for four different levels of job modules based on the definitions given in Chapter 12. These were summarized to make them immediately understandable to the workforce, as shown in Table 13.4.

COMPETENCE LEVEL DEFINITIONS	
LEVEL	**DEFINITIONS**
LEVEL 1 Awareness	Knows what is involved and when to ask for help
LEVEL 2 Working knowledge	Can explain the knowledge and use it with some help
LEVEL 3 Can do (skilled)	Competent, able to work with minimum help
LEVEL 4 Troubleshoot, adapt	Expert in his area and can diagnose & resolve significant and unusual problems

Table 13.4 Competence Definitions

Once operators or technicians had learned the skills relevant to each level, they would be assessed, as discussed in the next section. If successful, they would then be awarded a certain number of credits which would increase their salary. The credits for each level are shown below:

- Level 1 – 1 credit
- Level 2 – 2 credits
- Level 3 – 3 credits
- Level 4 – 5 credits

There were three salary bands:

- Operator or technician having up to 3 credits
- Experienced operator or technician having 3–9 credits
- Senior operator or technician with 9 or more credits

13.6 Assessment

The aim was to provide an auditable process that enabled operators and technicians to be assessed in the workplace against established performance criteria in an effective and consistent way. The assessment system was designed to:

- Accurately assess the candidate's performance against agreed criteria
- Be easily understood by all, use time effectively, and avoid unnecessary bureaucracy
- Be applied only by adequately trained assessors

The assessment process is shown in a simplified way in Figure 13.2.

As far as possible, we based the assessment of competence on practical demonstrations of the skills involved. At each level, we used the performance elements of each job, which enabled assessment to be done in stages if necessary. Samples of two competence descriptions are given in Tables 13.5 and 13.6.

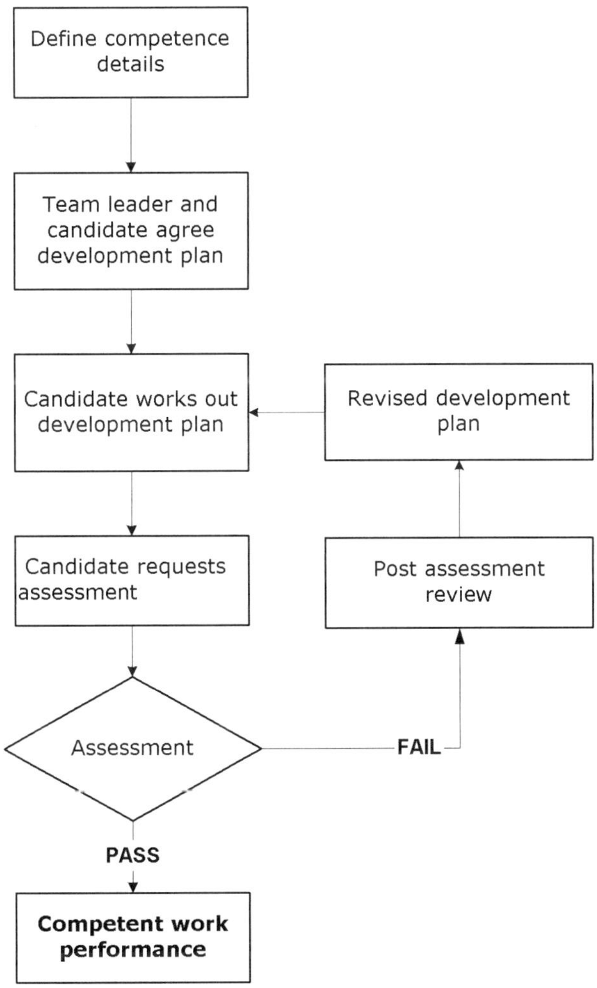

Figure 13.2 Assessment Process

MECHANICAL WORKSHOP LEVEL 1	
Operatives working at this level will under direct guidance demonstrate the ability to safely, correctly, and effectively perform the following tasks and those of lower levels as a minimum requirement.	
01. Follows company permit to work system	○ Completes and applies work permits
02. Follows site safety regulations and is aware of common refinery hazards	○ Safety regulations ○ Awareness of common hazards i.e. H_2S, fire, and inert gas
03. Follows safety procedures when operating workshop machinery	○ Uses guards as per workshop safety rules ○ Maintains a clean work area ○ Uses all required P.P.E.
04. Performs basic machining	○ Uses lathes for basic turning ○ Uses milling machine (basic) ○ Uses guillotine ○ Uses plate folder
05. Identifies piping systems	○ Types, valves, material, etc.
06. Identifies rotating equipment and their basic operation	○ Pumps ○ Compressors (reciprocating) ○ Compressors (rotary) ○ Turbines ○ Diesel engines
07. Identifies parts and functions of rotating equipment	○ Rotating elements, pistons, valves, seals, bearings, etc.
08. Assists in overhaul of rotating equipment	○ As part of a team, assist in the overhaul of equipment
09. Performs single flange work	○ Opening and closing flanges
10. Performs minor maintenance on manually operated valves	○ Gate, Globe, Ball, Butterfly, Plug,Needle ○ Non-return ○ Relief valve ○ Diaphragm
11. Performs basic lifting and rigging procedures	○ Slings and lifts loads to 1 ton
12. Uses CMMS at basic level	○ Picks up work and closes jobs

Table 13.5 Mechanical Workshop: Level 1 Competence Description

MECHANICAL PLANT LEVEL 4	
UNDER AN ENGINEER'S SUPERVISION Operatives working at this level will demonstrate the ability to safely, correctly, and effectively perform the following tasks and those of lower levels as a minimum requirement.	
01. Troubleshoots	• Identifies / Rectifies faults in basic cross discipline work areas
02. Co-ordinates	• Liaises and pre-plans jobs with plant maintenance people
03. Prepares specifications and costings	• Prepares job specification and cost estimates for minor work • Small project maintenance work and necessary site inspection work
04. Recommends spare parts	• Works out spares recommendations for Site equipment
05. Schedules	• Organizes and arranges day-to-day maintenance or project work activities
06. Organizes	• Organizes minor project work
07. Maintains lifting and rigging appliances	• Work with inspection group on maintaining the standard of lifting and rigging equipment
08. Understands basic process operations	• Has basic knowledge of refinery process systems
09. Interprets standards	• Interprets relevant standards selecting the correct equipment

Table 13.6 Mechanical Plant: Level 4 Competence Description

13.7 Lessons

1. It is important when carrying through an initiative that we explain it properly to the workforce and that key leaders in the organization support it.

2. Many old-time supervisors can be uncomfortable with anything different from the traditional craft roles and training methods. They may only be prepared to give lukewarm support (if any) to the new ways. As an example, it was difficult getting them to accept even a minor inconvenience of releasing staff for scheduled training.

3. Many of the engineering supervisors had come through a training route in which progression depended on residence time in specified positions, e.g., a four-year apprenticeship split into six-month modules. They found it difficult to accept a system with no residence time hurdles.

4. In Operations, Shift Managers are the key figures in bringing success. As far as their people are concerned, they are management. They must lead, explain, and sell management concepts and principles to their shift workforces.
5. Temporary staff may not be prepared to confront and solve issues. Many were content to push problems into the future for someone else to solve.
6. When work is urgent—for example, when there has been a plant stoppage or during critical aspects of a start-up—empowerment and mentoring of new staff tended to take second place to operational needs.
7. The grading system must have rigorous progression hurdles. (All too often, we fell into traps where we did the equivalent of giving one competence credit for using a shovel and one for using a pick. Therefore, people who used both a pick and a shovel could get promoted. In one absurd situation, they could get promoted again if they used these in an other area.)
8. Operators and technicians were enthusiastic to learn, attain, and demonstrate competencies which would result in salary increases. The demand on training resources was much higher than we had anticipated. We had bought inherently reliable equipment and were operating and maintaining it well. The reliability and achieved availability was, therefore, high with a low manpower demand.
9. Workers prefer to be members of a cohesive group working for one supervisor. In that way, they get to know each others' styles, strengths, and weaknesses. In our system, the workers could easily end up working for many different supervisors; this result brought relationship difficulties.

13.8 Principles

The traditional style of organization works reasonably well. It has known problems, but moving to a different style of organization brings new and unknown problems. For a business to achieve the results it wants in a slimly-staffed organization, we can delegate work successfully only when we know the competence of individuals.

Consistency and fairness have to be demonstrable in any grading system for it to gain acceptance and credibility.

Breadth of skills is important. However, be cautious in giving credit for it, as it might result in a poor competence profile.

Chapter 14

Building a Reliability Culture

I keep six honest serving-men (They taught me all I knew); Their names are What and Why and When and How and Where and Who.
Rudyard Kipling, The Elephant's Child (1902).

Author: V. Narayan

Location: 2.1.4 Large Petroleum Refinery

14.1 Background

At the time of these events, the refinery had only been in operation for about two years. Local trainees were recruited to work in the maintenance department. They were fresh out of school, with limited or no practical experience. A contractor supplied an expatriate workforce. Hence, there was not much ownership or pride, as the contract staff knew their tenure was limited. Short-term contracts and poor living conditions meant that motivation levels were quite low. The contract staff skill levels were reasonable, but we were employing 'hands,' not whole people who had brains as well. The attitudes and behaviors of the contract crew soon rubbed off on the local trainees.

14.2 Long-Term Issues

Poor attitudes and behaviors could potentially be serious problems, as they affect reliability. Habits die hard, so any behavior patterns established in the early stages would be hard to change in later years. It was, therefore, very important for us to get the workforce, especially the local trainees, to think reliability.

14.3 A Driver of Behavior

We had to find something to galvanize the expatriate workers into a reliability mindset. Monetary incentives were not possible; besides, I had limited faith in their sustainability. In this location, expatriate workers were not paid very well, lived in poor conditions, and had very little opportunity to play or entertain themselves. The fun element was missing in their life. They saved whatever money they could to send back to their families in the far east or in south Asia. That was their sole objective.

One way to bring some fun into their lives was to make work more interesting. Easier said than done, of course, but I had an idea, based on the experience I had with my own children when they were 2–3 years old. They were always asking the question why, and each answer invariably led to a further question. This curiosity led to rapid learning, as most readers with small children will recognize from their own experience. Quizzes and puzzles capture their imagination, and they always enjoy the thrill of getting the right answers themselves.

Adults, of course, are more circumspect; they do not ask many questions lest people think they are dumb. Their learning curve levels out and they stay within safe comfort zones. They enjoy fewer pleasures of problem-solving than they did as children.

14.4 Executing the Plan

Every morning, during my walkabout, I asked each team working on any equipment two or three questions. Typical questions were, what are you doing?, why are you doing it?, why did it fail?, etc. Initially their response was "I don't know." In fact, my supervisors were curious as to what I was doing, harassing their lads. I simply asked them to be patient. After a couple of weeks of this sort of pestering, the workers tried some tentative answers. Initially I made no comment, and tried to keep a passive expression. As the days went by, they became more adventurous and asked me for feedback.

14.5 Results: The Moment of Truth

Soon they started looking at drawings and manuals on their own initiative. I had never asked any of them to do so, but they figured out that they needed documents if they had to explain things to me. The supervisors, who kept the manuals and drawings in their offices, were surprised by the number of requests for these documents. Within about six weeks, they started answering my usual questions even before I asked, trying to check if they had the right answers.

They discovered that operators were not following start up and shutdown procedures properly. This started an initiative; they now insisted on being there at start-ups, and advising operators on the right procedures. Their own dismantling and assembly procedures became better aligned to those in the manuals. During shop overhauls, they began to visit the workshop to find out what exactly was found and whether they could do anything about it. They started thinking reliability without much prompting. Within a year, they began to challenge me, with drawings and manuals to support their case. In some cases, their thinking was much better than mine. When I told them that, they went away beaming with pride.

14.6 Rubbing Off the Right Way

The local trainees also enjoyed the process. They started doing the same things the contract workers did, and soon they were active, sometimes too active. All this was good, as we were getting them to stop accepting failures. Getting reliability between the ears was the objective, and I felt we had taken the first steps forward.

14.7 Lessons

1. Achieving high reliability requires people at all levels to behave appropriately. They have to stop accepting failure as normal and question any that occur.

2. Workers who think reliability is important often achieve more than large Reliability Departments. This mindset does not need either a high level of skill or a large knowledge base.

3. Operators are often more willing to listen to technicians than to implement reports from Reliability Departments.

14.8 Principles

Reliability issues can all be traced to human errors at some stage in an asset's life cycle. Human reliability is the key to achieving good reliability performance.

Work can be a great source of fun. That itself can be a better motivator than many of the conventional methods. Human beings love solving problems, but the scientific management theories of F.W. Taylor in the early 20th century removed most of the fun elements. Using his theories, people like Henry Ford made each worker an expert in one micro-element of the production process, thus converting them into high-speed robots. In the last quarter of the 20th century, Toyota reversed this philosophy, bringing in teams to work on complete automobile assembly, not just in putting on wheel nuts. Their commercial success and leadership in automobile manufacturing showed that people who enjoy their work are also more productive.

Chapter **15**

Managing Surplus Staff

Do not let what you cannot do interfere with what you can do.
John Wooden, US basketball coach.

Author: V. Narayan

Locations: 2.1.3 Petroleum Refinery

15.1 Background

In poor countries, unemployment levels are often high, as it was in this case. Prevailing wages were low, so it became easy to hire more people than necessary. We had many layers of staff in maintenance; the actual proportion of useful workers was perhaps 40% or less. The non-productive workers affected output in two ways: first, by producing little themselves and second, by slowing down the productive workers. They were not inherently bad or unskilled. Under different circumstances they could have been quite productive themselves.

The company was part of a very large multi-national oil company. The message from corporate headquarters was to de-layer and downsize. That part was hard to swallow but do-able. The difficult part was what to do with the surpluses. Social legislation in the country made it close to impossible to fire them. Also, the damage to the industrial climate was perceived to be more serious than any economic gain.

I did not play a role in these events. What follows is based on my observation of the Managing Director's leadership.

15.2 Empire Building

Many people see manpower reductions as almost suicidal. The more people one has:

- the higher one's job title is likely to be
- the more levels people can put between the problem and themselves, hence, the less likely they are to have to face the problem personally
- the better one can manage uncertainties; you never know what will

hit you next year or the one after, so it is wise to keep some spare capacity to meet those eventualities
- the easier it is to manage capacity shortfalls—why give up those you have?

An analogy can be made between managers and dairy farmers in Thailand. The more cows a farmer has in his fields, the wealthier and more important he becomes in the eyes of others.

Staff reductions require a lot of ground work, which busy managers find hard to fit into their schedules. It is painful, not glamorous. As C. Northcote Parkinson[i] said many years ago, "work expands so as to fill the time available for its completion."

So everybody 'is busy' doing 'something'. Whether that something is useful is not always questioned.

15.3 Leadership

We had a European Managing Director (MD). The local staff (including me) did not expect him, as someone from a different culture, to understand the prevailing limitations, both cultural and social. To our surprise, he proved us wrong and did appreciate the nuances of the local attitudes and culture. He may not have agreed with these values, but he systematically went about the major changes required, without upsetting staff or the trade union. Some members of the management team privately disagreed with the MD's position. However, he was not going to be deterred from his chosen path.

15.4 Identifying Surpluses

Within maintenance, we had to identify all those people who met the skills profile and competency standards of the identified and approved positions. The numbers were determined by the principal's corporate norms after making some allowance for local variations. We justified the need for these allowances for each case, in writing. As a result, we were allowed to enlarge the maintenance numbers by about 10% over the norm. The rest, whose strength was around 50% of the existing maintenance manpower numbers, were physically separated from the main team, and moved to a new department.

15.5 Opportunities

Over the years, quite a large volume of maintenance and minor project work had been done by contractors. In the reporting system, the headcount of those doing maintenance work such as tank repairs was averaged over the year and added to the maintenance strength. This nearly doubled the equivalent manpower numbers. Small projects required a significant number of 'equivalent' people as well.

The MD's plan was to use the new department to carry out small projects or significant maintenance work traditionally done by contractors, thereby displacing a portion of the 'equivalent' numbers that had to be added on earlier. Thus, instead of firing many of the company employees, we could issue less contract work.

15.6 Implementation Challenges

Many of the surplus staff had skills, but these were not necessarily suited to doing project work. Old habits die hard, and the MD realized that if they were allowed to mingle freely with the maintenance crews, they would gradually be re-absorbed, as people began to find work for them. Then there was the question of morale. People could feel rejected and, hence, dejected. Poor morale and safety don't go together; sooner or later we could expect accidents.

15.7 Managing These Challenges

The company offered training in skills to people in the new department. Many took up the offer and were trained in skills such as fitting, welding, scaffolding, cable jointing, or installation of process control instruments. Some took up refresher training in their own skills while others learned additional skills. A few did not take up the offer.

People in the new department were physically relocated. They were not allowed to take part in any routine maintenance activity. The company enforced this policy strictly and treated deviations as violations of discipline. In the initial period, there were a few instances when this occurred, but it soon became clear that the company was serious about the policy.

The new group was called the Development Department, giving it respectability. The majority of the people in it were surplus staff, but they knew that they were to execute minor projects, and were thus gainfully employed. In an interesting move, a few excellent technicians and two well-respected supervisors were also assigned to the department. The new department was strengthened with these skilled people, and thus would be able to deliver results. Their presence also served as a morale booster. Once the department was up and running for a year, they returned to the maintenance department.

15.8 Results

Over a period of two years, the new department progressively increased the proportion of minor project work that they handled to 80% of the total. A small group who were assigned to tank maintenance work did very well, and became quite enthusiastic. As there was a large backlog in this area, there was no problem finding work for them. Of those who had acquired new skills or accepted refresher training, several found better jobs elsewhere. Others accepted a voluntary severance package. The overall (equivalent) maintenance manpower number decreased by 50% within three years.

During this period, the maintenance work output remained quite steady, with backlog well under control. Overtime work fell slightly, and reliability improved.

15.9 Lessons

Too much manpower is always a significant problem. So it is right to avoid that problem by having barely enough own staff (but with enough competence) to do the required work. Additional resources can be contracted to cover peak labor or skill requirements.

We should keep the manpower situation under continuous review. This should cover not only labor, but also supervisory and managerial grades. It is much easier to make gradual reductions rather than to make step changes.

The MD demonstrated that he was not merely a manager, but also a leader. By recognizing the local cultural and social scenario, he adapted the implementation process, thereby making it more acceptable. This alone would not have sufficed in the prevailing situation. He also saw a potential win-win possibility and capitalized on it effectively. He handled the implementation challenges thoughtfully and was sensitive to social issues.

15.10 Principles

Many companies state publicly that people are their greatest assets. Under pressure, they often liquidate these assets, without exploring alternative solutions. Loss of employment obviously affects the person concerned, both materially and emotionally. Those who are fortunate to retain their jobs may also be adversely affected. In some situations, imaginative solutions can be found, but it requires somebody who actively seeks them.

Reference

i. Parkinson, C. Northcote. 2002. *Parkinson's Law: The Pursuit of Progress.* Penguin. ISBN-13: 978-0141186856

Appendix 15-A

Author: Jim Wardhaugh

Managing Overstaffing—Different Approaches from Around the World

15-A.1 General

It is amazing how common is the problem of overstaffing. The problem has been seen and managed in different ways in different locations on every continent. Almost always the situation has arisen because of lack of managerial foresight (or courage). Eventually, however, action is forced on the location. Some of the different ways this has been handled are shown below. In this case, we are considering the removal of a significant number of the workforce in a step-change. Slow attrition is hardly ever a problem.

15-A.2 Reduction Strategies

Voluntary Redundancy

Voluntary redundancy involves nothing more than a straightforward call for volunteers to leave. They get a severance payment and maybe a pension. Generally this is more attractive to the older personnel. It is easy to administer.

However, it can be costly and there is no control over who goes. Quite often the best will leave as they are the ones who can more readily get new jobs. Often the severance payments are more attractive to the older, more experienced people. They will leave and a drain on experience results.

Selective Voluntary Severance

Selective voluntary severance is similar to voluntary redundancy, but only those selected by the company can apply for severance. The advantage is that the company can choose who they would like to lose.

Often this flies in the face of union fairness rules, so it can be difficult to negotiate.

Another issue is that this approach is seen as rewarding the incompetent and disloyal, who walk out with barrow loads of money while the loyal and competent stay behind, working ever harder to make up for those who have left.

Straight Redundancy

Straight redundancy puts the company in total control of all the factors, but it is rarely used because of the amount of pain caused. It is only really applicable if the operation is closing completely.

15-A.3 Pain Reduction Strategies

"If the thing has to be done, it is best if it is done quickly." So where do these people go (and what do they do)?

Additional Plant

Opportunity can be taken of the building and putting into service of a new plant. This scenario can be explained as a temporary job on construction, safety, quality, etc., followed by separation. Alternatively, there could be a job in the permanent organization.

Transfer to Site Contractors

Company staff of the right caliber are often attractive to contractors who service the refineries. These contractors rarely train apprentices these days; they just "poach" competent personnel from other companies.

Set Up Own Contractor Company

If some competent people are leaving, it might be opportune to retain access to them by encouraging them to set up a firm which provides services to the company.

Company resources could give guidance on setting up the contracting firm, help with bridging finance, and provide or facilitate some training, and even arrange some temporary finance or loan guarantees.

The company could provide guaranteed work at a reasonable rate of return to the new firm for a buffer period before facing the cold wind of competition. Two years might be a reasonable period.

Construction Group

The affected people can be formed into a construction group which carries out small works such as plant changes. Good supervision and organization is essential. This approach is rarely seen as attractive for the following reasons:

- The group needs to set up separate facilities and buy or hire tools and vehicles.
- Productivity is often poor compared with contractors.
- Without tendering, there is no pressure to be efficient; but if they tender, they may not get the work.
- Unions tend to claim this is as normal work and make the group permanent.

Design Office

Competent, well-educated technicians, fitters, electricians, and instrument mechanics can readily be absorbed into the design office. Draftsmen once needed a long apprenticeship during which they learned drawing board skills over a lengthy learning curve. Computer Aided Drafting (Design) has largely done away with this; now, technicians who are relatively computer literate can be productive within weeks.

Transfer to Other Groups in the Company

Too often, possibilities are not fully explored in an open-minded way. Personnel from the maintenance force have background knowledge and skills which would stand them in good stead in activities such as:

- Firefighting
- Operating
- Safety
- Clerical work

Train for Different Work Outside the Company

Many of the surplus workers have transportable skills which can be used in many other fields outside the company. The company can provide job-seeking facilities and sponsor short courses to enhance any skills which would make them more attractive in the job market.

Chapter **16**

Retraining Surplus Staff

Should not shepherds take care of the flock?

Prophet Ezekiel.

Author: V. Narayan

Location: 2.1.3 Petroleum Refinery

16.1 Background

Our company sold bitumen in steel barrels, not in tankers, as most of the users in the country did not have suitable facilities to receive bulk deliveries. We fabricated barrels in a captive barrel-fabrication factory. The steel mills supplied us sheets in the form of coils. A contractor operated a de-coiling plant adjacent to the barrel factory, and sheared the sheets into blanks, suitable either for rolling into shells or punching and forming into end covers.

One of my responsibilities was to manage this facility. We made 10,000 barrels a day in three shifts. Each shift had a supervisor with an experienced superintendent in charge, who was my direct report.

Just prior to my taking over this responsibility, the company had reorganized the Utilities department, merging its operations with that of the Fluid Catalytic Cracking department. The Utilities shift supervisors were re-assigned and one of them joined the barrel factory. He was in his early 50s, a lifelong boiler operator with no knowledge of fabricating barrels. A modest and humble person, he was known to be a competent boiler operator. The company had not provided him any special training prior to his re-assignment, and he was expected to learn on the job. I had to make do with this situation, which I inherited.

16.2 Outline of Barrel Production Process

We had two identical production lines for the barrel shells and assembly. First we rolled the sheets into cylindrical shells and welded the seam in a seam-welding machine. We then provided corrugations on the shells to give them stiffness. Next we bent the ends and flanged them in a flanging machine. We punched out and formed the circular top and bottom covers separately, using heavy duty presses. The end covers also had flanges to match those on the shells. In the assembly line, we fitted the end covers on the shells. The next step was to bend the flanges together in a seaming machine. Finally, we tested the barrels in a testing machine. For this, we used air at 1" water pressure and soap solution. The operator filled the barrel with air and brushed on the soap solution on the welds and folded seams. Leaks were unacceptable, and leaking barrels (called leakers) had to be repaired by gas welding.

16.3 Quality Problems

On a normal shift, we experienced 10–15 leakers, or less than 0.5%. These were often due to sheet thickness variations, machine setting errors, etc. Prompt action could keep the leaker rates quite low. But production workers and supervisors had to be vigilant and quick to react when a set of leakers was detected. Without a clear understanding of the process and the machinery, the ex-Utilities supervisor was unable to manage the production line in his shift. As a result, sometimes there were leak rates as high as 3%. This meant that in a single shift, 150 or more empty barrels had to be stored inside the production area till they were repaired. Extra welders had to be brought in, sometimes at the cost of delaying maintenance work in the refinery. On occasion, there was no moving space left inside the barrel factory.

16.4 An Unfortunate Set of Circumstances

Within a month of taking on this responsibility, I went away on a planned vacation for two weeks. During this period, at the end of a particularly bad shift, the refinery manager happened to visit the barrel factory. He was not at all pleased at what he saw; a factory overflowing with leakers. As a result of this incident, the supervisor was formally warned about his performance. While prevailing social legislation made it difficult to fire blue collar workers, supervisory staff had less protection. Even so, he could not ordinarily have been fired. Instead, taking advantage of his timid nature, the company applied pressure on him to resign his position. I came back from vacation to find him serving out his notice period.

16.5 The Challenge

I objected strongly to this treatment from the company and requested he be allowed to stay on, provided I could train him to do useful work. I offered to mentor and rehabilitate him in three months. If this failed, I agreed to withdraw my objection. This request was granted, so I had to justify my brash action by proving my claim.

I had a plan in mind. We had a serious backlog of work on atmospheric storage tank maintenance. They were on a 15-year cycle of cleaning and repair, and many were overdue. Contractors did the actual physical work, but our supervisor managed the program. Work in enclosed spaces like tanks can be quite hazardous, so the quality of supervision matters. Each tank could be out of service for several months at a time. Scheduling their outage was a tricky affair, as we had to ensure that crude and product handling capacity shortages did not create loss of production. Currently we had one supervisor dedicated to this work. My plan was to train a second supervisor to handle the backlog, and I had a candidate.

16.6 Tank Construction and Repair Information

Most tank bottom plates are only 1/4″ or 5/16″ thick, except where they meet the shell, or where drains or other appurtenances are attached. These plates are laid out in a linear arrangement and fillet welded. The outermost set of bottom plates, called annulars, are thicker, 3/8″ or more. These are laid around the circumference, where the floor meets the tank shell. The welding detail at the point where two annulars meet the bottom plate is somewhat complex. The entire weight of the heavy shell is borne by the annulars and transferred to the supporting foundation pad. These pads are made of compacted earth with a bituminous tarmac-like surface. Tank shells have relatively thin plates at the top, sometimes just 1/4″ thick. The hoop stresses are higher at the bottom, so they need thicker plates. Hence lower strakes are thicker, to suit the increasing liquid pressure.

Tank shell plates are prone to corrosion damage on the product side and, if the exterior painting is in poor condition, on the outside as well. Rain water tends to collect under the annular floor plates if the tank foundation pad is damaged by weathering; therefore, we sometimes find underside corrosion damage of the annulars.

Floating roof tanks have seals to minimize leakage of product vapors to the atmosphere. These seals are made of thin and suitably formed (curved) sheet steel, in sections, each covering a part of the circumference about 20′ long. The seals are pressed against the shell using a pantograph arrangement loaded by weights or, in some designs, by springs. This allows each section to follow the shape of the shell, which could be slightly oval or uneven. There may also be slight differences in diameter from strake to strake. At the top of the floating roof, the moving sheet forming the seal is connected to the fixed circumferential edge of the floating roof with a rubber sheet seal. This allows vapor sealing in spite of the radial movement of the steel sheet seals. There are many pins and bearings in the pantographs which wear out with use and need replacement. The thin-sheet steel seals wear out and corrode, so they too need replacement.

16.7 Training Program

The company had an excellent set of procedures relating to tank cleaning and maintenance, complete with detailed drawings. These described various processes one could use for cleaning tanks, the equipment required, safety practices to follow, and other relevant information. Similarly there were procedures for repair work on fixed roof and floating roof tanks. These procedures were bound in two volumes, each about 11/2″ thick.

For the first two weeks I asked my new trainee, the supervisor, to read about 20–30 marked pages every day. I would spend 30–45 minutes quizzing him every afternoon. He was a dedicated student, and soon knew the procedures quite well. After that initial period, he accompanied me whenever I visited a tank cleaning or repair site. Initially he just listened while I spoke to our site supervisor or contract workers. Meanwhile, he continued to read and answer my quizzes for two more weeks.

In the second month, he started assisting the first supervisor in tank cleaning activities. By the middle of the second month, he started assisting in supervising repair work as well. By the third month, he took independent responsibility for one tank and did fairly well. Throughout, I was with him for at least 30 minutes every day. At the end of three months, I was satisfied he was competent enough to work independently.

16.8 Results

The company noted his performance, attitude, and sincerity. Other supervisors assisted him as they did not want their colleague fired. He learned rapidly and soon became the in-house technical expert on tank maintenance. I was allowed to retain him as a tank maintenance supervisor. The planning supervisor taught him project planning and management. His personality developed and he was able to assert himself when required. He learned to work with contractors and handle commercial issues within a few more months and became a very useful member of the team.

16.9 Lessons

1. It is easy to blame somebody for poor performance, even when it is not his fault in any way. No effort had been put in to do a gap analysis and training needs before this person was reassigned, so the fault lay with the company.

2. People can pick up new skills if they are given the training and an opportunity to shine.

3. It is very important to look after people. The outcome can help the person involved and be very satisfying to the mentor.

16.10 Principles

Companies have to remain lean and healthy. They cannot carry non-performing staff as baggage. But manpower reductions in high unemployment economies can cause a lot of pain, not just to the persons involved and their families, but also to those fortunate enough to retain their jobs. Sometimes, an opportunity to re-deploy people presents itself, but their line supervisors have to make an effort to seek this out. The rest will fall in place, once the will is there.

LEADERSHIP

PART 4: PLAN

Chapter 17

Integrated Planning

Teamwork is the ability to work together toward a common vision, the ability to direct individual accomplishments toward organizational objectives. It is the fuel that allows common people to attain uncommon results.

Andrew Carnegie, Industrialist.

Author: Mahen Das

Location: 2.4.1 Medium-Sized Semi-Complex Petroleum Refinery

17.1 Background

Prior to my arrival, the refinery already had a fairly advanced work planning process in place. For major projects and plant shutdowns (called turnarounds in North America), they used Critical Path or Network Planning. Commercially-available software with resource leveling capability was used for this purpose. Once project execution began, however, there was little or no progress monitoring and updating of the plan. The critical path charts remained as decorations on the wall.

Preparation for a shutdown generally meant pulling out the last worklist, adding the current wishes of the operating and inspection departments, estimating the contract work volume, and converting the list into a critical path plan. Operators added the operational tasks of shutting down, gas-freeing, etc., at the front end of the plan and, separately, starting up the plants at the back end. They also added the requirements of Technologists to their plan. Project engineers made their own separate mini-plans for new construction work and appended them parallel to the main plan. There was little coordination in the preparation activities between these departments.

With poor overall coordination and lacking a milestone chart, the time available for proper award of contract work was invariably inadequate. And contract work was definitely required. The lack of time meant that proper competitive bidding was not possible and the price was invariably higher than necessary. Local contractors maintained a skeleton work force of skilled craftsmen. Temporarily, they hired whoever was willing to work during large projects, such as shutdowns, without much regard to skills. Contractors and their personnel were viewed with suspicion by the refinery and always kept at arm's length.

Thus the shutdown "plan" was a collection of at least three diverse plans,

1. Activities of all the maintenance disciplines
2. Operational and technological activities
3. New project activities

These three plans ran in parallel, but were disjointed; their interfaces were managed poorly.

During execution of a shutdown, the maintenance engineer was supposed to be the coordinator. People in the other participating departments did not accept this because top management never announced it formally. As a result, execution looked like one done by a mixture of football teams on one side rather than one united team.

17.2 Revised Model

In Chapter 5, Changing Paradigms, we discussed revised work models which incorporated features based on an agreed maintenance philosophy. The new shutdown work model with these features is illustrated in Figure 17.1.

At the kick-off meeting, we set the premises for the shutdown. Thereafter, we named the leader and key members of the team, who then began the compilation of the preliminary work list. Contributions to this list came from:

- Last shutdown inspection report
- Latest inspection findings
- Operators' and technologists' fault lists
- All maintenance disciplines' fault lists
- Projects' requirements
- Various wish lists

A cross-functional team carried out a rigorous risk-based challenge of this preliminary list. Team members included:

- The shutdown leader
- The planner
- The inspector
- The member whose work contribution was being challenged

This process ensured that the final integrated list contained only that work which was in line with the premise and justified from a business perspective. At the end of this stage, we froze the work list and began the process of planning and contracting. The purpose of freezing the work list was to avoid costly disruptions to the planning and contracting process.

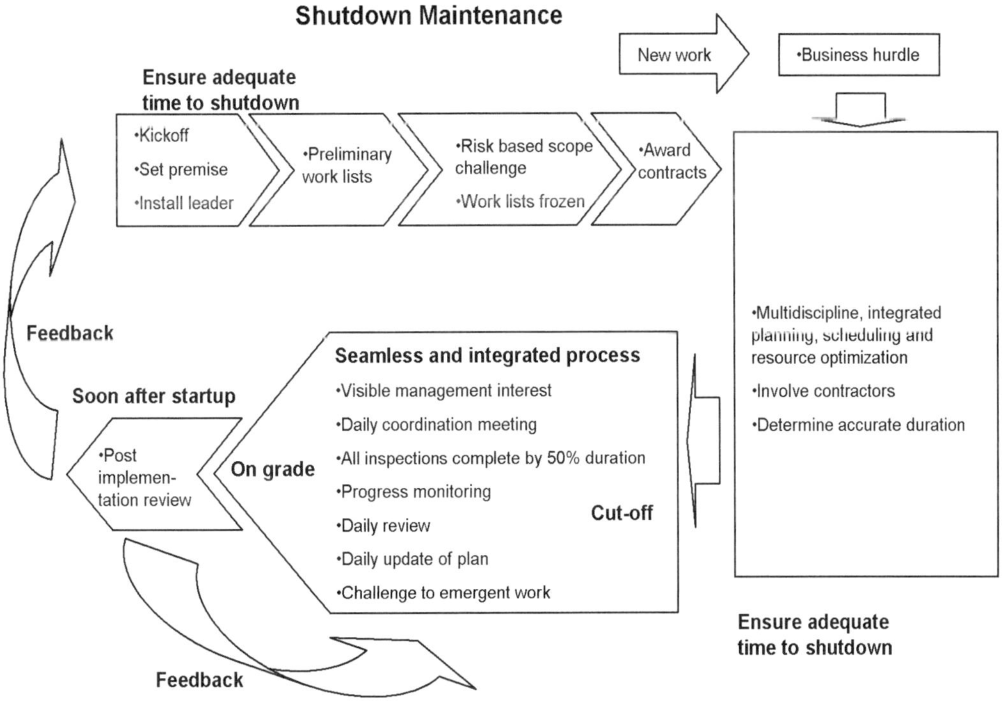

Figure 17.1 ShutdownMaintenance Process

Any additional work identified after this point in time had to cross a tough business hurdle set by the management team. For example, a plant change proposal by technologists would need to have a payback of at most 2 years and bring at least $500k benefit to the business in the first year. Also, if the execution caused an extension of the shutdown duration, the business cost of this delay would have to be added to the cost of the change before computing the economic justification.

17.3 Planning Stage

The integrated planning process then began using this final work list. It was carried out by a cross-functional team. The planner led, helped by the discipline engineer whose activity was being planned.

What we mean by integrated planning and scheduling is that all tasks, irrespective of the department executing them, are planned and scheduled in their logical sequence and with their logical interlinks in a single network. For example, the front-end operational activities in the network of the shutdown of a Crude Distillation Unit could be:

1. Switch intake from crude oil to gas oil
2. Divert products to slops tank
3. Trip to close all fuel to furnaces

4. Insert 3″, 150# spade in fuel oil line
5. Insert 6″, 150# spade in fuel gas line
6. Circulate gas oil until furnace outlet temperature drops to 95°C
7. Remove following spades:
   a. 6″ naphtha to slops
   b. 6″ kerosene to slops
   c. 8″ light gas oil to slops
   d. Etc.
8. Drain following accumulators
   a. Naphtha
   b. Kerosene
   c. Light Gas Oil
   d. Etc.
9. Drain following heat exchangers from both sides:
   a. Naphtha/Crude Oil
   b. Kerosene/Crude Oil
   c. Gas Oil/Crude Oil
   d. Etc.
10. Make following temporary connections:
    a. 1″ Steam to primary column bottom
    b. 1″ Steam to Gas Oil Splitter bottom
    c. 1″ Steam to Furnace Coil
    d. 2″ water to flushing system bottom
    e. Etc.
11. Steam out as follows:
    a. Primary Column for 10 hours or until hydrocarbon-free
    b. Furnace coils for 10 hours or until hydrocarbon-free
    c. Etc.
12. Bring in crane, remove the following Safety Relief Valves (SRV), send to workshop, and carry out gas test at nozzle outlet every 2 hours:
    a. SRV 001 Primary Column Top
    b. SRV 004 Stabilizer Column Top
    c. Etc.
13. Check flushing water samples for hydrocarbons every 2 hours at following points:
    a. Naphtha/Crude Oil heat exchanger shell and tube sides
    b. Gas oil splitter bottom
    c. Etc.
14. Open the following manway covers when hydrocarbon testing is negative:
    a. Primary Column, all
    b. Gas oil splitter top and bottom
    c. Etc.

15. Carry out gas test and, if negative, issue entry permit for the following:
    a. Primary Column
    b. Etc.
16. Stop steam to plant, reverse 8″, 300# spectacle blind at battery limit
17. Stop cooling water to plant, insert 24″, 150# spectacle blind at battery limit

Similarly, for a new project, the activities in the integrated network could read as follows:

1. Build scaffolding under the 8″ furnace inlet line (prohibited during operation)
2. Bring in crane and string up prefabricated pipe sections (prohibited during operation); Note: try to combine crane work with safety valve removal
3. Remove block valve from 6″ tie-in stub hot tapped during operation
4. Complete field welds
5. De-spade
6. Make electrical connection to the new pump
7. Make water flushing connection
8. Water flush new system with the pump using fine suction strainer
9. Remove water flushing connection
10. Box-up system and hand over for commissioning
11. Etc.

All other disciplines integrated their activities into the network in a similar way.

17.4 Scheduling Stage

Scheduling was carried out together with the operations day assistant and the relevant discipline engineer. Time and resource estimation was carried out by the scheduler together with the relevant discipline engineer. Resources were leveled for the total integrated plan. This optimized the resources for maximum overall benefit. Contractors participated fully wherever they are involved.

17.5 Results

The final critical path was 24 hours shorter than was earlier provided for in the production plan. The actual completion beat this plan by another 12 hours. The actual direct cost was 10% lower than the budget, which already had been 10% lower than the previous comparable shutdown.

17.6 Lessons

During the preparation and execution of a multi-discipline project, it is necessary to have one overall leader; this results in maximum efficiency.

Rigorously challenging shutdown work lists to confirm they are in harmony with the premise, and then freezing them, are essential control steps. Any additional work thereafter has to meet stricter acceptance criteria.

Using a team to plan and schedule adds different perspectives and improves the quality of the process.

Integration of the network plan results in overall optimization. In a major project, this approach delivers significant benefits.

Keeping the same project team from concept to completion enhances ownership.

Turnarounds and shutdowns are projects with compressed time scales. Some specifics, providing a recipe for success, are given in Chapter 19, Shutdown Maintenance.

17.7 Principles

A framework is necessary to enable people to perform well. Good leaders set clear objectives, install and empower teams, and enable work flow by ensuring discipline in the way work is generated and controlled. They communicate these steps to the relevant people in a timely manner. Once this framework is in place, people at the working level will be able to produce good results.

The framework alone is not enough. We need in addition, competent and motivated people. Training, mentoring, and experience help develop competence. People who enjoy their work are usually well motivated, so every effort must be made to make their work interesting and challenging. We can then expect good results consistently.

Chapter 18

Critical Path Planning Capability

For the man who knows not what harbor he sails,
no wind is the right wind.
Lucius Annaeus Seneca, Roman dramatist, philosopher, politician

Author: Mahen Das

Location: 2.2.5 Oil Company Operating a Group of Petroleum Refineries

18.1 Background

In my capacity as a Maintenance and Reliability Consultant from the head office, I visited the refineries and regional offices of this company a few times. The aim of these visits was to help them make improvements in maintenance and reliability of assets.

At the time of these events, although the region was technically advanced, the companies in this group had little computerization for maintenance management. They also did not have sufficient in-house capability for estimating and planning maintenance work. Hence they contracted out all such work fully, including the planning and execution of shutdowns. The company only provided overview supervision at a high level.

Project managers all over the world knew the benefits of Critical Path Planning (CPP), also called Network Planning, and this company was no exception. Their shutdown contracts required CPPs to be prepared and applied. Strangely, the contractors were not familiar with commercially-available software for this purpose, so they would produce CPPs manually to meet contractual requirements.

18.2 Critical Path Planning

One of the major benefits of critical path planning is that if progress is monitored periodically, typically once a day, we can get a regular update of the plan. This way, we can get the latest estimated completion date and associated critical paths, thus enabling suitable measures to be taken to realize the full benefit of an earlier completion or mitigate the consequences of a later completion. It is not practical to manually re-process all the data involved in such an exercise. The solution is, of course, to computerize the plan.

Therefore, the plans produced by the contractors manually were worthless. They served little more than being decorations on the walls of the shutdown headquarters on the first day of the shutdown. After that, they became outdated even for that purpose. Although the companies paid the contractors for preparing the CPPs, the shutdowns were in fact executed with the help of simple Gantt charts. They thus lost the most important benefit from the CPP, that of updating the plan regularly.

Only a couple of junior engineers in the contractors' organizations understood how to use a CPP. Neither the client nor contractors' organizations realized this situation at the level of the decision makers. The client was satisfied that his shutdowns were being planned in the state-of-the-art manner. The contractor was convinced he met his contractual obligation adequately. Neither of them realized that the client was not getting the benefit for a technique he was paying for, because nether party knew about the real benefits.

Understanding all this took a considerable amount of discreet probing through discussions with people in the clients' and contractors' organizations. In the local culture, people did not appreciate "loss of face." This made it difficult to communicate these issues to senior management. Corrective action could only commence once they knew the economic consequences of their current practices.

18.3 Treading on Eggshells

I organized an impromptu presentation on "Appreciating Benefits of Critical Path Planning," and made sure that the movers and shakers in the operations and engineering organizations attended it. We had a very lively question-and-answer session after the presentation, lasting an hour. I got a clear signal that all attendees had grasped the essence of CPP and its benefits. They had also grasped the lack of substance in their shutdown contracts. It dawned upon them that they had been losing the return on their investment in CPP.

The difficult question was how to redeem the situation. We decided to train a number of staff in computerized CPP, from their own organization as well as from those of the contractors. I offered assistance in organizing such training.

The idea of combining their own staff with contractors took some more selling. Eventually, the proposal was accepted. It was agreed that a selected number of staff and contractor personnel would be trained in time for the forthcoming series of shutdowns. They finally identified a group of 12 people; 6 of them were their own staff and 6 were contractors' personnel.

18.4 Training Course

At the head office, we put together a suitable training course. It consisted of a 3-day classroom course/workshop on the techniques. Thereafter, there was a fortnight long hands-on session in a group refinery which happened to be working on a real-life shutdown plan, using an off-the-shelf software package. This hands-on session included all aspects of CPP, namely, task breakdown, estimation of resources, logical sequencing, preparing the network, leveling of resources, optimizing resources, and the critical path by using what-if scenarios. As the actual shutdown of this refinery was some time away, the exercise included mock progress reviews in order to practice updating of the plan. At the end of three weeks, the participants felt confident that they could do this in their own situation.

18.5 Results

The results were impressive. We sent a seasoned planner to help them during the planning effort as a confidence-building measure. The plan was completed on time and put into practice. Once the shutdown commenced, the outside planner visited the site for a few days to review the actual progress monitoring and plan updating. For the first time in their history, the shutdown was executed in accordance with a live and dynamic plan with daily progress monitoring and plan updating. A senior manager commented that the ownership of the plan and, indeed, for the entire shutdown exercise was such as never witnessed before—both in their own staff as well as those of the contractors.

The shutdown was completed in record time, 25 days instead of the historical 30, and at a considerably lower cost than originally planned.

Critical Path Planning with daily progress monitoring and updating has now become an essential feature of the shutdown and major projects work of this group of companies.

18.6 Lessons

1. Computerized Critical Path Planning enables efficient execution of plant shutdowns and new projects. In order to draw the maximum benefit:
   - Actual progress must be reviewed periodically during the execution of the plan.
   - The plan must be updated on the basis of this periodic review.
2. There should be in-house capability for computerized CPP. Without this, the client will be fully dependent on the contractor. The result is that the full benefits of CPP cannot be realized.
3. When contractors and the company's own staff work as one team for a project, the feeling of ownership is greatly enhanced.

18.7 Principles

All investments have to be justified by their business benefits. Resources will be wasted by pursuing technology for its own sake rather than its business benefits. If this link is lost, performance will drop.

Individuals and companies can slide into using work processes and technologies without appreciating their true value. It is often easier for outsiders to recognize the situation, so external reviews are valuable.

Cultural issues are important, and reviewers have to be able to understand these before suggesting corrective actions.

Promoting teamwork between competing interests can result in a win-win situation.

Chapter 19

Shutdown Management

...Keep the money machine running.

Perfection is achieved, not when there is nothing more to add, but when there is nothing left to take away

Antoine de Saint-Exupery, French poet, writer.

Author: Jim Wardhaugh

Location: 2.2.3 New Medium-Sized Complex Refinery

19.1 Background

We aimed to make our brand-new refinery a top performer in every conceivable aspect. When we contemplated our first shutdown, we decided to learn from both poor performers and top performers those practices that brought poor performance or made them top performers.

I had been collecting good practices on shutdowns over many years, both while working in various types of plants and while visiting them on consulting visits from the technical headquarters. Many shutdowns of all sorts were done each year all over the world. Therefore, I found it surprising that there was no standard recipe which all these locations followed. All over the world, shutdowns were handled in completely different ways. Some of these ways brought success, others mediocrity, and still others brought abject failure in performance terms.

I think I have a good handle on what should bring success (or failure). But let us start by looking at what we are trying to achieve.

19.2 Shutdown Aims

Why do we have shutdowns? Ideally we would like a facility to operate for 20–30 years without interruption. We would like to design it, build it, run it, make product, sell product, and take the money to the bank. That is the sort of business model your bank manager can understand and love.

The reality is unfortunately quite different. Many factors reduce possible plant run lengths, among them:

- Loss of plant integrity or reliability
- Loss of production efficiency due to fouling, catalyst degradation etc.
- Need for modification to meet changing market demands

We do need to have shutdowns, but it would obviously be sensible to avoid unnecessary ones. Clearly our shutdown aims should be business driven. I suggest that, stated simply, the aims should be as follows:

- Maintain production performance into the future
- Promote high plant availability
- Reduce the probability of breakdowns between shutdowns
- Do the needed tie-ins for plant modifications

I like to consider the plant as a money machine. In that mental model, what we are trying to do, in financial terms, is to improve the profit stream now and into the future. So what brings success? I suggest:

- An entire series of effective business-driven decisions
- Each decision backed up by a solid business case
- A business-driven partnership among the facility's core departments (e.g., Production, Technology, Maintenance, Inspection)

Let us have a look at the relative performances of some different sites so we can get a feel for what is good performance.

19.3 Performance Numbers

In the first place, you need to be absolutely sure what you are talking about. Are you looking at an apple or is it a pear? Definitions or lack of clarity of definitions can cause a lot of confusion. Even a simple thing like shutdown duration is open to different interpretations. Conventionally, in benchmarking, shutdown duration is measured from the time the plant feed is cut to the time the plant is back in production with the product on specification. You will hear some engineers boast of achieving incredibly short durations, but these are usually pure engineering time, i.e., the time that the plant is in the engineer's hands. To this time you still need to add the time to shut down the plant, make it safe, and then start it up again to make product. These are very different numbers. Figure 19.1 shows the definition of plant shutdown duration and shutdown intervals. Always remember: we are in the business of making saleable product.

In Tables 19.1, 19.2, and 19.3, we compare the achieved performance numbers for poor and top performers (or pacesetters).

In these tables, the value shown for a top performer is the average of the top quartile of performance (if you rank performances from best to worst, the top quartile is the highest quarter of the group). A poor performer is the average of the fourth (and lowest) quartile. This implies that for the top performer there are sites that can be significantly better than the quartile average. Meanwhile, for the poor performer, there are sites that can be significantly worse than the bottom quartile average. What is clear from these tables is that there is a significant gap between the worst and the best.

If the economic drive for plant availability is so obvious we would expect to find both the worst and the best driving for availability (or uptime or whatever other name your particular facility calls it). Furthermore, we would expect

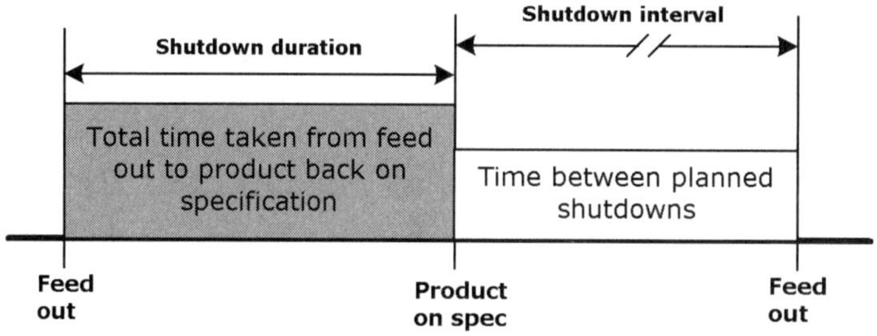

Figure 19.1 Shutdown Definitions

Table 19.1 Comparison of Shutdown intervals

Comparison Poor / Pacesetter Shutdown Intervals (years)		
Performance	**Poor**	**Pace setter**
Crude Distillation Unit CDU	3	4.5
Fluid Catalytic Cracker FCC	3	4.5
Hydro Cracker HC	2	4
Platformer	3	4.5

Table 19.2 Comparison of Shutdown Durations

Comparison Poor / Pacesetter Shutdown Durations (weeks)		
Performance	**Poor**	**Pace setter**
Crude Distillation Unit CDU	6	2
Fluid Catalytic Cracker FCC	12	4
Hydro Cracker HC	5	3.5
Platformer	5	3

Table 19.3 Comparison of Annualized Availability

Comparison Poor / Pacesetter Plant Annualized % Unavailability		
Performance	**Poor**	**Pace setter**
Crude Distillation Unit CDU	4-10	2
Fluid Catalytic Cracker FCC	5	3
Hydro Cracker HC	6	4
Platformer	6	4

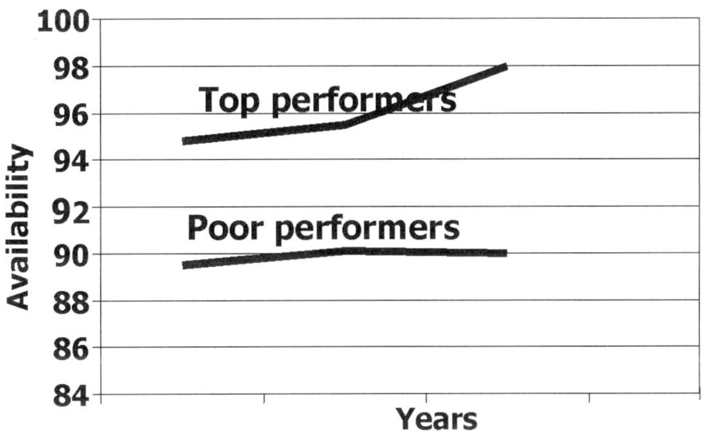

Figure 19.2 Availability Trends

them to be converging on some perceived optimum. This is not what we find when we compare performances. See Figure 19.2 Availability Trends, which shows that the gap is actually getting bigger.

19.4 Practices of Poor Performers

We won't spend too much time on these because I assume that you do not want to learn how to fail. Furthermore, if you do, I am sure you can conjure up many more ways of ensuring failure than I could! The four approaches we found, which seemed to lead inexorably to poor performance were as follows:

1. Having frequent shutdowns in obedience to some real or perceived regulatory regime, which insisted on pointless opening up of much of the equipment. The large scope, driven largely by an inspection

group acting as integrity policemen, is costly (see Chapter 6); it inevitably brings scope creep and lengthy shutdowns.
2. Habitual shutdowns in which Operations grabs the opportunity to have a look inside their plants and open up lots of equipment. Maintenance would see this as providing a move away from the boredom of the daily routine, some visibility and, with some luck, a moment of glory for themselves.
3. A universal opportunity used to hide and bury mistakes of all sorts (Operations, Maintenance, and Projects).
4. Occasionally, the shutdown is project driven, done to modify the plant. This is another opportunity to have some glory. It is hardly ever realized that if the downtime costs of lost production were taken into the project payback calculation, very few of these projects would be viable.

The above approaches are distinguished by bringing reduced plant availability, high costs, and little apparent business benefit. All too often, each shutdown is done in a traditional way with a firm belief that this way has stood the test of time and is the best. If benchmarking shows someone else to be better, you can always go into denial. See Chapter 6 for a list of excuses for poor performers. What we found in our studies was that the following aspects did not matter much when considering shutdown performance:

- Extent of use of contractors
- Extent of unionization
- Size of the facility
- Age of facility
- Geography

There were top and poor performers at each end of these spectra.

Trying to summarize the things that really did bring poor performance, we found that the following were the real factors:

- Too frequent shutdowns, driven by habit, conservatism, regulators
- Shutdowns which are too big because of too much unnecessary work (often because of over-conservative inspection regimes)
- Shutdowns which take too long to do because of:
  - Too much work
  - Overstretch of local labor competence pools
  - Poor planning, scheduling and organization of work
  - Shutdown and start-up problems

19.5 Practices of Top Performers

Top performers do not replicate themselves in every way, but they do all the important things in a consistently similar way. They treat their plant as a money machine. It is only going to make money when the plant is running well. Their prime aim is to attain high plant availability at the right price. Each proposal to shut down the plant or increase the frequency or duration of shutdowns is looked at from a hard-nosed business perspective.

This is their recipe.

Minimize the Need for Shutdowns

- They shut down only for good business reasons.
- No unthinking routines, no bad habits
- Flexibility of scope and timing is agreed with regulators
- They use unplanned stoppages as opportunities to do work.
- They pre-plan opportunity packages for fast activation
- Less tendency towards conservatism
- Reduced work scope of main shutdown
- They manage degradation.
- They buy decent equipment, configure it sensibly, operate and maintain it well
- Run the plant in a way that makes failure modes and rates predictable so inspection intervals can be confidently determined
- Eliminate failure causes by use of techniques like Root Cause Analysis
- Reliability is institutionalized (like safety)

Manage Regulators

- In the past, regulators often insisted on frequent shutdowns for internal inspections, which made decisions easy, but life expensive.
- Top performers demonstrate to the regulators that they manage a professional operation and gain control of all the factors.
- Decisions can then be business driven

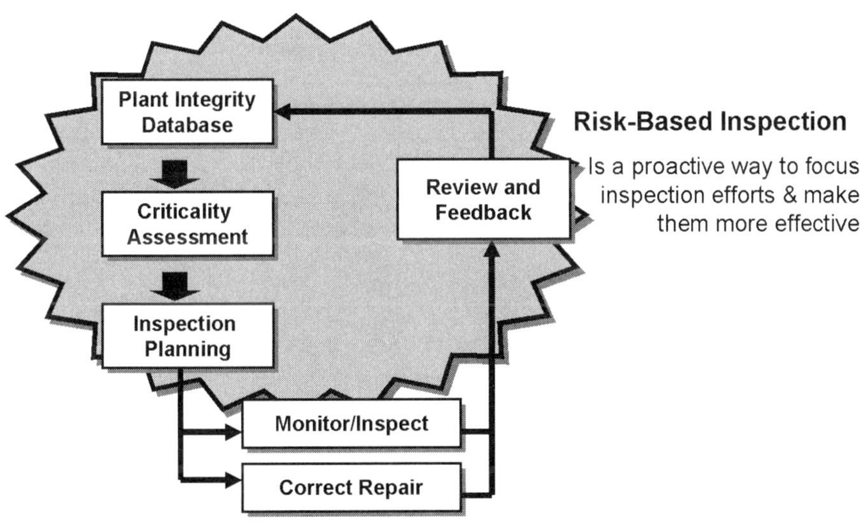

Figure 19.3 Risk-Based Inspection

Effective Inspection Regimes

- The role of the inspector has changed from policeman to business partner. See Chapter 6.
- Use of Risk-Based Inspection to bring cost-effective inspection plans. See Figure 19.3.
- Production operates within agreed parameters and there is a joint management of degradation.
- Grasp opportunity windows to do necessary inspection work; increase use of non-intrusive inspection.

Avoid Expensive Shutdown Work

- It was once considered sensible to do work on a shutdown, as work could be planned, and skills, cranes, etc., made available.
- Studies now show that doing work on a shutdown is nearly 50% more expensive than single-focused jobs done on the run.
- Productivity also degrades when manpower is more than 700.

Have Small Shutdowns

- Do most work on the run.
- Use non-invasive inspection techniques.
- Focus on the critical business aspects.
- Have clear shutdown objectives as an effective work-filter.

Months before shutdown						
	24	18	12	6	0	1
Ongoing Business plan	S/D concept	Preliminary preparation	Detail planning	Final preparation	Execute	Post S/D
○ Form steering groups ○ 10/5 year plan ○ Annual plan ○ Integrate with projects ○ Optimize timings ○ Capital and revenue budgets ○ Why, what, when and business expecta-tions	○ Form steering team ○ Establish S/D purpose & objectives ○ Make high level work scope ○ Establish one page milestone plan ○ Make 30% estimate & staffing ○ Decide organization structure ○ Initiate long delivery procure-ments	○ Establish organization ○ Define/ freeze 90% scope ○ Preliminary critical path ○ 20% estimate ○ Agree contract type ○ Start contract negotiations ○ Develop overall S/D plan ○ Establish detailed milestones: ○ QM plan	○ Challenge / Review risk ○ Establish S/D work package ○ Prepare S/D schedule ○ Make detail mini plans ○ Make 10% budget ○ Prepare management instructions ○ Establish HSE instructions ○ Order materials / services ○ Award contracts ○ Arrange manuals / drawings	○ Infrastruc-ture in place ○ Scaffolding ○ Get permits ○ Contractor induction ○ Contractor training ○ Allocate tasks ○ Mobilize supervision ○ Audit material supply ○ Final inspections confirm scope	○ Manage changes ○ Close out tasks ○ Report job status ○ Monitor progress / Update plan ○ Monitor costs ○ Write repair history ○ Identify non conformances/ Incidents ○ Commission plant equipment	○ Identify lessons learned ○ Adjust strategy & tactics ○ Adjust procedures & instruc-tions

Table 19.4 Business Processes of a Shutdown

These steps minimize shutdown complexity and competence demands. A small competent workforce brings:

- Quality work
- Equipment reliability
- Good communications
- Ownership and pride in workmanship

19.6 Effective Work Planning, Organization, and Scheduling

This area is characterized by:

- Clearly-stated objectives for the shutdown, focusing efforts appropriately
- Site departments' alignment on these objectives
- Agreed framework for action, see Table 19.4
- Effective and visible business-driven leadership
- Integrated planning (Production, Maintenance, Inspection, etc.) for shutting down, work processes, starting up, etc.
- Detailed plans for critical issues (heavy lifts, permits, productivity, competent, waste, etc.)
- Challenging, but realistic targets
- Effective learning-capture processes
- All aspects are critically reviewed shortly after the end of the shutdown
- Improvements are put in place

19.7 Results

For our first shutdown, we followed the recipe given above in detail and it paid dividends:

- No significant accidents
- Minimal rework
- Shutdown duration was a day less than planned, bringing increased sales of about US $400,000
- Cost was less than budget
- Start-up and next two year's run were trouble free

19.8 Lessons

- Have a shutdown only if it will improve the ability of the plant to make on-specification product or retain integrity now and into the future.
- Delay is usually a good thing, but do not run the plant into the ground. What matters is attained overall availability, not macho run lengths.
- Have a clear purpose and derived objectives along with visible man agement commitment to these.

- Have a small scope of work driven by business, use RBI, and modern practices.
- Use a small, competent workforce.
- Emphasize a cohesive team working an integrated plan.
- Do it better every time.

19.9 Principles

Production plants are machines for making money, so don't shut them down unless there is a good business reason. Recipes for success are available, but the will to succeed must be there in the first place.

Chapter **20**

Electrical Maintenance Strategies

...Doing unnecessary work won't stop failures

The greatest danger for most of us is not that we aim too high and miss, but that it is too low and we reach it

Michaelangelo.

Author: Jim Wardhaugh

Location: 2.3.4 Small Complex Refinery in Europe

20.1 Background

I had been working in the Projects and Technical groups in this refinery for a couple of years. My leader decided that I should move from there to take over the electrical maintenance section. He did not think that the fact that I knew nothing about maintenance was a problem. I remember him saying that, as we don't have significant problems there, I should just carry on doing what has been done in the past. Maybe do a bit of fine tuning?

I had three engineers, a host of supervisors, and a horde of electricians, the number of which would have made Genghis Khan swell with pride. I had no idea what these electricians did to fill their day. What was a bit worrying was that neither did the engineers and supervisors. Although I knew nothing about maintenance, the size of my electrician horde made me suspect that they were either busy doing unnecessary work or not busy at all. Eventually I found that it was a lot of both.

20.2 Maintenance Theory

I read a few books on maintenance. I must have chosen the wrong books as intellectual bite was totally missing. The message coming from these books was that, in the past, maintenance consisted of waiting until the equipment broke down and then fixing it.

The books argued that modern equipment was less robust than in the past, more complex, and needed more care. They advocated time-based preventive maintenance as their solution. Their thesis in the books was that we could prevent all failures. Failures were themselves a symptom of some shortcoming or failure in the maintenance systems. The books suggested as best practice targets, a profile of 90% time-based preventive work and 10% breakdown work. This thesis and the preventive maintenance approach had been adopted in the electrical maintenance group. In this chapter, we will show how ineffective this approach can be by looking at a couple of case studies.

20.3 Preventive Maintenance of Motor Starters

My initial questions about what my electrician horde did show that a team of a supervisor and six electricians spent every day of the working week checking the condition of 415-volt motor starters. Their goal was early identification of overheating due to loose connections. Any problems could then be rectified and failure prevented.

A significant planning and scheduling effort went into this activity, with a planner working almost full time on it. He spent his day arranging with production for the release of the equipment and making sure spares were available in case a problem was found. I think that the intention was for each starter to be checked every two years.

I questioned the planner and supervisor on these activities. They were unembarrassed when they revealed that over the last three years of checking they had:

- found one loose connection that could have become a fault
- found two issues which needed some non-urgent attention
- caused four faults by fiddling inside the starters; these had caused a loss of production.

They argued that this showed that our maintenance systems were robust and we had excellent strategies in place. I suggested to the supervisor and planner that perhaps the payback was not commensurate with the checking effort and asked whether it was sensible to continue. Such talk was clearly heresy and they looked stunned. Obviously I didn't understand maintenance. It took almost a week of debate to convince them that this effort should be stopped. The moment of truth came when I asked them if they would do this

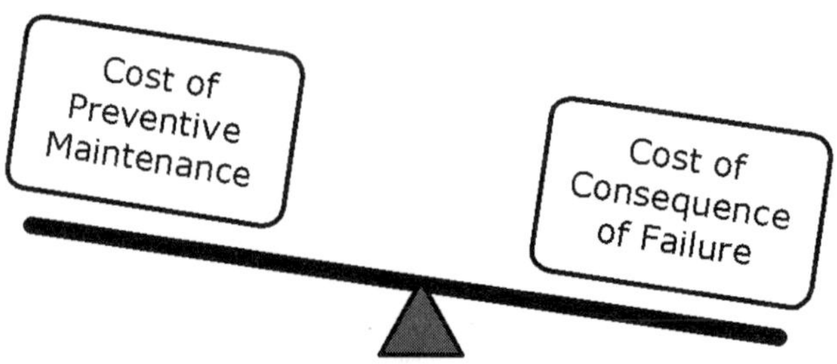

Figure 20.1 Cost-Benefit Balance

work if it were with their own money.

We stopped this activity and replaced it with a check of the starters at the three yearly shutdowns when the motors were not in use. Only rarely was anything ever found. However, as the check consisted only of opening a starter door and looking inside, we carried on; see Figure 20.1.

20.4 Preventive Maintenance of Batteries

Each of our electrical substations had lead-acid batteries to operate the high-voltage switchgear. There were also quite a number used as backups for instrument systems in control rooms, the telephone exchange, etc. The batteries were defined as "critical" to the business because failure could cause instrument or switchgear mal-operation. In line with the prevention of failure-at-all-costs concept, a large effort was being put in to ensure reliable operation.

There were about six electricians working on this, and four contractors doing basic work which the electricians felt was beneath them. My challenge to the approach on switchgear checks was still fresh in people's minds. Questioning what the battery people did to ensure business benefit brought paroxysms of anguish and accusations of near blasphemy. I knew then that I was on to something.

I was swamped with jargon and justifications. But cutting through all the technical verbiage, these workers filled their days topping up the batteries with distilled water, cleaning off acid deposits, and greasing battery terminals. This seemed to involve a weekly visit to every battery. There were occasional battery changes so that batteries were brought back to the shop for some mysterious re-energizing process which I didn't understand at all. But as I am a mere electrical engineer, what would I know!

I asked one of my young new engineers to investigate why the batteries needed topping up so frequently. He discovered that it was because the chargers were set at too high a charge rate. Therefore, they overcharged the batteries and thus boiled off the water. Overcharging had become the norm many years before, to ensure batteries did not go flat.

Setting the chargers correctly stopped the boiling off of electrolyte; it also stopped the need for a team doing regular weekly top ups. During normal duties, electricians visited most substations almost every day. During these visits, we asked electricians to keep a lookout for unusual sounds or smells. In addition, however, we set up a system for formal monthly substation visits when particular aspects were scrutinized against a check list. In essence, inspectors used their human senses to look for symptoms of overheating, etc. They checked batteries visually during these monthly visits.

In line with the company's purchasing specification, all battery chargers had low voltage alarms. As a back-up, we connected these alarms to the existing substation alarm panel at minimal cost. Any problems would then be alarmed in the control room and an electrician sent to investigate.

20.5 Modern Maintenance Theory

In retrospect, it was clear that the group I had taken over were trapped in a mind set time-warp. I will not discuss Reliability Centered Maintenance and all the other current three letter acronyms as I wish to emphasize the common sense approach. Modern maintenance management puts a significant focus on the cost effective elimination of failures and then balances the cost of failures against the cost of prevention and/or detection (See Figure 20.1). In this approach:

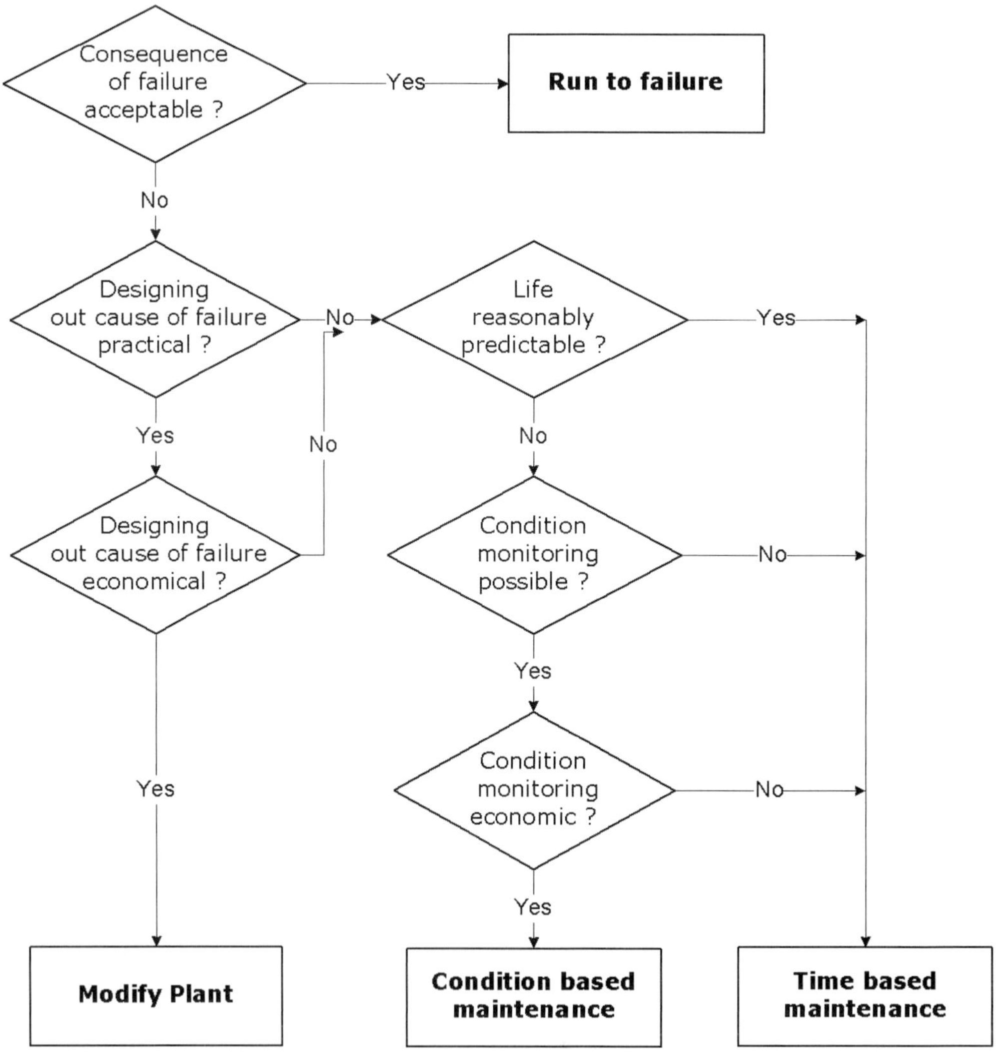

Figure 20.2 Maintenance algorithm

- Efforts are focused where the consequences of failure are high
- Repeat failures are investigated and the root cause eliminated where economic
- Run to failure is a respectable strategy for low consequence events, e.g., where the equipment is spared

Figure 20.2 summarizes the modern approach.

We often hear some gurus promoting time-based preventive and condition-based approaches to maintenance as best practice. They recommend we should put our emphasis in applying these practices. The reality is that in most top performers, what drives a decision to do work is the consequence of failure. If the failure consequences are low—i.e., there is little impact on production, safety, environment or the equipment itself—a top performer is unlikely to spend significant amounts of money on preventive or condition monitoring measures. In industries where there is a high level of sparing, this can lead to a work profile approaching 70% based on a run to failure approach and 30% based on a condition/ preventive approach.

20.6 Electrical Maintenance Strategy

Where does this take us in the electrical field? We'll leave electric motors to one side for the moment as we are going to look at these in Chapter 39. Most electrical equipment, e.g., switchgear, is an assembly of electrical, electronic, and mechanical components. So what should we do?

Most of the mechanical components of electrical equipment tend to suffer from lack of use rather than use. Many items of production switchgear, for example, would only be operated at plant shutdowns or when a production drive was changed over to a spare. The mechanical components, contacts, and mechanical relays are not subject to significant wear. Therefore, only occasional operation and lubrication would be needed to prevent them sticking. The exception to the above is with very high use switchgear, which should be viewed as a specific item with specific problems. The types of applications where you find these are gasoline pumps at filling stations, pumps on level control, inching drives on cranes, etc.

Electrical components like coils, electronic relays, rectifiers, transistors, and integrated circuits tend to fail randomly. They exhibit almost no wear-out. Very densely packed integrated circuits can exhibit wear-out if they are so densely packed that they are operating almost at the molecular level and subject to friction wear from electron movement. It is unlikely that this problem will cause you many sleepless nights.

When you have little wear out, and random failure is common, it is difficult to establish a frequency for inspection or condition monitoring. Quite often the bulk of the equipment is in a type of duty where the consequences of failure are low. Only minimal monitoring effort can then be justified. For example, a regime of occasional external visual inspections with an internal scrutiny with minimal strip down at shutdowns might be appropriate. If the fault could have significant consequences, some kind of testing is required. These tend to be

Risk Assessment Matrix

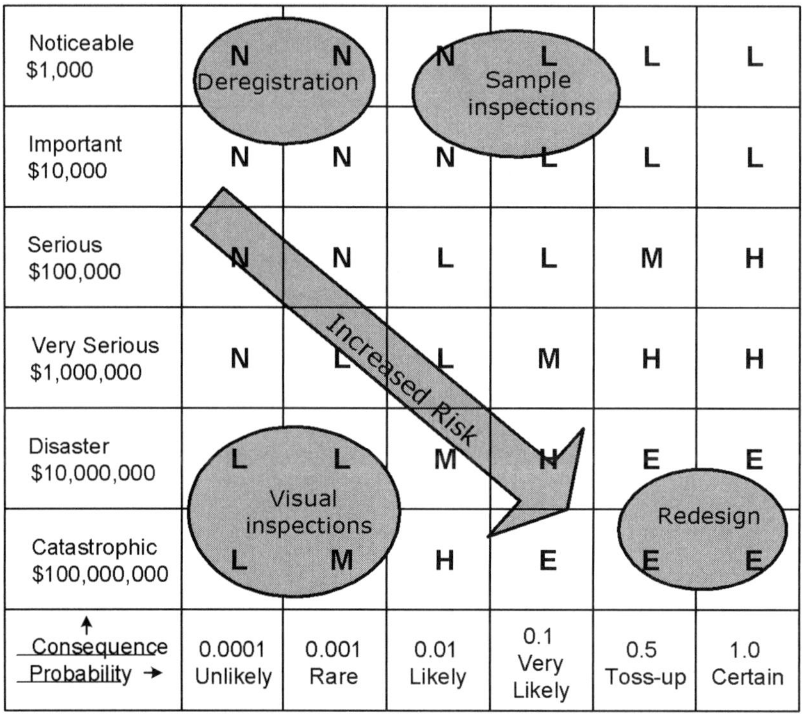

N = Negligible L = Low M = Medium H = High E = Extreme

Figure 20.3 Risk Matrix for Electrical Inspection

less common, so they are usually treated as special cases.

While it might be difficult to devise a credible inspection frequency, 'never' seems unacceptable. What is clear, however, is that at the other end of the spectrum, poor performers tend to inspect equipment excessively. They do a lot of preventive and condition monitoring activities, find little, but keep on doing it anyway.

To put some logic into the decision-making process, it has become popular to use a risk-based approach. API RP580 and API RP581 have a very commonly used approach for the inspection of pressurized systems. API RP580 specifically excludes electrical (and other) systems, but the approach is sound and valid for electrical equipment as well. A risk-based approach using a simple API type matrix, as shown in Figure 20.3, can focus the knowledge of a few capable individuals and prompt a cost effective and defensible regime. Some further aspects of the use of this type of matrix is given in Chapters 24 and 26.

The bubbles give guidance on the inspection approach to be considered, depending on where the system falls on the criticality matrix. Where failure is inconsequential, you might consider not registering the equipment at all. This is a "do nothing" strategy with a vengeance. Note the use of sampling tech-

Typical Inspection Intervals for Electrical Equipment		
Type of Equipment	Top Perfomer Interval	Average Interval
Non-hazardous Area	5 years	1 year
Motors/Plant Lights	4 years	2 years
LV Switchgear	4/5 years	2 years
HV Switchgear	4/5 years	2 years
LV Protection	4/5 years	2 years
HV Protection	4/5 years	2 years
Transformers	4 years	2 years
UPS Battery Systems	2 years	2 years
Earthing	4/5 years	1 year
Moveable Equipment	Continuous (user)	1 year
Portable Tools (Battery)	No inspections	3 months
Portable Tools (Mains)	Continuous (user)	3 months

Table 20.1 Typical Inspection Regimes for Electrical Equipment

niques for large inventories of similar equipment items. In the extreme, redesign may be necessary. Local rules and regulations may impose more onerous conditions, but these should always be challenged in an appropriate way in the most appropriate forums.

It is possible to go for a more detailed and a semi-quantitative approach based on the hardware protection strategies used in IEC61508 and 61511, but these are labor intensive and require some expertise. I would suggest that these are kept for the clearly special occasions or where you want to verify the answer from your simplified approach.

Top performers tend to have a run-to-failure strategy for most equipment. That is the default, but it would be backed up by an inspection regime, as shown in Table 20.1. Top performers would also consider sampling techniques for a large number of equipment items. A limited amount of testing would be done if safety considerations required it. However, it should be noted that most electrical protection systems are backed up and duplicated in some way.

20.7 Electrical Inspection Strategy—Our Choices

We decided to adopt the top performer approach although prophesies of doom came from the old hands. Somehow, balancing risk and cost had never been part of their mental model.

We believed that many potential problems could be avoided by the simplest of checks. People visit or pass by much of the equipment frequently in the course of their normal work. We encouraged them to use all their senses to determine if there were any problems. For example, we asked them to pay attention to the equipment's normal operating characteristics, e.g., load, sound, temperature, smell, vibration. They recorded any unusual conditions in the CMMS and thus triggered action. To achieve the ownership necessary for this, we allocated electricians to specific geographic areas of the plant and made them responsible for condition. This was the essence of our inspection regime.

We backed this up with a formally-scheduled inspection and test effort at the intervals shown in Table 20.1. Generally we scheduled an external visual inspection by a competent person using his senses. We arranged these whenever access was available, e.g., during plant shutdowns. He would look for damage, accumulations of dirt, disconnected earth, unauthorized modifications, etc. Our experience of the switchgear preventive maintenance had brought us to believe that opening up equipment only brought problems. Testing was only done where there was a real justification, and then often on a sampling basis. If problems were found, we solved them. If no problems were found, we extended the interval between inspections.

An issue which generated considerable heat was the routine inspection of portable equipment. Note that here we are talking about the small stuff, not the big moveable equipment like three-phase welding machines. We had been routinely tagging each piece of portable equipment, calling it into the electrical shop from the site, carrying out a formal inspection, labeling it tested, and sending it back to the users. When we thought about this, the regime seemed increasingly absurd. Think about this. Modern portable equipment is either:

- battery-driven (at a voltage which is inherently unable to cause danger to personnel)
- double-insulated (plastic with no earth), so again, it is inherently safe unless mechanical damage exposes the conductors

In both these cases, a competent user can easily identify that a potential problem exists and initiate maintenance action. A formal inspection by an electrician adds little value when we have frequent scrutiny by the user. Sadly, all too often logic flies out the door and regulators unthinkingly write this nonsense into rules.

We could choose self inspection as this work was not covered by legislation.

20.8 Excess Staff

By now you are probably wondering how many surplus men this threw up in total and what happened to them. The answer is that, by asking logical questions and balancing effort against business benefit, we identified that some 40% of the maintenance work-force was employed in unnecessary jobs. We had to handle this sensitively because we were aware of the explosive nature of this information. Over a period of three years we redeployed about 40% of the electricians as follows:

- Opportunities at a new plant being built
- Commissioning work for the site project department
- Replacement of contractors doing work inside the department
- Normal retirements
- Early retirements
- Draftsmen positions
- Left to go to college full time

20.9 Lessons

- Just because something has been done for twenty years doesn't make it correct.
- We do not need to search for self-revealing failures, which are random and infrequent.
- Spending effort on condition monitoring which would cost more than the failure consequences is poor business economics.
- If you need workers for a new plant, examine your existing workforce to check for spare capacity before considering recruiting additional staff.

20.10 Principles

The maintenance strategy we develop must balance effort and business benefits.

Common sense questioning can eliminate a large amount of unnecessary work. Do not be driven by fashion or the latest three-letter acronym.

Chapter 21

Minor Maintenance by Operators

...Do operators have useable spare time?

Author: Jim Wardhaugh
Locations: 2.3.4 Small Complex Refinery in Europe
2.2.2 Large Complex Refinery in Asia

21.1 Background

There is a widely-held belief that production shifts are inherently overstaffed. They contain spare people to cope with no-shows due to sickness or to cope with emergencies, which often need more people than steady-state operation. We know that operators also have skills that would enable an enlargement of their role. By utilizing their skills and spare time, we could increase operators' job satisfaction and improve business performance—a true win-win situation.

If the reader wants stories of triumphant implementations, stop reading now. But if you can accept a story of a failure and another story which might be defined as a partial success, please read on.

Both companies described here are part of a large multi-national group of companies, with corporate headquarters in Europe.

21.2 Introduction to Minor Maintenance by Operators (MMBO)

Several companies within this multi-national group had implemented Minor Maintenance by Operators (MMBO) with variable levels of success. New plants had successfully recruited operators with proven technical skills to become operator-maintainers (see Chapters 13 and 31 for related information). In some plants, MMBO had been introduced successfully; this had brought significant benefits. However, in the older, often unionized locations, the implementation of MMBO has been more difficult with poor results.

21.3 Benefits of MMBO

The benefits of MMBO are considerable. Independent benchmarking studies show that top performers with relatively slimly-staffed shift work forces can get each operator to do up to 1.5 hours of maintenance-related activities during an 8-hour shift. (See also Chapter 31; in that study, they identified 25% of an operator's time being available for maintenance). These man-hour inputs can be directly offset against maintenance resources, thus reducing the size of the maintenance work-force. In addition maintenance becomes more effective because of earlier interventions and faster repairs avoiding escalation of problems.

An example of this is where operators notice a small steam leak from a flange. In a traditional organization, operators would put in a job request to maintenance. Depending on the time of occurrence, it could take 2–12 hours before maintenance gets there. By that time, the flange gasket might have been destroyed by the steam flow. If the response time is longer, e.g., over a weekend, the flange gasket face itself might have been damaged. The job has now become large and might require a shutdown of the steam system. In the operator-maintainer situation, operators would pre-empt these big problems. They would get out their spanner, tighten up the flange, and stop the leak.

Reduction in staffing numbers is always a seductive argument as far as management is concerned, with immediate benefits accruing. In my view, a more powerful argument for using MMBO is that it teaches operators the maintenance needs of equipment, and blurs the boundaries between maintenance and production groups. This is an excellent first step, albeit small, in aligning both groups on business objectives. Joint reliability initiatives can then become a feasible and practical proposition. These can bring huge long-term benefits, even if nothing happens immediately.

21.4 Maintenance Skills Available in Production Operations

We should not underestimate the level of technical skills already available within the operator ranks. We can determine these skill levels by checking individuals' previous employment. Some of the operators may have had craft-type training in the past. Hobbies are a rich source of expertise. Those citing do-it-yourself activities as a hobby can do most of the simpler maintenance work without a lot of training. People change light bulbs, wire up plugs for domestic equipment, mend a sticking lock, do minor repairs on their car, etc. These skills are transportable to the workplace.

I am always amazed at how little we value some of our people's attributes. I remember a fitter who played the trumpet in a brass band. The members were all unpaid volunteers although there was some sponsorship from a local company. He was not the conductor, but he had taken on the role of organizer of the rehearsals. Over the years this had been extended so he organized an annual tour of some European country. This year it might be France, next year Holland, and the year after Sweden. He handled travel arrangements and accommodations, and arranged concert venues and all the minutiae of the tour. Yet at work, he was given rather mundane tasks because he supposedly had little organizational abilities. I rest my case.

21.5 Case Study of an MMBO Implementation in Europe

This was a very overstaffed and high-cost location. They had a strong unionized workforce who defended strict demarcations. Any changes in work practices needed to be negotiated with a number of different union groups. Additional payment was a pre-requisite for any action. In this situation, we decided to implement MMBO.

We set up a working group to select the job tasks that operators would be expected to perform. We believed that maximum benefit would come from the routine, frequent, repetitive, simple tasks and these should represent the core of the work. This would ensure that they could practice any skills they acquired regularly. We tried not to be over-ambitious, in order to minimize training needs. Indeed we felt that we were setting our ambition and skill level requirements too low. Appendix 21-A has a sample list of tasks.

However, there was a gap between the available skills and the skills required to complete the defined job tasks safely and correctly. We needed to bridge this gap by a number of practical training programs. We devised a simple scheme to identify skill gaps and the training programs needed. By introducing a certification process to recognize skills and competence formally, we would provide confidence and satisfaction, perhaps even pride, to the individuals. Supervisors gained confidence with the certification scheme.

By the end of the year, we obtained agreement with the union; we paid a lot of money in salary increases to enable MMBO to take off. So what was the result? Zilch! Absolutely nothing! We had spent a lot of time, effort, and money, and we got virtually no maintenance activities done on shift. The operators thought that the maintenance group was dumping all the low-skill work that they didn't want onto the Production group. There was no glory in doing this work and, indeed, it was demeaning. The Production supervision clearly held this view and undermined the management line. The maintenance workforce saw this as giving away their work, which would imply a reduction in numbers, creating a problem both now and in the future when their children sought employment. The maintenance supervisors agreed with this. Although they gave lukewarm support, they also undermined the management line.

21.6 Case Study of an Implementation in Asia

Having learned from the European experience, when I moved to Asia and heard that the MMBO was advocated, I was determined to stay clear, but failed. The company set up a small group (including me) to decide on our approach. We decided to adopt a totally different approach. It was:

- to define the additional work we wanted operators to do as the normal work of operators. The analogy we made was with car

ownership; a driver would not contemplate taking his car to a garage to top up oils, change a light bulb, etc.
- to define the work with one simple criterion: Would it be easier for operators to do the job themselves than to create a job request and put it in the system for the maintenance force to do it?
- to avoid negotiation, as the union agreement would not be altered.
- to reveal only a part of our hand to a few trusted (formal and informal) leaders in the operator group.
- to utilize a number of frequently-occurring problems in Production. Here, it was easier for the operators to solve these themselves rather than go through the effort of calling upon maintenance. We would help operators to do this, but keep it quiet! This might be slightly disingenuous, but not much.

We were now ready to go.
- The first step was to get the Production supervisors some very simple tools.
- Next, we placed a box of light bulbs and fluorescent tubes in the control rooms for free use.
- Operators found it easier to change tubes and bulbs themselves.
- Then we moved on to putting a box of different-sized pressure gauges in the control room. When operators were doubtful of a pressure reading, they would change the gauge and put in a box marked faulty gauges.

We carried on like this in a slow and steady manner, introducing something new, usually at the request of the operators. The sort of things included printer paper and cartridges, panel bulbs, gas testers, small filters, etc. We didn't negotiate or pay a cent. But we got the work done!

21.7 The Control Issue

As a rule, work done by maintenance is captured in a CMMS where it can be evaluated, action taken, and maintenance tactics altered if needed. Work done by operators is normally not captured in any formal way. It might be in a note book or a log sheet, but this is not suitable for analysis. Hence the work of operators is inherently invisible.

This creates a conundrum. Some work, indeed much work does not need to be captured for analysis. But the rest does need analysis. So how do we handle that? Do we put it in the CMMS and accept the administrative effort? If we insist that operators raise job requests for their own work, it may then become unattractive for them to do the work. Just to make life harder, a number of CMMS systems will not allow you to be both an operator and a maintainer. Can we have a simpler parallel system? None of the options are particularly attractive, especially to engineers who are obsessed with capture and analysis of activities.

Many years ago, a Dutch friend of mine expounded the Law of Remaining Misery. What this said was that in every activity there is a certain amount of misery which you can't reduce. The only thing you can do is to decide where you are going to take it so that you minimize its impact. The control issue may be a case in point.

21.8 New Opportunities to Consider

Advances in technology have made it possible for operators to do many tasks which only recently were considered too complex to be carried out in house. The use of hand-held data collectors which can be programmed to collect specific data such as temperature, sound, vibration, etc., is an example of this approach. Some production facilities are now using these technologies to:

- Identify what the operators do
- Ensure that particular plant aspects are regularly scrutinized
- Compare plant readings with predetermined standards
- Capture plant data for post-mortem reviews

All this data can be captured, recorded, and analyzed.

21.9 Lessons

1. Set reasonable limits on MMBO, taking into account the time available on each shift and the skills of its members. It is easy for individuals or teams to become over-ambitious.
2. Unions see their role as negotiating good deals for their members, but enforcing compliance is always the company's problem. Just because it has been paid for, do not expect that MMBO will be accepted.
3. Compliance cannot be enforced if it is counter-culture.
4. It is difficult to capture and control MMBO activities without becoming bureaucratic and, therefore, losing some of the benefits.
5. A slow introduction of some MMBO works if it is easier for operators to do the activities themselves rather than use formal maintenance channels.

21.10 Principles

1. Handle the concept and its impact on emotions and culture before being trapped by logic and detailed development.
2. The principle of MMBO must be fully supported and encouraged by man agement, supervision, and the informal site leaders.
3. Success will come only if all stakeholders win to some extent.

Appendix 21-A

List of typical tasks for implementation in European location:

- Lubricator adjustments
- Luboil changes and sampling
- Small valve changes
- Clearing of blockages
- Screwed valves
- Caps and plugs
- Blanks and spading
- Adjustment of pump glands
- Pump isolation for engineering
- Replacing gaskets
- Temporary fitting of vent and drain lines
- Fitter assistance on appropriate jobs
- Filter cleaning and changing
- Minor safety jobs
- Lighting bulb changes
- Replacing local pressure gauges
- Thermocouple replacements
- Indicator replacements
- Temporary lighting
- Electrician assistance, including isolations
- Motor isolations
- Main switch isolations of LV equipment
- Heat exchanger bolting
- Making/breaking and pulling up flanges, replacing gaskets
- Removal of relief valves
- Simple workshop practice
- Valve overhauls
- Simple lifts

Chapter 22

Relocating Machine Tools

There is nothing more powerful than an idea whose time has come.
Victor Hugo, French dramatist, novelist and poet

Author: V. Narayan

Location: 2.1.2 Automobile Parts Manufacturer

22.1 Introduction

The company had three large factory buildings at this site. Building No.1 was 150,000 sq. ft., Building No.2 was 90,000 sq. ft., and Building No.3 was 60,000 sq. ft. in area. The production facilities for fuel injection bodies and spark plugs were in the first building, where the main store was also located. Critical component parts of the pumps—namely the barrel and plunger—were produced in the second factory building. A layout of the factory is shown in Figure 22.1.

There were wide fluctuations in market demand for the company's range of products. Flexible production processes were used and batches ran for a few months at a time. It was necessary to change the layout of the machinery to suit the production process for the specific item being produced. This meant that we had to relocate some machines to suit the process. The production planning department had a dedicated layout planning section to design these changes.

Each machine tool required several services. These included power supply, air supply for operating the pneumatic controls and actuators, water supply, and cutting oil or emulsion. In order to facilitate rapid layout changes, the design of the factory allowed the services to be provided on a plug-in basis (see Figure 22.2). There were overhead bus-bars suitably encased in sheet metal ducts, with power sockets located about 10 feet apart. This enabled the ma-

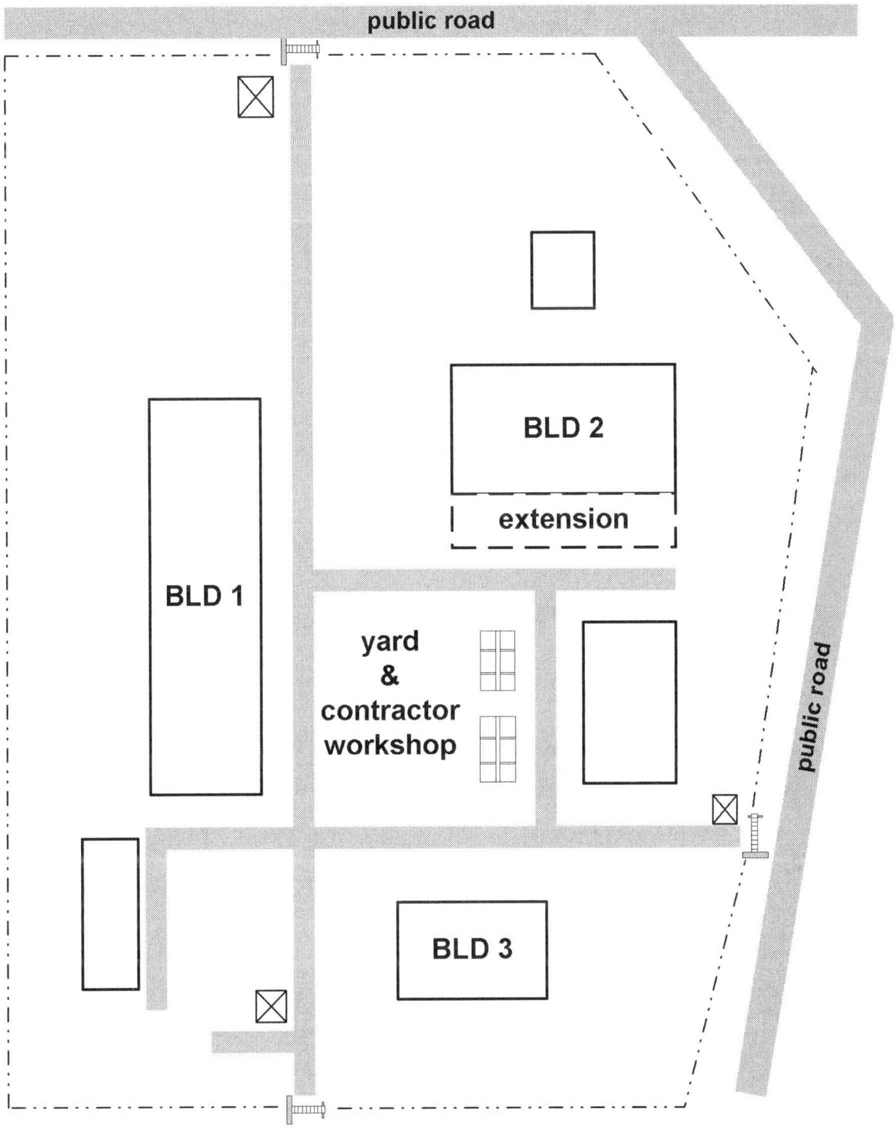

Figure 22.1 Layout of factory

chines to be wired up in a new location fairly quickly. Similarly, there were overhead water and air pipelines along the bus bars about 15 feet above the floor. These had outlets with valves, also located about 10 feet apart (see Figure 22.3). This design enabled us to make utility connections safely and quickly.

To move the machines, we had to empty the cutting-oil/emulsion tanks. Once transported, we refilled them at the new location. Most of the machines did not have permanent foundations, just vibration damping pads. A few items like presses had heavy concrete foundations and, hence, remained in one place. The production planning layout section identified the machine tools that

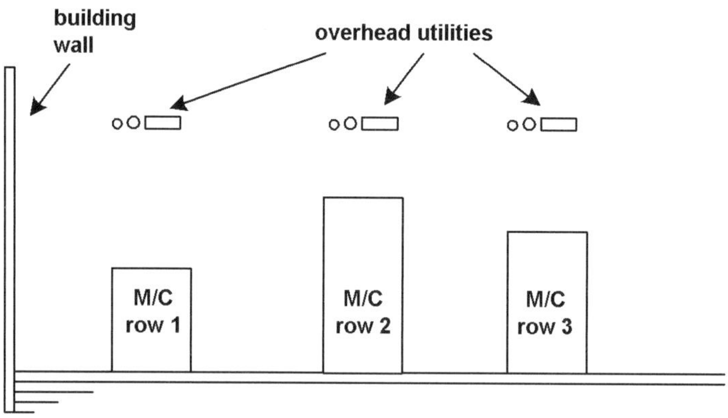

Figure 22.2 Section showing overhead utilities

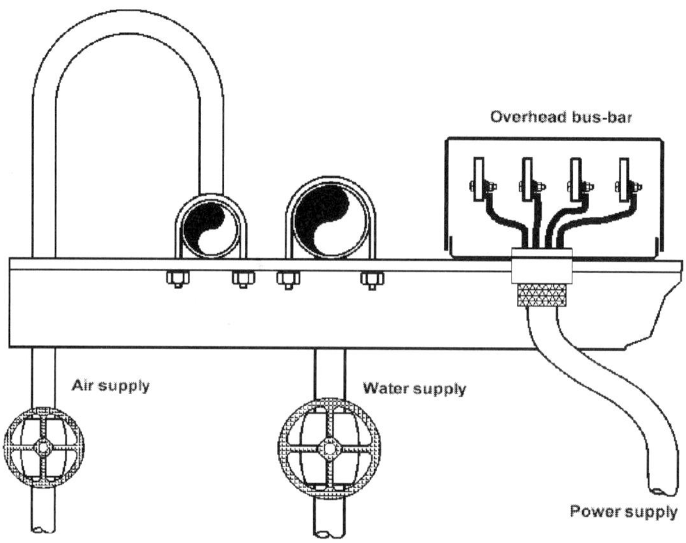

Figure 22.3 Details of service connections

required shifting. Most of the movements were within a building, but some were between buildings.

The company designed and built many of the special purpose machine tools (SPMs) they needed for manufacturing their product range. This work was done in Building 3.

In this company, the work ethic was very strong. Staff accepted flexible working practices. People were expected to meet their targets—this was part of the company culture. The "not-invented-here" disease had not taken hold in this environment. People were used to drawing conclusions after proper analysis. The company was nearly 20 years old, and their staff had seen many changes during its rapid growth. The concept of moving precision machine tools from one location to another was not common practice elsewhere. In this company, it was the norm and the factory itself was designed to facilitate such movement.

22.2 Background

At the time of these events, Building 2 was being extended by 35,000 square feet. This was due for completion in July. Thereafter we had to install about 500 machine tools in this extension. Most of the movements were from Building 1. A few new machines were on order and some SPMs being built in Building 3 would also be located in this extension.

The manufacturing process in Building 2 was very sensitive to dust. The precision grinding and polishing operations were to very fine tolerances. For example, the clearance between barrel bore and plunger was 0.3 to 0.4 micron or 0.000012″ to 0.000016″. In this situation, fine dust particles could result in serious production defects.

22.3 Existing Procedure for Machine Relocation

Over the years, the company had built up a machine-shifting crew. One of my supervisors led this crew. A machine moving contractor's workers removed the item from its old site and loaded it on a special heavy-duty, low-slung trolley. They then manually towed the machine across the factory and internal roads to the new location. Company electricians disconnected and reconnected the machines. An electrical contractor dismantled the conduit and cabling from the overhead bus-bar and installed new or modified conduits and cables in the new location. A plumbing contractor did the piping work for the air and water supply. Company workmen handled the cutting-oil (or emulsion) tank draining and refilling. On average, this crew was able to handle 3 machine relocations per day. The machines were restarted in the new location within 12–14 hours.

New piping and conduits were fabricated in the contractors' workshops, located centrally between the factory buildings (see Figure 22.1). Once the machine was placed in the new location, they could measure the pipe/conduit dimensions. The plumbing and electrical contractors then went away to their workshops to cut and thread the new piping. Meanwhile, company workmen installed the cutting-oil tank, control panels, and other accessories. While the piping and cabling was being installed, the cutting oil (or emulsion) was refilled, so the whole process was reasonably streamlined and efficient.

The crew had to set up ladders to connect the conduits and pipes to the overhead services. This activity and that of locating the machines themselves produced dust. The crew took care to keep airborne dust levels to a minimum.

22.4 The Challenge

In a planning meeting in February, the Production General Manager (GM) who was chairing it wanted a plan to complete the installation of the machine tools in the new extension within 30 days. At the current rate of relocations, the activity would take over six months; he declared this to be unacceptable. The dust level over a prolonged period was likely to cause quality problems.

Furthermore, the inability to run the new lines at one site would result in unnecessary parts' movement, resulting in production scheduling problems.

Most of the people at the meeting were taken aback by this demand. The production process planning engineer and my mechanical department head said that raising the tempo from 3 to 17 machines a day appeared impractical. We had three low-slung trolleys and it was not possible to get 12 additional trolleys by August. The time available to train people to do the work was inadequate.

I sought time to study the issues involved before taking a stance.

22.5 Analysis

About this time, a young engineer joined my team as an intern for a year. He seemed an intelligent and enthusiastic person, so I assigned him to do the analysis. He knew nothing at all about Method Study, which was the process I wanted to use. We discussed the project and drafted a plan of action. The first job was to train him. As a student, I had myself used a good book entitled Introduction to Work Study published by The International Labour Office (ILO) in Geneva[i]. I gave him my copy and asked him to read the chapter on Method Study dealing with Flow Process Charts and String Diagrams. We spent an hour every day discussing his queries and after three days he was ready to start.

He spent a week following the work crew from the time that machines were disconnected to the time they were restarted. He charted the process carefully, noting the time for each activity. All the machine movements were recorded, so at the end of the week we had 16 charts (see Appendix 22-A for an explanation of process flow charts and string diagrams). During the next week, he recorded people-movements, using string diagrams. By the end of the second week, he had identified many weak spots in the current system. He reported that if we were able to make the necessary changes to the current process, the existing crews could move 5–6 machines per day. That was good, but not good enough.

Next he studied the string diagrams and looked for improvements (see Figure 22.4). One work practice caused significant delays, adding 40–60 minutes to the cycle time. This delay was the result of the plumbers and electricians walking to their workshops to fabricate the pipes and conduits. His plan was to move the light-weight pipe threading and bending machines to within 50–100 feet of the new location of the machines. This was acceptable, as long as dust levels were controlled. Reducing cycle time meant that 7–8 machines could be moved per day. This was progress, but still not good enough.

Our first impulse was to consider doubling the resources. We resisted that knee-jerk reaction, and decided to analyze the bottleneck resources and their effective working time.

1. The company electricians isolated the machines before anybody worked on them. Similarly, the company electricians waited for all other work to be completed before reconnecting the machine. They were thus

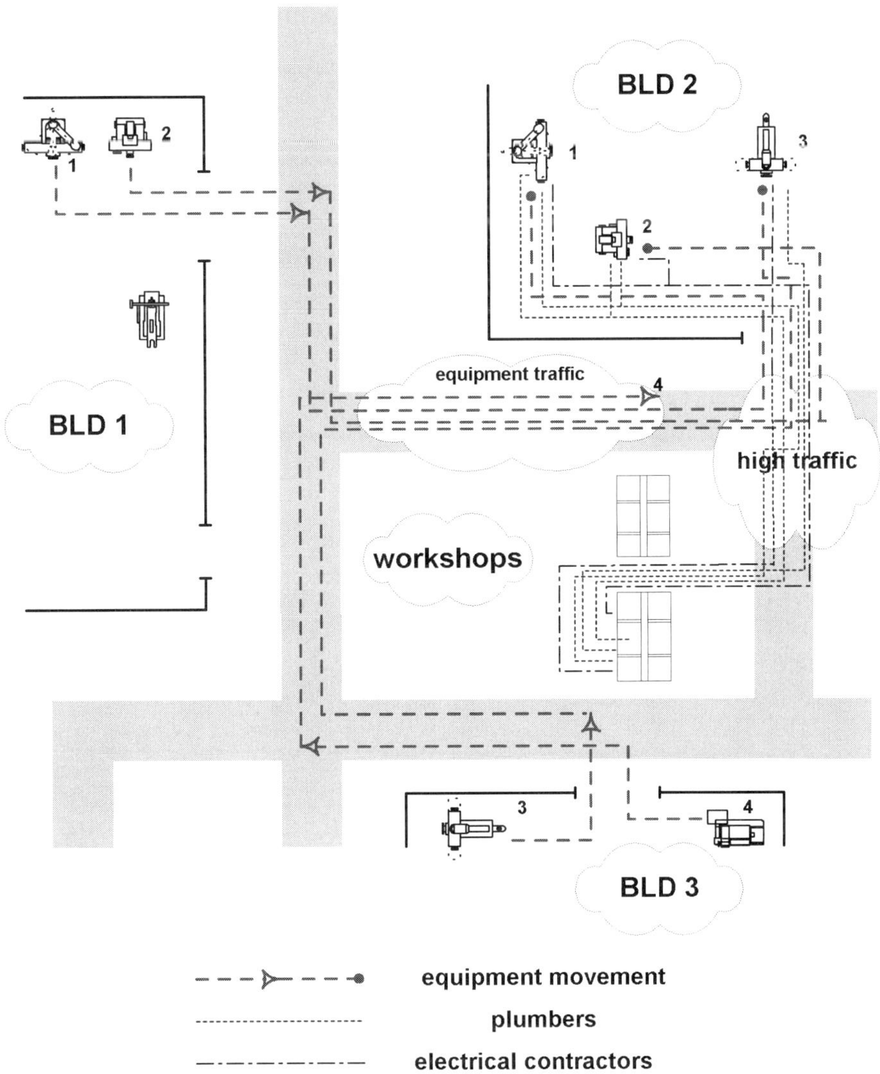

Figure 22.4 String diagram of present method

the first and the last people to work on any machine.

2. Company electricians worked 2 hours overtime per day for machine shifting activities.
3. The contractors took about 2–3 hours per machine to measure, fabricate, and install the conduits and cabling.
4. Each trolley was in use for 2–4 hours per machine movement.
5. The 3 plumbers were busy for about 4 hours per day.

The main players in the team, including the production process planning engineer, met with the intern to review these findings and hear his proposals. He suggested the following actions.

a. We should schedule the work in two shifts of 8 hours each in place of the current 8-hour day work.
b. We start work with one electrician arriving 30 minutes before the rest, so that machines could be worked on as soon as the other crew members came in to work. The second shift electrician worked an extra 30 minutes at the end of the shift to reconnect and energize the last machine that was ready.
c. We would electrically isolate and disconnect 3 machines every 1 1/2 hours and prepare them for shifting.
d. The machine operators along with their tools, parts, etc., would move with the machines. They would help with the cutting oil refilling and in arranging the tool cabinets, etc., at the new location.
e. We would temporarily locate the conduit and pipe fabrication machinery and tools in the new building extension, between 50 and 100 feet of the machine tools being installed.
f. We would increase the contract crew size by two electricians, two plumbers, and four additional workers in the machine shifting crew. With these resources, we could form two shifts.

While there were doubts expressed with these proposals, especially about the resources and lack of additional trolleys, the participants agreed it was worth a trial. The only change was that we decided to order one additional trolley for delivery by end of July. We agreed to test out the new scheme during normal day shift movements in the existing buildings. This would help us evaluate the risks involved in attempting a step change in the rate of machine movements.

22.6 First Shop Floor Trial

The intern joined the machine shifting crew, starting work along with the first electrician. Within a week of commencing the trial, the team was able to handle 6–7 machines per day shift, without additional resources. They required less overtime work and we expected we could reach a rate of 12–13 machines in two shifts. I was of the view that we would climb up the learning curve and reach the final target rate quite quickly. The remaining team members were not too happy with my position and urged me to be cautious while making any commitments to the GM.

In the May planning meeting, I told the GM that it was feasible to complete the work in 35 to 40 days and that a 30-day target may just be achievable. He accepted this evaluation, so we started working on the detailed plans.

22.7 Building Extension Delays

Meanwhile, there were problems with the building construction work. Our civil engineer assured me that we would meet the scheduled completion target, but I was not convinced. The project progress charts (S-curves) indicated that we would overshoot the target completion date by six weeks. As the person responsible for the building extension and machine shifting, I was concerned that the whole program would go awry. After discussions with the architect, the civil engineer, and the building contractor, we concluded that additional resources were required. In the event, the building construction was only finished on July 30, in spite of a 30-percent increase in resource levels.

22.8 Second Shop Floor Trial

In early June, we attempted a second trial for a week, with two shifts working. By this time, the contractors had recruited and trained additional workers. On the first day, 11 machines were shifted. Thereafter, it improved steadily, till by the end of the week we were able to move 14 machines. The second trial indicated that the target of 17–18 machines per day was within reach. From a position where we needed to double the resources to get six or seven machines moved a day, the team had achieved double that number with a minor increase in resources.

22.9 Results

In August, we put the new machine shifting plan into action. The crew members understood the new system with the experience gained in the two earlier trials. They felt challenged and became very motivated. By now they were used to working with the intern, and started seeking his guidance. As before, he kept accurate records, with process charts for each machine movement. On the first day, we moved 13 machines, but this rate improved during the week to 15 machines per day. Careful analysis showed the weak points, which were mainly in coordination and communication.

By day 8, the team managed 17 machines per day. By the end of the second week, they were moving 18–19 machines per day. The overtime work reduced dramatically. Eventually, the shifting program invariably ended by the middle of the second shift. The entire operation of shifting 500 machine tools was completed in 28 days.

22.10 Lessons

People tend to accept the current status and ways of working. An outsider is better able to challenge these practices. Analysis and understanding are the first steps in any improvement effort, but each situation is different. Blindly copying practices from elsewhere does not always work. Adapting methods that blend with the prevailing company culture improves the chance of success.

The availability of an intelligent and motivated analyst was a lucky break. The fact that he was able to pick up concepts quickly contributed greatly to the success of the project. He worked long hours, but obviously enjoyed the challenge. It was a bonus that he was able to get along with people, and was able to earn their respect quickly.

In making significant changes to existing work practices, it is always a good idea to test them out first. The two trials and their relative success boosted the morale of the entire crew. They were eager to try out their skills and beat the new challenge with the guidance and coordination of the intern. The relatively long time available for preparation and training helped reduce some uncertainties.

At the beginning of the project, I was confident we would find a way, though at that stage, I did not have any idea of what exactly we would do. Nearly every other person associated with the project thought I was over-optimistic and would land in deep trouble.

Method Study was a suitable process for the analysis, and the intern collected hard data carefully. He had the right qualities of inquisitiveness and industry. He was innovative and had analytical and people skills.

We did not commit ourselves to any targets till we had carried out field trials in normal working conditions. This reduced the risks involved. The team had a say in the decisions made, as we had weekly meetings to discuss the data, analysis results, and progress. Everybody had a role in the successful outcome, and the "feel-good" levels were quite high.

The one decision that was questionable was the ordering of the fourth trolley. Though it was used occasionally, we could have managed without this trolley. Its role was in risk mitigation, i.e., as an insurance spare, which it fulfilled adequately.

22.11 Principles

1. The role of a leader is to challenge the status quo. The GM did this by making a seemingly impossible demand.

2. Knee-jerk reactions are poor substitutes for careful analysis and understanding.

3. "Look before you leap" is an apt saying; the risks can be high when attempting new ways of working. It is necessary to take proper risk control or mitigation measures, and these have to be thought through in advance. A field trial is one way to evaluate risks.

4. The tools we use must be carefully selected. In this case, Method Study was appropriate, but there is a wide range of techniques available for use in other situations.

Reference

i. Kanawaty, G., ed. 1992. Introduction to Work Study. 4th ed. Geneva: ILO Publications. ISBN: 92-2-107108-1.

Appendix 22-A

Process Flow Diagrams, Charts, and String Diagrams

This explanation and the figures are based on the book, reference (i) above, Introduction to Work Study, published by the International Labour Office (ILO). The figures in this Appendix are reproduced courtesy of the International Labour Office.

A flow diagram is a plan of the factory to scale, with the machines and work

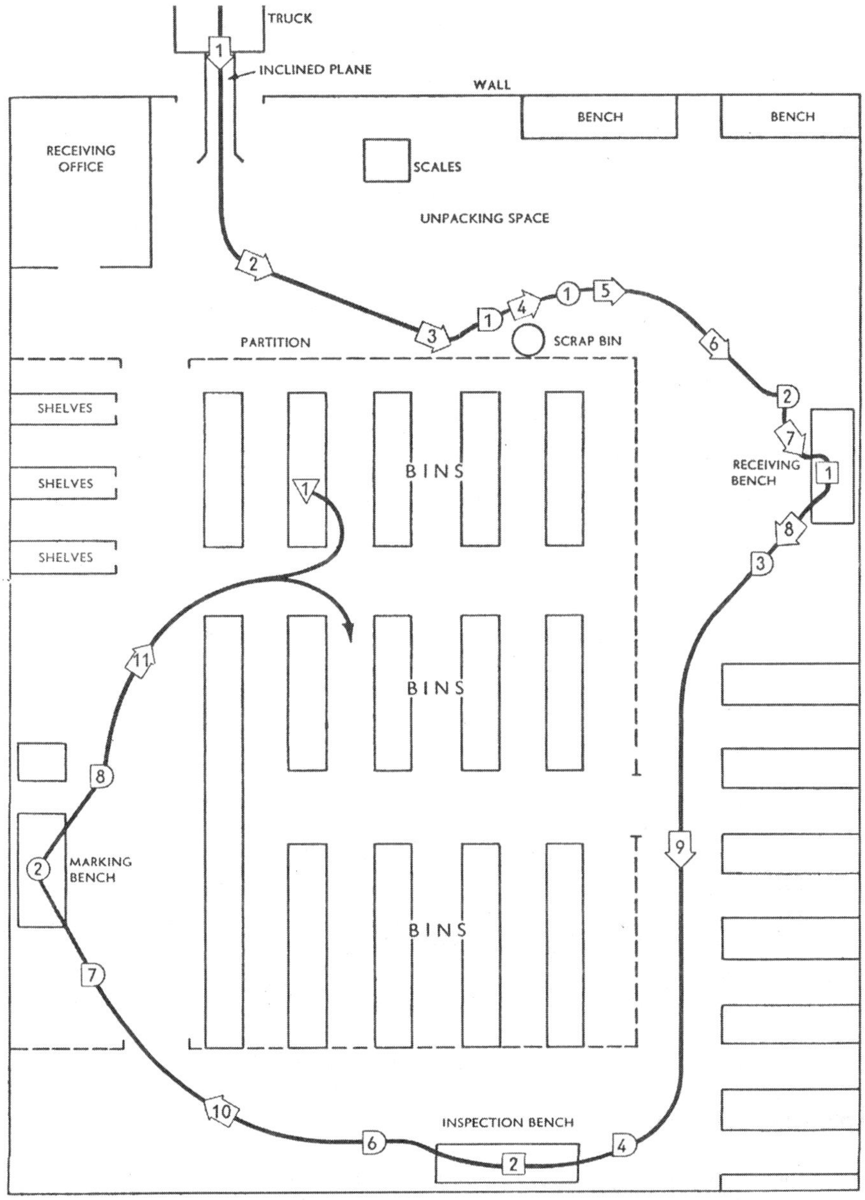

Figure 22-A.1 Flow Diagram—Old Method

areas marked in their correct locations, and showing the path of the movement of materials or components between the work areas and machines. In a production flow diagram, the flow will start with the arrival of raw materials at the store; through the various stages of production and assembly, till the product reaches the dispatch area. A flow diagram for inspecting incoming parts is shown in Figure 22-A.1.

A flow process chart is a tabular record of the movements, showing the activity types, time, and distance for each movement. It shows the relevant data relating to the process shown in the flow diagram. A chart relating to Figure 22-A.1 is shown in Figure 22-A.2

CHART No. 3 SHEET No. 1 OF 1	SUMMARY			
PRODUCT/~~MATERIAL/MAN~~ *Case of BX 487 Tee-pieces (10 per case in cartons)*	ACTIVITY	PRESENT	PROPOSED	SAVING
ACTIVITY: *Receive, check, inspect and number tee-pieces and store in case*	OPERATION ○	2		
	TRANSPORT ⇨	11		
	DELAY D	7		
	INSPECTION □	2		
	STORAGE ▽	1		
METHOD: PRESENT/~~PROPOSED~~	DISTANCE *(ft.)*	185		
LOCATION: *Receiving Dept.*	TIME *(man-hrs.)*	1.96		
OPERATOR(S): CLOCK No. *See Remarks column*	COST LABOUR MATERIAL	 $3.24 —		
CHARTED BY: *B.C.* DATE: *4.11.48* APPROVED BY: *T.H.* DATE: *5.11.48*	TOTAL	$3.24		

DESCRIPTION	QTY. *1 case*	DISTANCE *(ft.)*	TIME *(mins.)*	○	⇨	D	□	▽	REMARKS
Lift from truck: place on inclined plane		4							2 labourers
Slide on inclined plane		20	10						2 "
Slide to storage and stack		20							2 "
Await unpacking		—	30						
Unstack case		—							
Remove lid and take out delivery note		—	5						2 "
Place on hand truck		3							
Truck to reception bench		30	5						2 "
Await discharge from truck		—	10						
Place case on bench		3	2						2 "
Take cartons from case: open: check contents: replace		—	15						Storekeeper
Load case on hand truck		3	2						2 labourers
Delay awaiting transport		—	5						
Truck to inspection bench		54	10						1 labourer
Await inspection.		—	10						Case on truck
Remove tee-pieces from case and cartons: inspect to drawing: replace		3	20						Inspector
Await transport labourer		—	5						Case on truck
Truck to numbering bench		30	5						1 labourer
Await numbering		—	15						Case on truck
Withdraw tee-pieces from case and cartons: number on bench and replace		—	15						Stores labourer
Await transport labourer		—	5						Case on truck
Transport to distribution point		15	5						1 labourer
Store									
TOTAL		185	174	2	11	7	2	1	

Figure 22-A.2 Flow Chart—Old Method

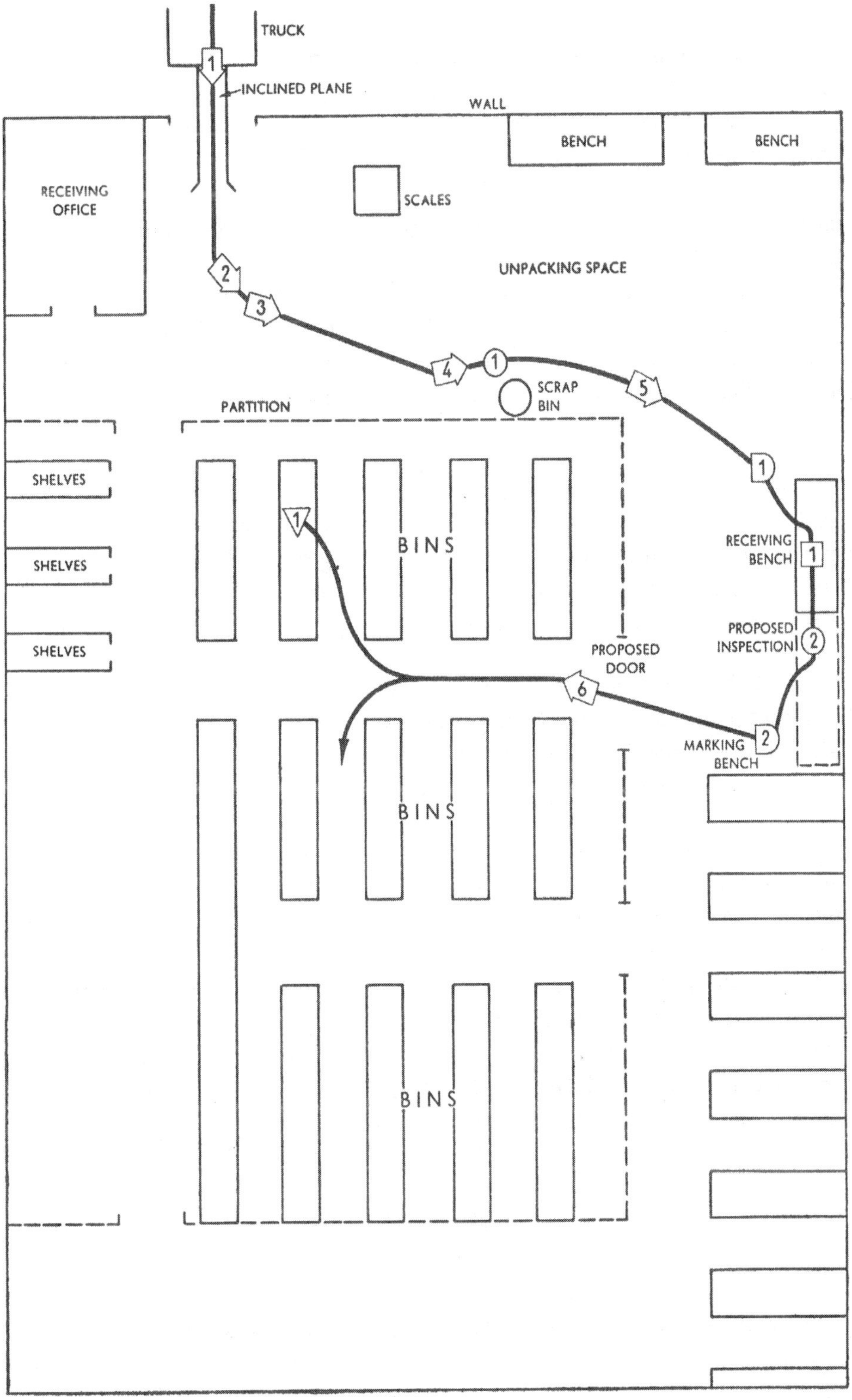

Figure 22-A.3 Flow Diagram—New Method

CHART No. 4 SHEET No. 1 OF 1	SUMMARY			
PRODUCT/~~MATERIAL~~/~~MAN~~ *Case of BX 487 tee-pieces (10 per case in cartons)*	ACTIVITY	PRESENT	PROPOSED	SAVING
ACTIVITY: *Receive, check, inspect and number tee-pieces: store in case*	OPERATION ○	2	2	—
	TRANSPORT ⇨	11	6	5
	DELAY D	7	2	5
	INSPECTION □	2	1	1
METHOD: ~~PRESENT~~/PROPOSED	STORAGE ▽	1	1	—
	DISTANCE *(ft.)*	185	107	78
LOCATION: *Receiving Dept.*	TIME *(man-hrs.)*	1.96	1.16	.80
OPERATOR(S) CLOCK No *. See Remarks column*	COST *per case* LABOUR	$3.24	$1.97	$1.27
CHARTED BY: *B.C* DATE: *6.11.48*	MATERIAL	—	—	—
APPROVED BY: *T.H.* DATE: *7.11.48*	TOTAL	$3.24	$1.97	$1.27

DESCRIPTION	QTY. 1 case	DISTANCE (ft.)	TIME (mins.)	○	⇨	D	□	▽	REMARKS
Lift from truck: place on inclined plane		4			•				2 labourers
Slide on inclined plane		20	5		•				2 "
Place on hand truck		3			•				2 "
Truck to unpacking space		20	5		•				1 labourer
Take lid off case		—	5	•					1 "
Truck to receiving bench		30	5		•				1 "
Await unloading		—	5			•			
Take cartons from case: open and place tee-pieces on bench: count and inspect to drawing		—	20				•		Inspector
Number and replace in case				•					Stores labourer
Await transport labourer		—	5			•			
Truck to distribution point		30	5		•				1 labourer
Store		—	—					•	
TOTAL		107	55	2	6	2	1	1	

Figure 22-A.4 Flow Chart—New Method

Examining the flow diagram and analyzing the flow process chart allows us to identify the weaknesses in the present method and design a better method. We ask the following type of questions:

1. Why do we do things in this manner?
2. What else could we do?
3. Why are some work areas or machines where they are at present?

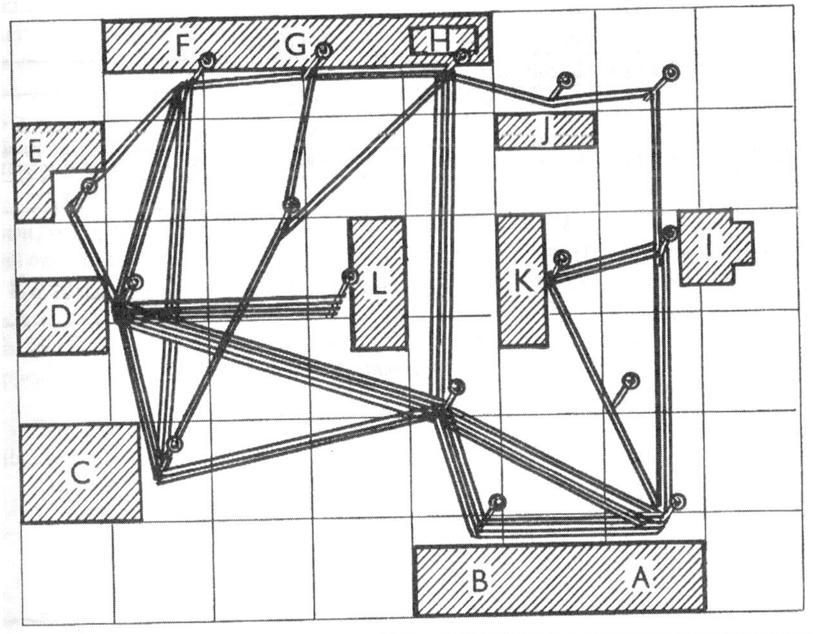

Figure 22-A.5 String Diagram

4. Where else could they be?
5. Where should they be?
6. Should the path be as long as it is? What makes it so long?

As a result of such analysis, some rearrangements are designed and an improved flow path becomes possible. This is shown in the flow diagram in Figure 22-A.3, and the corresponding flow chart in Figure 22-A.4.

A string diagram is a drawing of the workplace or factory to scale, with the relevant machines and work areas shown in their correct locations, and showing the movement of a worker, using a continuous thread. Using a string diagram, we can see the density of the travel paths and distances.

We start with a scale drawing of the relevant factory area with the machines, work tables, doorways, aisles, partition walls, and pillars correctly located. This drawing is mounted on a softwood board. Drawing pins are fixed at every stopping or turning point in the path that affects the movement of the worker. The pin-heads are left clear of the board by about 1/2″ so that the thread can be maneuvered around the shaft of the pin.

We take a measured length of thread, tie it to the pin at the starting point, and trace the worker's movement through the activity, using the pins on the way as guides. The result is a picture showing the frequency of travel along each section of the path, the thickness of the thread showing the frequency. An example of such a chart is shown in Figure 22-A.5. The length of the thread remaining unused helps us compute the total travel distance.

As you can see, movements between points A and D, A and H, and D and L are quite frequent. These are also far apart, so we examine whether the work areas A, D, H, and L should be relocated. All such relocations may not be feasible, but we will try to minimize the worker's travel path as far as possible. Movement adds costs but not value. A string diagram made after the work area relocation shows the improvement in travel distance. Such before-and-after string diagrams are useful communication tools when explaining proposed changes to workers and management.

Chapter 23

Painting Contract Strategy

...Enumerate a few basic principles and then permit great amounts of autonomy. People support what they create.
Margaret J Wheatley, Writer and Management Consultant.

Author: V. Narayan
Location: 2.2.1 Liquefied Natural Gas Plant

23.1 Background

Paint work on several structures, vessels, and pipelines in the plant had started degrading even at the time of commissioning and hand-over. The main contractor obtained a release from his painting repair obligations by agreeing to a lump sum settlement.

The technical issues of selecting the right paint, surface preparation, coating thickness, etc., were manageable. It was more difficult to award watertight contracts that would guarantee an acceptable coating life and were legally enforceable. The prevailing practice was to micromanage the painting contractors by providing detailed specifications on surface preparation, paint application, and quality control, followed by close supervision from the company. We supplied the paints, but this led to a situation where the contractor would blame the paint quality and the paint vendor would blame the application quality.

We identified three levels of scope of painting work, as follows:

1. General corrosion damage over large areas of the plant, typically with over 100,000 square feet of painted surface. We had time to plan and award contracts for such work.

2. Corrosion damage caused by service conditions in the plant area, typically involving less than 10,000 square feet of painted surface.

3. Localized corrosion in clusters or individual spots, which were difficult to define and usually needed urgent attention. Generally these involved surface areas less than 1000 square feet.

23.2 Overall policy direction

As the total volume of work was quite large, we needed a clear plan of action, suitably budgeted and one which permitted a steady workload. Large packages of work needed high levels of technical and management skills, but the work could be specified in large packages suitable for the company's major tendering process. Where possible, we wanted to engage local contractors who were capable of executing medium-sized work packages. This policy would help support the local economy and improve the company's image. Small but urgent repair works should not have to wait unduly for long tendering exercises.

We intended to award the large work packages to international contractors and paint manufacturers, if we could design a watertight contracting system. They would have the resources and experience to manage such work. Medium-sized packages could be awarded to local contractors, using the faster tendering process for minor contracts. The small, often urgent repair works could be done by our own civil engineering staff. We could thus ensure that such work was done with speed, flexibility, and quality. These proposals were formally approved by management and became company policy.

23.3 Commercial Issues

There were serious challenges to resolve before we could award the large work packages. In order to minimize the life cycle costs of maintenance painting, we had to ensure that:

a. We could identify the level of degradation and match the repair to suit.
b. Both parties could agree with this assessment easily.
c. The durability of the painted surfaces improved at a reasonable cost.
d. We helped contractors to minimize their overheads.

We could then award a contract where the paint supplier also applied the paint (or the contractor provided the paint). These are called supply-and-apply contracts. We needed a clear and unambiguous definition of the scope of work to award such contracts. The bidders will not normally be willing to take responsibility for the condition of the existing paint work. They would only agree to guarantee the work if they were allowed to strip the existing paint work to bare metal if required. This was not practical in an operating scenario, and we were at an impasse. If we could find a method of defining the current and required conditions against a standard, then a solution was possible.

23.4 The Standard

Our civil engineer located a Swedish Standard that categorized in-service coating defects photographically. This standard is different from that used for specifying surface preparation quality (SIS O5 59 00:19237); it showed a set of nine photographs ranging from a perfect finish to one with a severely degraded paint condition. The size and number of defects progressed logarithmically, so that the range of damage displayed was quite considerable. Using this standard, one could monitor the paint condition visually, allowing judgment of the paint condition from ground level. We could match the current condition to one of the photographs and specify the improvements required with another photograph. These photographic standards were definitive and the bidders were willing to use them.

At present, this Swedish standard appears to be unavailable. There is, however, an excellent publication called Fitz's Atlas[i] with a similar set of graphical standards. These standards define localized and scattered coating breakdown and are used in conjunction with the European Scale of Corrosion ISO 4628 and ASTM 610. The Fitz scale, kindly provided by the publishers MPI Publications UK, is shown in Appendix 23-A.

23.5 Large Painting Works

We drafted a master plan to cover the painting program for the whole plant over a five-year period. Next we divided the main plant areas into blocks, using the internal roads as boundary markers. Each of these blocks had over 100,000 square feet of painted area to inspect and repair as necessary. There were 49 such blocks in all, so 10 blocks were planned per annum.

We put these large contracts out to tender, asking the bidders to inspect and decide the scope of work themselves. The goal was to bring the entire paint work within that block to a specified photographic standard, e.g., 1% localized breakdown (see Appendix 23-A). We would help them inspect the current condition and provide a limited amount of scaffolding material as well as our hydraulic man-lift. Each bidder could take up to three days to do the initial inspection. We left it to the bidder to decide the specification for surface preparation and paint application. We specified only the end results, using the photographic standard, which we could monitor visually.

Successful bidders had to carry out three further inspections after the initial work. The first was at the end of 6 months, the second after 18 months, and the third and final inspection was after three years. After each inspection, they had to repair any damage found, but this was to a less demanding photographic standard than originally required, e.g., 3% localized breakdown. The bidder had to guarantee the paint work for a period of three years.

We would pay in installments. On completion of the initial paint work, we paid 75% of the contract sum. The first inspection and repair entitled them to a further 5%, the second inspection 10%, and the final inspection 10%. If only minor touch-up work was required, we agreed to provide, free of cost, a hydraulic man-lift and/or scaffolding materials for a period of 10 days on each occasion. Our view was that after the final inspection, the paint work would last for a further two years when a new round of repainting would commence.

We selected the prospective bidders from a list of international painting contractors and paint suppliers, most of whom formed alliances. We were able to attract enough bidders, as the work volume was steady and large. The bidders received free facilities for access during inspections and minor repair work and did not need to mobilize these services at their cost. Because the plant was remote and short of infrastructure, such mobilizations could be quite expensive. This policy brought down their costs significantly; being in a competitive tendering situation, they were inclined to pass on some or all of these savings to us.

23.6 Intermediate Work Packages

Within the plant area, some structures, equipment, and piping suffered severe corrosion damage, due to the local service conditions. In general, we were dealing with painted surfaces less than 10,000 square feet in area. We prepared work packages for these, with detailed specifications. These were for surface preparation, primer, and finish coat details and dry film thickness. They had to apply paint during dry weather conditions. We provided scaffolding material and paints free of charge, so the contractors only needed to cover labor costs and consumables. This limited their capital expenditure and overheads.

We used the minor work tendering procedure to process these packages, through small local contractors. Processing time was relatively short, so these did not need as much forward planning as the large contracts discussed above.

23.7 Small Work Packages

Our own staff painted some special items such as control valves or pumps. These items suffered coating damage through use or corrosion due to service conditions. The painted surface areas were usually less than 1000 square feet. Our staff helped manage urgent work with reduced supervision.

23.8 Results

It took some time for all the parties to adapt to this new way of working, especially with the large contracts. Some contractors had misgivings about the guarantee clause, and one dropped out of the bidding process. The others adapted fairly well, and we were able to implement the plan successfully. Costs dropped steadily every year. The physical appearance of the plant improved dramatically, and the contractual inspections proved valuable.

Local contractors coped very well with the intermediate packages, and a strong link was established with the local community.

The ability to tackle urgent work with our own staff proved a great advantage and it helped retain competencies. We used our workers to monitor the work quality of the local contractors as well, reducing the workload on the supervisor.

23.9 Lessons

Clear strategies help manage external corrosion issues effectively. A life-cycle approach is necessary, and long-term planning is well justified.

Maintenance painting can account for a significant portion of the budget. When the environment is aggressive and if the initial application is marginal, the problems are accentuated and costs can easily spiral out of control.

Commercial considerations are important. Contracts must be enforceable.

Under the right conditions, they can be designed to provide benefits for both parties.

23.10 Principles

1. Specifying end results rather than intermediate results can help harness the knowledge, experience, and ingenuity of service providers.

2. Providing limited infrastructure support to contractors can reduce costs for both contractors and the company.

Reference

i. Weatherhead, Roger and Peter Morgan. Lithgow & Associates. *Fitz's Atlas™ of Coating Defects.* MPI Publications. ISBN: 0 9513940 2 9. URL: http://www.mpigroup.co.uk/fitzs-atlas.asp

Appendix 23-A

Coating Breakdown Standards (based on Fitz's Atlas)

In Section 23.4, we discussed the use of a graphical standard for monitoring coating breakdowns visually. These Fitz standards are illustrated in Figure 23-A.1. on the following page.

Contents of Fitz's Atlas

The first two sections identify Welding Faults and Surface Conditions, which may be encountered and need to be addressed prior to the application of any coating system.

Two further sections deal with surface preparation, giving notes and visual guides to the standard preparation grades achieved through both Dry Abrasive Blasting and Water Jetting.

104 photographs in the Coating & Application Defects section enable the user to identify a range of coating failures. Advice is given on probable causes, prevention, and repair.

The last section deals with various types of Marine Fouling.

Finally, an Appendix gives breakdown scales, to help determine the degree of coating failure, and quick reference guides to specific paint characteristics and paint compatibility charts.

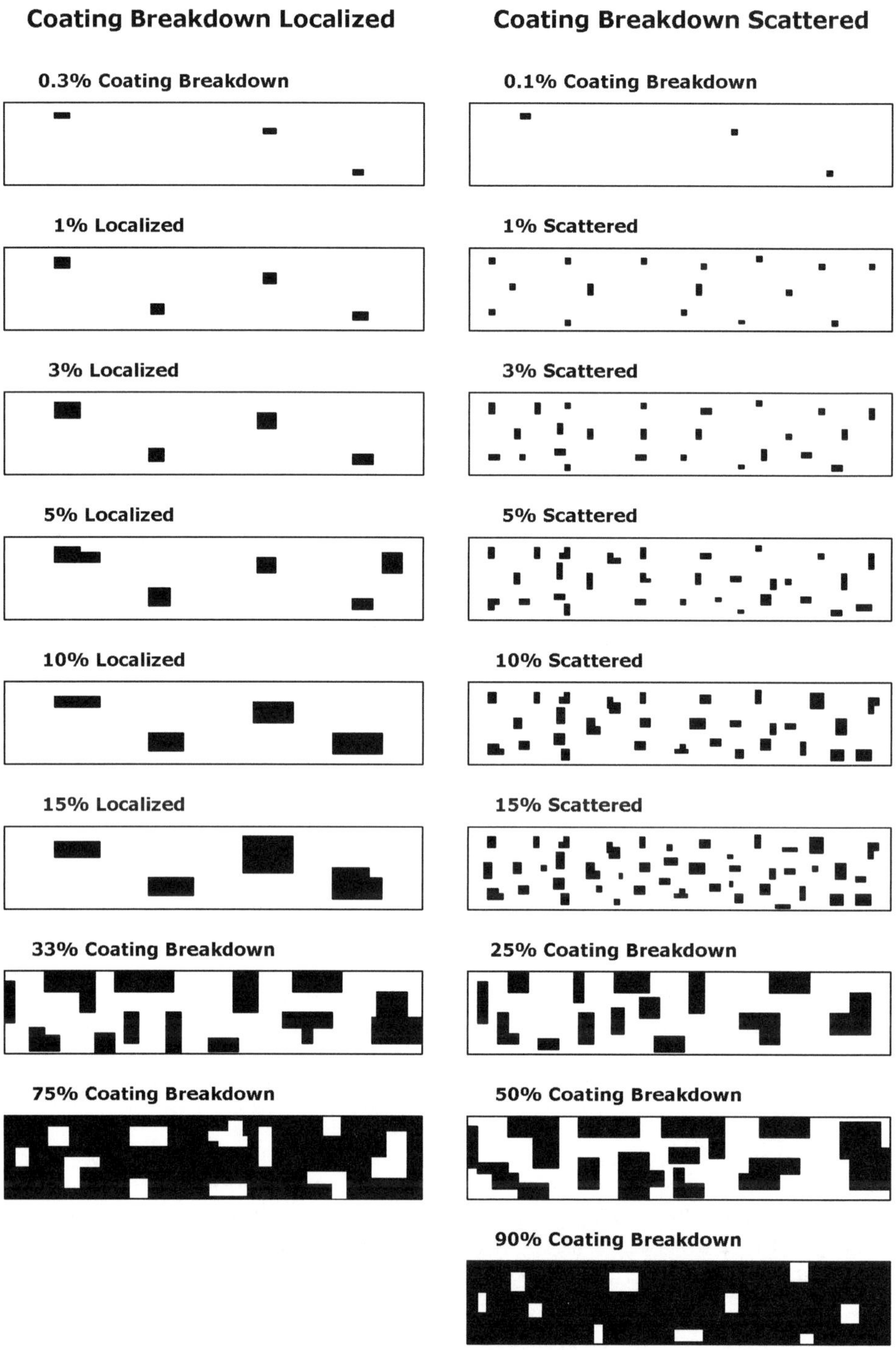

Figure 23-A.1 Coating Breakdown (Localized and Scattered) (From the Fitz Atlas)

PART 5: SCHEDULE

Chapter 24

Long Look-Ahead Plan

...where no plan is laid, where the disposal of time is surrendered merely to the chance of incidence, chaos will soon reign.
Victor Hugo, French dramatist, novelist, and poet

Author: Mahen Das
Location: 2.5.1 Small Petroleum Refinery

24.1 Background

The refinery adopted risk-based techniques such as RCM (Reliability Centered Maintenance) and RBI (Risk Based Inspection) as soon as they became available in the early 1990s. At the time of these events, they had just installed a CMMS (Computerized Maintenance Management System). With a strong tradition of good manual systems to plan, prioritize, schedule, and optimize resources, their transition to the CMMS was quite smooth.

Plant reliability was high and there were few breakdowns. With an effective prioritization system in place, there were few jobs which demanded urgent action.

24.2 The Day-to-Day Maintenance Process

Soon after the introduction of the CMMS, I visited the refinery as an internal maintenance and reliability consultant. As a step in their continuous improvement program, they agreed to try out the day-to-day maintenance management process, previously described in Chapter 5 and illustrated in Figure 24.1. Refineries that consistently perform well, or top performers, consider this as best practice and use this business process.

The main features of this process are:

1. The key participants in the maintenance process work as a team. Priorities are clearly defined and understood by all. All work is screened by this team.

2. Proactive work is determined with the help of risk-based methodologies. It is planned and scheduled over a long period. This is the long look-ahead plan of known work.
3. Emergent work is subjected to daily scrutiny and appropriately prioritized.
4. Backlog is used as a repository of work. It is managed within defined parameters, e.g., ceilings on total volume and residence time for each item.
5. The current week's work-plan is firm. It consists of proactive work and any high priority emergent work which was known before the issue of the plan the previous week.
6. Work on this plan will be displaced by new emergent work only if the team decides that the new work has high enough priority. Otherwise it will be put in the back log repository.
7. The following weeks' work-plan consists of proactive work and appropriately scheduled emergent work from the repository.
8. Every week the work executed in the earlier week is reviewed. Any lessons learned can be extracted and applied in the future.

Most elements of this process—e.g., the challenge of day-to-day emergent work by the joint maintenance/operations team, risk-based prioritization of work, and backlog management—were already in place. Tasks arising out of

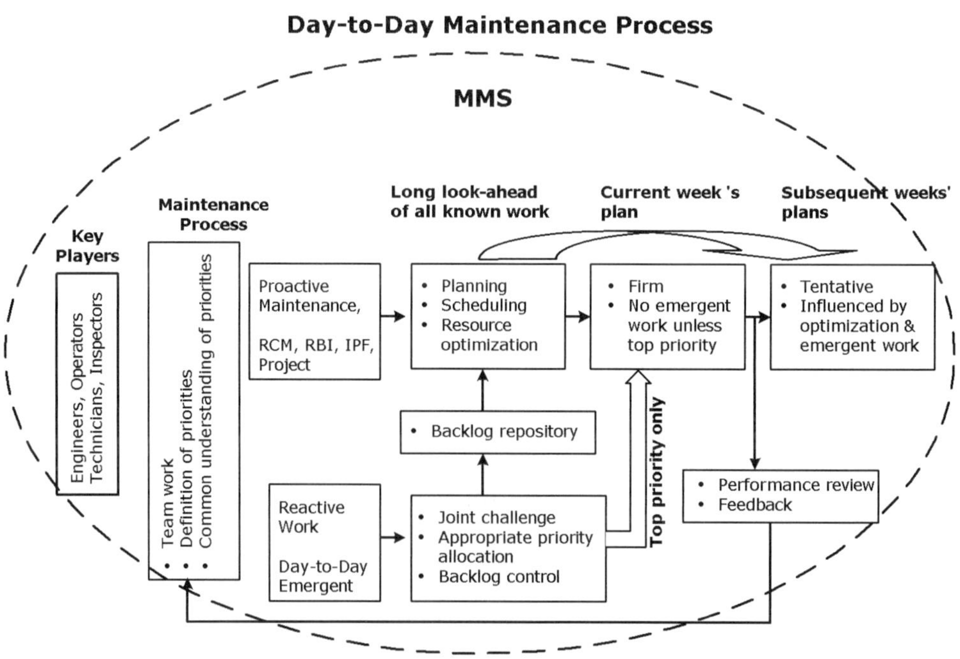

Figure 24.1 Day to Day Maintenance Process

Actual or potential consequence (LOW → HIGH)	CONSEQUENCES: People (Risk to Safety)	Assets Production (Financial Risk)	Environment (Risk to Environment)	Rating	← LOW PROBABILITY HIGH →: Unlikely within 12 months / Improbable / A	Likely in <12 months / Possible / B	Likely in <3 months / Medium / C	Likely in <2 weeks / Significant / D	Likely in<1 day or now occurring / High / E
LOW	Minor Risk of Injury	Damage/ Loss <$1k	Slight Effect, On-Site Impact	1	4	4	4	3	3
	Slight Injury First Aid	Damage/ Loss <$10k	Minor Effect, Single Breach	2	4	4	3	2	2
	Minor Injury Record-able	Damage/ Loss <$100k	Local Effect, Multiple Breach	3	3	3	3	2	1
	Major Injury LTI	Damage/ Loss <$1M	Major Effect, External Interven-tion	4	3	3	2	1	1
HIGH	Permanent Disability or Fatality	Damage/ Loss >$1M	Massive Effect	5	3	2	1	1	1

Explanation of Priority Definition	
4	CMMS Prio. 4 - Longer Term Backlog, after considering "Do Nothing" option
3	CMMS Prio. 3 - General Backlog, for completion within 3 months
2	CMMS Prio. 2 - Priority Over "2 week Plan" Work
1	CMMS Prio. 1 - Priority Over Today's Planned Work
1	CMMS Priority "RUSH" - Justification for Call-In if no mitigating ways with resources already on site

Figure 24.2 RAM for Maintenance Priority Setting

the recently completed RCM and RBI exercises were stored in a data base. They would be reviewed after two years.

As part of the day-to-day maintenance process, a look-ahead plan has to be made. This is a series of weekly schedules starting from the current week and going as far as possible into the future. These weekly plans are assigned optimized resources for all known work, such as periodic routines arising from RCM and RBI analyses and known project-related work. This is the base-load plan, to which the day-to-day emergent tasks are added once they have been screened and prioritized. The process described in feature #6 above defines how emergent work is handled. The following weeks' schedule follows the process in feature #7 above.

24.3 Using the Process

The company agreed to create a look-ahead plan for 104 weeks. At the end of this period, the original RCM and RBI results were due for review. They transferred all these tasks from the RCM/RBI output data base into this schedule. To this, they added any known project-related work.

Thereafter, they estimated the resource requirements for these tasks. The next step was to optimize the schedule so as to level the resource requirements. The start and end dates of most tasks had some flexibility. By using the available slack and shuffling the tasks, resource requirements could be leveled fairly well. The optimized 104 weeks look-ahead plan was now ready for use.

24.4 Rush Work

Rush work is hard to plan and costs at least twice as much as planned work. Moreover, rush work displaces planned work from that day's schedule. This wastes effort already spent on planning the displaced activity. Therefore, rush work has to be properly justified. It is a major drain on the efficiency of a work management process. What we need is an effective work prioritization system.

24.5 Prioritization of Work

The refinery already had in place a prioritization method for maintenance work. This was based on an estimation of the business risk if a task was NOT carried out. It had been in use for some time but it was open to individual interpretation of consequence and probability. This led to emotive discussions at times. Another refinery had developed and used a risk matrix for this purpose (see Chapter 26, Figure 26.1). Using a similar process, they developed a risk assessment matrix (RAM). This is, illustrated in Figure 24.2.

24.6 Launch of the New System

It took three weeks to complete all work relating to the preparation of the 104 week look-ahead plan. Once they launched it, daily scrutiny and challenge of all emergent work began in right earnest with the help of the newly-adopted matrix. During the first two weeks, none of the emergent tasks were assigned the top priority. In the first year, they needed to add 5 jobs a week on average to the next week's schedule and just 1 job a week to the current week's schedule.

24.7 Results

During two years of observation, there were a total of 11 jobs for which the current day's schedule had to be interrupted. During the same period, there were a total of 59 call-outs; i.e., an average of just over one a fortnight. These were for jobs with an average duration of 30 minutes per call-out, but the responder was paid for 4 hours in accordance with the trade agreement.

With fixed and dependable schedules published in advance, all technicians knew what was expected of them. The operators also knew when and which equipment to prepare for maintenance. The supervisor became more of a facilitator. Technicians did not have to wait for instructions or job hand outs in the morning, unless there was an emergent top priority work. All these contributed to increased efficiency of execution.

In the first two years, they achieved a plan compliance rate of nearly 90%. The ratio of top priority jobs to total jobs was less than 3% (see also Chapter 25)

The company reduced overtime work by 20%. The new system helped lower contract labor costs. In international benchmarking exercises, they continued to occupy top quartile maintenance and reliability performance.

24.8 Lessons

1. Top management can create the environment necessary for consistent high performance.
2. The techniques of Reliability Centered Maintenance and Risk Based Inspection enable creation of weekly schedules a long time into the future.
3. This level of planning with optimized long-term schedules gives significant benefits.
4. An effective rational system for prioritization of work is essential for the success of a planning system.
5. With fixed and dependable schedules which are published well ahead, workers need not wait for task allocation. This eliminates delay and wastage of time.

24.9 Principles

Adopting the right business process can bring a step change in performance. Implementing such changes requires careful planning and preparation, and they must blend with the local social and cultural environment. The use of the best available technologies, tools, and software support can enhance this process significantly. Before using these, we need a stable environment where proper systems are already in place and working well.

Chapter **25**

Workload Management ...from chaos to order

Simply by arranging the next day—defining on paper what I want to accomplish—I feel that I have a head start.

Mark McCormack, Founder and Chairman of International Management Group.

Author: V. Narayan

Location: 2.1.3 Petroleum Refinery

25.1 Background

In this refinery, I was responsible for the mechanical maintenance of the process plants and in charge of the mechanical workshops. The maintenance areas were aligned to the production sections, with a supervisor in charge of each area.

Overtime work had risen steadily due to the high level of breakdowns, and was running at about 15 % at the time of these events. Operations staff realized that those who shouted loudest got the attention of maintenance; the result was a high noise level. Maintenance craftsmen were constantly being moved from job to job, resulting in low productivity and quality. Equipment downtime was high as craftsmen were unable to complete many jobs in time. Morale was low, both in maintenance and in operations.

A colleague and I had recently attended an optimum maintenance course conducted by a then well-known authority, Dr. Howard Finley. We both came back from the course full of enthusiasm and charged with ideas. These were the days when the bath-tub curve was the only known reliability model, and my colleague was obsessed with being able to find the knee of the curve. I had other ideas. What we had learnt from Dr. Finley was that setting work priorities was the first order of business. I convinced my colleagues that we should tackle that quickly.

25.2 Existing Foundation

There was a clear set of definitions of priorities in place. Jobs with an immediate effect on safety of people or equipment had the highest importance, denoted Priority A. Similarly, if there was an immediate and significant production loss, that was also Priority A. If the loss was not immediate but was considered significant, it was assigned a lower priority, denoted Priority B. All other jobs were assigned the lowest level, namely Priority C.

25.3 Corruption of the Priority System

However, the system deteriorated over the years to the point where most jobs were marked Priority A or B. We had not identified the persons who could authorize the assignment of these priorities. Maintenance had to respond to suit the assigned priorities, so operations marked up jobs to higher priority levels. This meant working overtime to clear the 'urgent' workload, and often meant interrupting jobs to attend to other breakdowns. Chaos reigned, but we had gotten used to this state of affairs for some time. We had 60–70 percent Priority A, 20–25 percent Priority B, and only 5–10 percent Priority C jobs in the system.

25.4 Analysis

Most people were aware of the misuse of the priority ranking system. They knew that if they wanted attention quickly, all they had to do was to raise the ranking. Maintenance did not have the authority to question these, and operations allowed practically anybody to assign the priority. Nobody bothered to read, understand, or apply the existing priority definitions. We proposed a new authorization system where only priority C jobs would be authorized by the operations shift foreman. Priority B jobs went up by one level to the operations supervisor, and Priority A jobs by two levels to the section manager. We expected that with this system, the need for a higher priority would be challenged at each stage. This would improve the quality of the ranking.

25.5 Communication and Training

The two of us decided that all the supervisors in operations and maintenance, section managers, and support department heads had to be brought on board to share our vision. We compressed the week-long session we had attended into one lasting two hours. We sold the idea to management, persuading most of them to attend the presentation as well. Because of the numbers involved, we conducted 10 presentations, with 10–12 people on each occasion.

Presented with the philosophy, examples of best practices, and opportunities, most attendees responded favorably. Toward the end of each session, we explained the effect of incorrect priority setting and our suggestions on how to tackle this issue. There were some people we could not convince, but there were also many who shared our enthusiasm and were eager to start the journey.

25.6 When the Rubber Hits the Road—Pilot Study

Thus far, we had completed the easy part of the exercise. Selling the idea to operators was a different ballgame. In our training sessions, we identified one section where the manager and supervisor were both very supportive. We decided to pilot it in this section.

I had the advantage of having the maintenance supervisors in the plant and in the workshop as direct reports. I explained to them the need to provide a high level of service to the pilot group. As far as the pilot group was concerned, shouting loud was no longer an option, so additional safeguards were required.

My colleague kept a close watch on the performance of both operations and maintenance. Whenever we felt that priority setting was incorrect, we brought it to the attention of the operations supervisor. From the maintenance side, if there were any delays, one of us intervened, so we retained a high service level.

The first two weeks were chaotic, and we thought that the experiment was a failure. By the third week, however, things started to improve. Operations started noticing that work was being completed on time and to better quality standards than they had seen before. This was mainly because maintenance craftsmen were able to complete jobs they had started without interruption. The positive comments from the operators helped raise the motivation among maintenance craftsmen. It took a further six weeks or so before we could see an improvement in the reliability of the pilot section processing units. At the same time, the overtime level dropped to less than 5 percent in the pilot area, thanks to the fewer breakdowns. The distribution of priorities in this area changed quite significantly. We now had only 5–10 percent Priority A, 20–25 percent Priority B, and about 70 percent Priority C jobs in the system. Scheduling of work became a lot easier. This was in line with the theory we had learned earlier. The before-and-after situations were similar to that of poor and best-in-class performers illustrated in Figure 25.1.

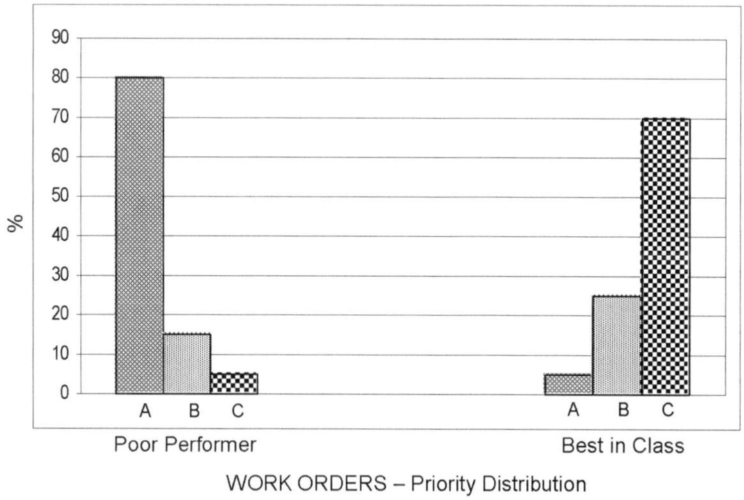

Figure 25.1 Comparison of Priority Distributions

25.7 Roll-Out to Other Sections

With this successful trial, it was relatively easy to introduce the authorization system to the other sections. We did this in a phased manner, taking six months more, so that people could adjust themselves to the revised ways of working.

In the past, the maintenance supervisors usually stayed back, mainly because some of their people were working overtime. They were themselves not paid any extra money, but working late became a custom. Once the new system was in place, they continued to stay late even though there was no overtime work in progress. It took quite some effort, including mild threats of disciplinary action, to break this habit.

In spite of the obvious success of the program, it was very difficult to introduce it in one section. Here the section manager did not agree with the change. In the weekly planning and scheduling meetings, chaired by co-author Mahen Das (who was then the planning engineer), the performance in this section stood out like a sore thumb. Eventually the reluctant manager gave in and allowed us to implement the new authorization system.

25.8 Results

The implementation was fairly successful. The improved reliability and reduced overtime raised the credibility of the maintenance department and made further improvements easier. We had managed to bring back some order from the chaos.

25.9 Lessons

1. Get all of the stakeholders on board from the beginning. Preparing the soil before planting is always a good approach. Sharing your vision with all those affected by the proposed change makes implementation easier.
2. Find at least one customer who is willing to try your proposal.
3. Your credibility is important; this means that your team may have to put in extra effort in the early stages.
4. Measure performance before commencing the program and at regular intervals during and after implementation.

25.10 Principles

1. A business process is often more important than technology.
2. Identification of business benefits before starting the program is quite important. Sell this idea to the customers.
3. In any program in which people are involved, introduce change gradually.

Chapter **26**

Infrastructure Maintenance

The key is to understand the processes that lead to superior performance, not to arbitrarily cut budgets with the hope that everything will turn out all right.
Ron Moore, Author

Author: Mahen Das

Location: 2.4.1 Medium Sized Semi-Complex Petroleum Refinery

26.1 Background

In Chapter 5, we discussed the reasons for the deteriorating business performance of the refinery. In a knee-jerk reaction, the company board simply passed a directive to "cut costs or else" down the line. When the crunch came, a maintenance philosophy was not in place. Cutting costs resulted in slashing the budget willy-nilly. As is often the case, the first victim of such cost cutting is the infrastructure. These assets tend to deteriorate slowly. As a result, neglect is less noticeable and, therefore, less disturbing in infrastructure than elsewhere.

26.2 Infrastructure Items

In this chapter, we include the following assets in the term infrastructure:

- Roads and culverts
- Ground level and overhead pipe-tracks
- Structures carrying overhead pipe-tracks
- Storage tank farms
- Outside plot pipelines
- Drainage channels
- Jetty supporting structures
- Boundary fences
- Outside-plot pump houses

26.3 A Trigger Event

The neglect due to cost cutting went on for three years. It would have gone on longer but for an incident which served as a wake-up call. A gasoline line from storage tanks to the loading jetty leaked and many gallons spread over the pipe track before the section could be isolated.

Some sections of ground-level pipe tracks were prone to flooding during rain showers. Also, as the accumulated water flowed away, small amounts of debris and dirt were carried with it and deposited where overgrown grass offered more resistance to the flow. One such point happened to be underneath the gasoline line in question. The accumulation built up and started deposit corrosion on the underside of the line. The leak was detected fairly quickly, but we had to stop loading operations immediately. Only after completing a temporary repair and making the ground around the spillage safe could we use this pipeline. A quick inspection of all pipe tracks revealed several other spots where the same hazard existed. We suffered significant losses due to shipping delays, product loss, and cleaning costs as a result of this incident. Potentially, we faced a serious public relations problem as well, due to high local sensitivity to environmental damage.

26.4 Inspection Results

A systematic and thorough inspection revealed the following:

- Sections of roads with potholes
- Culverts choked with debris, contributing to pipe track flooding
- Badly deteriorated storage tank foundation pads
- Partially choked drainage channels, contributing to pipe track flooding
- Cracks in the concrete jetty structure

26.5 Where Do We Start?

When funds are scarce we have to prioritize work so that resources may be used to maximum benefit. For this we needed a methodology to take rational decisions. But we did not have such a methodology. I was at my wit's end in trying to find a solution and sought help from my colleagues in the management team. The finance and commercial managers thought they could help. They were using what they described as decision matrices; one for deciding when to buy foreign currencies and another to decide when to buy crude oil or sell products on the spot market.

We set up a cross-functional team comprising a maintenance engineer, an operations day assistant, an inspector, and a young economist from the finance function. We asked them to design a simple decision matrix, similar to the ones already in use in Finance, which could help prioritize maintenance projects. They were to report back in two days.

26.6 Developing the Decision Matrix

The result of their work is illustrated in Figure 26.1 on the next page.

Risk Assessment Matrix

Noticeable $1,000	N	N	N	L	L	L
Important $10,000	N	N	N	L	L	L
Serious $100,000	N	N	L	L	M	H
Very Serious $1,000,000	N	L	L	M	H	H
Disaster $10,000,000	L	L	M	H	E	E
Catastrophic $100,000,000	L	M	H	E	E	E
↑ Consequence Probability →	0.0001 Unlikely	0.001 Rare	0.01 Likely	0.1 Very Likely	0.5 Toss-up	1.0 Certain

Increased Risk

N = Negligible L = Low M = Medium H = High E = Extreme

Figure 26.1 Decision Matrix

The matrix is based on the definition of risk, as described by the following equation:

Risk = (Consequence of an event) x (Probability of that event happening)

In this matrix, the team graded Probability into six categories—unlikely, rare, likely, very likely, toss-up, and certain—along the horizontal axis. Similarly, they graded Consequence into six categories, ranging from noticeable to catastrophic, along the vertical axis. The impact which each category of consequence had on the business was specified in real money terms.

Risk was graded into five categories, from negligible to extreme, depending upon the value of the product of an event's probability and its consequence. Each cell position represented a grade of risk, which was the product of consequence (vertical axis) and probability (horizontal axis). The management team helped the team finalize each risk category grade. For example, an event which was certain to happen and had a catastrophic business consequence ($100,000,000) was graded as an event of extreme risk for the business. Another event of the same consequence, but unlikely to happen (a probability of 1 in 10,000), was graded as an event of low risk to the business.

26.7 Using the Matrix

In order to determine the importance of any project, we set up other cross-functional teams (with at least 2 people) to assess the risk to the business if a project was not carried out. They estimated the likelihood and the consequence of occurrence. For example, a team comprising the maintenance engineer, a shift supervisor from oil-movements, and an external civil engineering consultant determined that repairing the cracks in the concrete jetty structure would cost $100,000 if carried out without delay. Each year of delay would increase the cost exponentially. A delay of 5 years could put the business in extreme risk category. This was a shock to us. We had always thought that concrete needed no maintenance, so it had become invisible.

Similarly, a team comprising the outside plot maintenance supervisor, an inspector, and a shift supervisor addressed another project with the following scope.

- Clean up the pipe-tracks, and keep them clean
- Inspect pipes
- Reinforce thickness if required
- Touch-up paint where required

They determined that the project would cost $80,000 if done without delay that year, with a recurring annual cost of $20,000. If postponed for a year, the business risk would move into the extreme category.

This process was applied to all visible manifestations of past neglect. They were ranked in order of the risk they represented if not carried out.

26.8 Finding the Money

The budget for the year included projects of various magnitudes, e.g., overhaul of a number of major items of rotating equipment, cleaning, and internal inspection of a number of storage tanks. These were justified by historical time-based logic. We reviewed these projects with the new matrix and ranked them in the same manner as the new infra-structure projects. This process helped displace some of the already-approved projects with some new ones. The jetty and pipe-track projects mentioned above were among the justified projects.

From that time onward, we used the decision matrix for all maintenance budgeting work.

26.9 Results

One more piece of the puzzle was in place to keep us on our improvement path. There were no further serious incidents, thanks to the diversion of funds into the infrastructure projects. Many people at working level applied the matrix to evaluate business risks. This helped us communicate the method to a larger section of the employees. We were also able to demonstrate to pressure groups that we had a sound, logical approach to manage risks.

26.10 Lessons

1. Top-down cost-cutting directives can lead to deterioration of assets, resulting in major risks to the business.

2. Using cross-functional teams to solve specific problems is very effective. Each function looks at the other from a lay-person's perspective and throws challenges without the risk of offence. The result is a well-balanced product.

3. A risk matrix brings objectivity to decision making.

4. The 10x rule for concrete:
   - Cost C to fill slight cracks
   - If left until the crack gets big: cost 10C
   - If left until spalling starts: cost 100C

5. As an engineer, it was a humbling experience to learn how to do my job better from the finance and commercial managers.

26.11 Principles

To manage any business well, we must manage risks effectively. We have to replace arbitrary decision-making regimes with logical processes that evaluate risks and help minimize them.

Cross-functional teams bring diversity and unconventional solutions; in general, we do not tap into their power fully.

Chapter 27

Workflow Management

McGregor's Theory X or Theory Y, authoritarian or participative management?

So much of what we call management consists in making it difficult for people to work.

Peter Drucker, Management Guru.

Author: Jim Wardhaugh

Location: 2.2.2 Large Complex Refinery in Asia

27.1 Background

A benchmarking exercise showed that our refinery was a poor performer, and while it was making lots of money, it had the potential to make much more.

There were many problems we could identify. The one that we look at here is how we managed maintenance and how we could change things for the better. We changed the organization to replicate the lean look of the top performers, and empowered our technicians to take a more active and decisive role. We also modified our CMMS to help with these changes.

Business process redesign had become fashionable about that time and we imported a few ideas from that field. These ideas were to:

- eliminate work if possible
- reduce number of people handling a transaction
- eliminate unnecessary authorization hurdles
- work in parallel, not in series
- eliminate duplication
- organize around results, not tasks

27.2 The existing organization

The organization was very traditional and functional in style. We saw nothing wrong with that. It might not be the fashionable business unit style, but the traditional and functional organization had stood the test of time. While there were problems with it, we knew how to handle these. There are aspects of this type of organization which bring benefits. In locations where workforce skill sets are not very broad, the compartmentalized approach brings clarity and confidence. Figure 27.1 shows the top level of the Engineering organization in a simplified way.

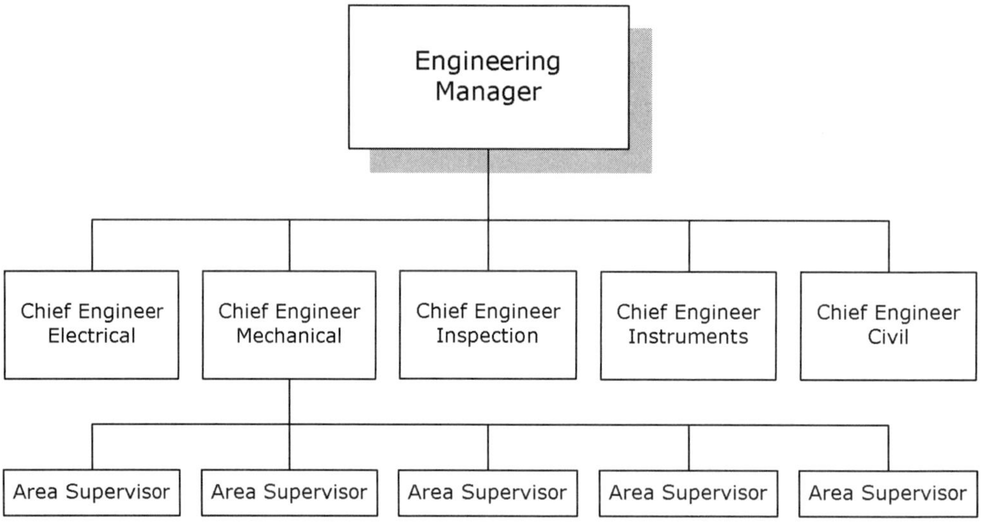

Figure 27.1 Simplified Engineering Organization Chart

Each discipline had a Chief Engineer who reported to the Engineering Manager. Under each of the Chief Engineers were a bunch of Area Supervisors each of whom ran a geographic maintenance area. Figure 27.2 shows the organization of an Area Mechanical Supervisor.

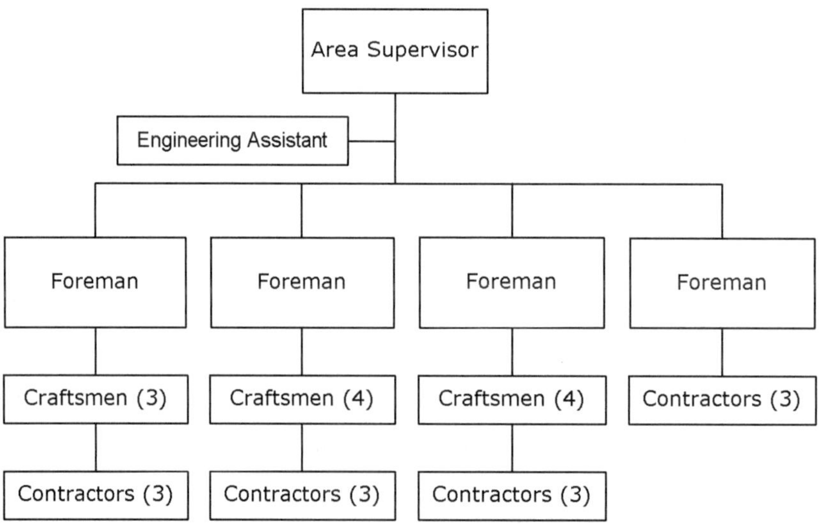

Figure 27.2 Basic Mechanical Element before Re-organization

In the element shown, we found that the Area Supervisor spent most of his time in meetings and the Engineering Assistant designed small changes and improvements. The foremen provided overall supervision, while the company craftsmen then gave detailed direction to the contractors, who did all the hands on tools work.

The mechanical grouping is shown. Instrument, Electrical, and Civil had similar, but separate groups. These operated in a similar way.

27.3 Performance of the Existing Organization

We looked at these groups and saw:

- High staffing levels compared with top performers
- Excessively high level of supervision
- Roles and responsibilities were unclear
- A slow response to problems
- A defensive compartmentalization of activities
- Change was difficult to achieve

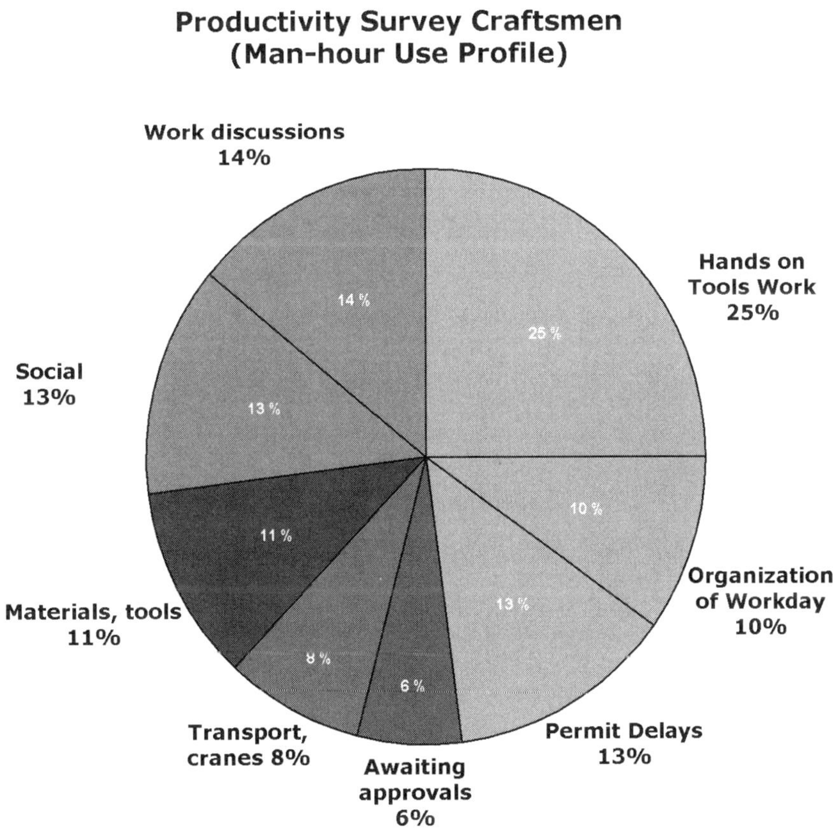

Figure 27.3 Time Utilization in Normal Work-Day (based on possible hands-on tools time)

We did a productivity survey by sampling representative maintenance activities over about a month. The results are shown in the pie chart shown in Figure 27.3. 100% of the pie represents the total time that was practically available for work in the normal day. We removed from consideration time for lunch and tea breaks.

What was striking was the amount of delay. The role of supervisors was to facilitate work so that delays were minimized. If supervisors were doing their jobs well, the delays shown in the figure would have been minimal. We had a huge number of people supervising formally and informally, so it was apparent that supervision was not effective.

We captured some more numbers to confirm this and found that:

- 80% of jobs were treated as rush
- 10% of jobs (only) were scheduled for next day
- 15% overtime for craft and contractors
- Backlog for Inst./Electr. was negligible
- Backlog for Mechanical was 1 week

Sounds familiar? Recall the discussion in Chapter 25—that related to events many years before the ones we are discussing now.

27.4 Some Management Theory

In 1960, in his book The Human Side of Enterprise, McGregor demonstrated that the way managers manage depends on assumptions about human behavior. He grouped these assumptions into two broad theories:

X where the worker needs (and indeed wants) to be directed and con trolled

Y where workers' and organizational goals can be beneficially aligned

I'm a Theory Y man, as are most other managers I have met. I have always resisted the belief that we should organize in such a way that the workers feel they should leave their brains hanging on the gate when they get to work.

When reviewing CMMSs for functionality and user friendliness, it became apparent that many CMMSs of the day were constructed along Theory X lines. It was assumed that all work would be planned and scheduled in the minutest of detail before releasing them for execution. Planning to this level is what economists call "Fatal Conceit" and is one of the reasons why the communist planned economy failed.

Our review of how top performers did their business suggested that we should go strongly for Theory Y. In the maintenance organization we were looking at, this implied:

- a very flat organization
- technicians would do the hands on maintenance and inspection work
- technicians would be given clear roles in geographic areas and respon sibilities
- technicians would pick up their own jobs from the CMMS, create a just-in-time backlog and schedule
- technicians were empowered to arrange permits, spares, drawings,

manuals, scaffolding, transport, cranes, etc.

- technicians would enter work history
- we have a CMMS that allows high visibility of actions and overview

27.5 Redesign of the Organization

Key members of the Engineering Management team spent time identifying how to organize a top performing operation and make it work. We decided to establish a concept of centralized direction (for best practices, rules, etc.) with small maintenance groups focused around individual production zones.

- Zones were to be geographical / production based in general, e.g., crude distillation, hydro-processes.
- People in the groups were to be accountable for zone performance.

In this concept each Zone would be:

- A business unit, as small as reasonably practical
- Part of the overall operation
- Operated by a team who controlled, maintained, and improved that zone against defined performance standards

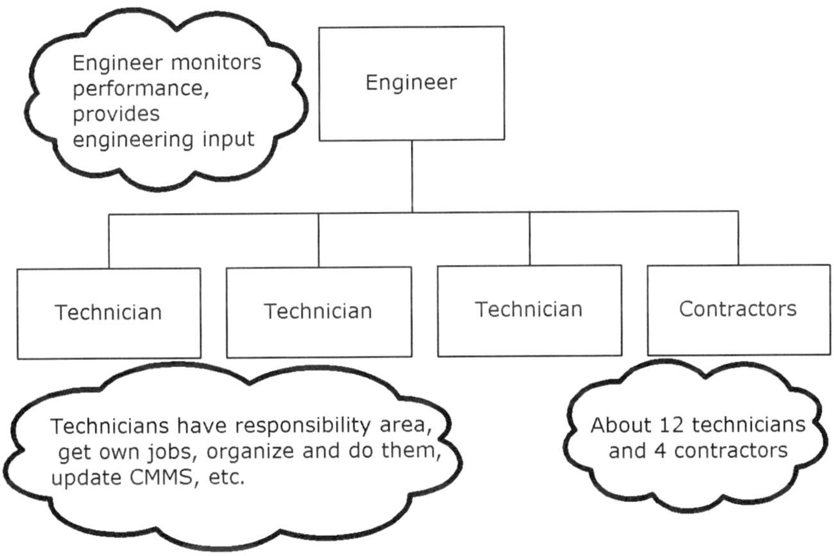

Figure 27.4 Revised Maintenance Element—Zone

Figure 27.4 shows the re-organized maintenance element, which we now called a Maintenance Zone. It covered the same work responsibility area as shown before but had significantly reduced levels of supervision and staffing. In the new mechanical zone there were about 11 company technicians with 4 or 5 contractors on routine activities. Roles had been changed substantially from those seen previously:

- Engineers monitored performance and provided the inputs needed by technicians or their operating counterparts. They also handled very large jobs and budget, appraisals, etc.

- All of the technicians had their own responsibility areas. They got their own jobs from the CMMS, organized them, and did them. They updated the CMMS. They were given significant delegated authority for issuing materials, arranging use of contractors, etc. However, their actions were transparent through the CMMS and finance systems. Thus we could monitor them and this helped prevent misuse of delegated authorities.

Driven by the Fatal Conceit argument, we chose to be very different from some of the received wisdom. We insisted that each technician receive, plan, schedule, and organize his own jobs. Many gurus suggest that all requests should go to a planner, then only when the work is ready should they go to the worker. We found that most jobs were rather repetitive, so the focus was on scheduling rather than planning. It seemed better to us that the person who was going to do the job should be able to interface directly with the scaffolders, crane operators, truck drivers, contractors, permit signatories, and the materials group. These short communications paths seemed inherently more error free than by using third party intermediaries. Detailed daily contact with other maintenance disciplines and operators ensured that the technicians

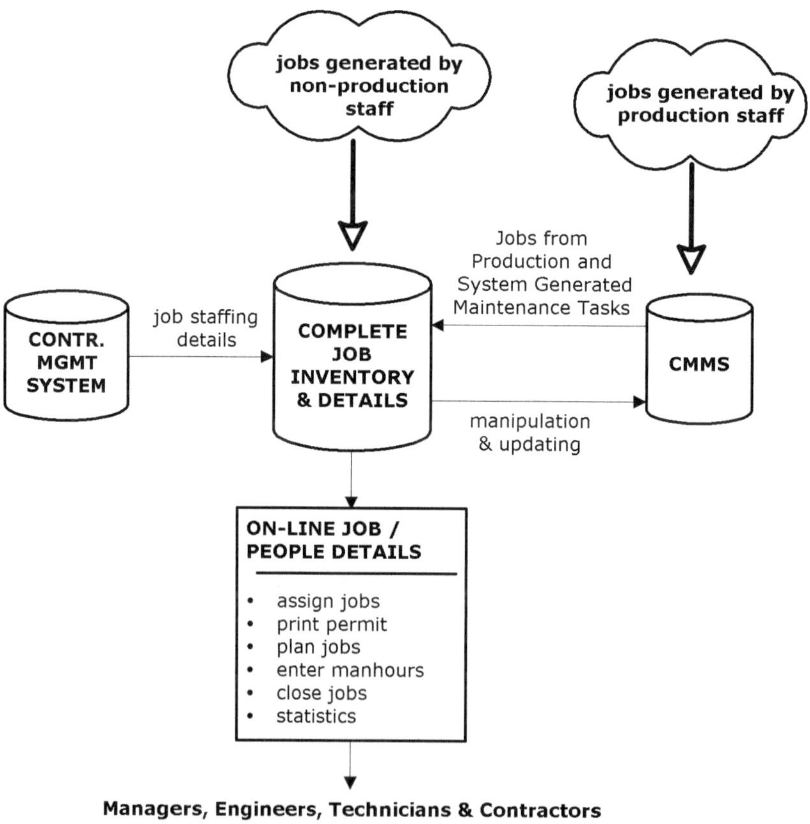

Figure 27.5 Zone Management System Overview

knew what jobs were important. They could then arrange to do these "just in time".

27.6 An IT Enabler

We agreed that an IT system needed to be implemented as an enabler for the new way of working. We wanted a single entry point to computer systems through which zone maintenance personnel could receive work, assign activities to personnel, plan the implementation, print permits, and enter job status, staffing, and man-hours. In a similar way the job could be closed and all relevant failure data entered.

We wanted a system which would make clearly visible what was going on in maintenance and provide full information on jobs, people, and contracts. We wanted the system structure to encourage greater delegation, job planning, and increased output per man. See Figures 27.5 and 27.6 for an overview of the system.

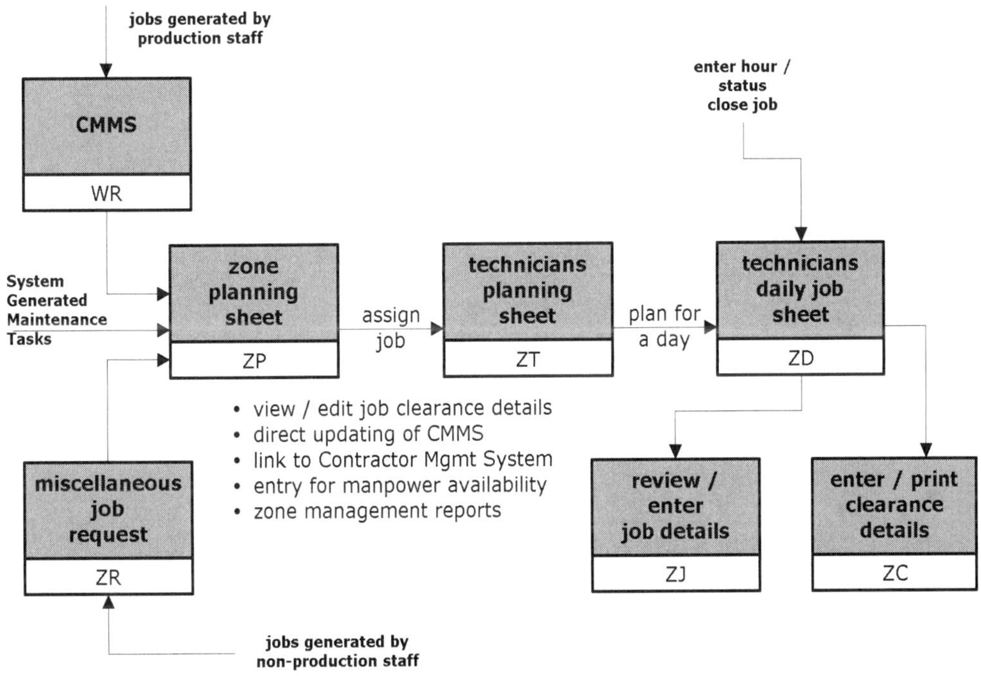

Figure 27.6 Zone Management Screen Structure

What we wanted did not seem to be available readily on the market so we had to build our own. The characteristics we wanted were:

- A user-friendly computer system (minimum key strokes and screens) focused on key users
- To capture work requirements (system generated jobs, requests, sub-tasks)
- To enable easy scheduling and job organization (labor, parts, permits)

- To enable easy entry of repair information
- To make visible details of work, timing of events, and individual productivity
- To provide a package of performance reports

We wanted it to be easy for the Engineering Manager and Chief Engineers to view the screens and question individuals on details of activities. In this way, they would show interest and mentor individuals.

The Zone Management software was to sit on top of the existing maintenance management system, which was very unfriendly to users. This would make the old system invisible as far as possible. The new system was created in house in a few weeks using prototyping methodology and Powerhouse®™, a fourth generation language.

27.7 Results

Very soon after putting the new systems in place we found that things had become a lot better. The characteristics of the organization were:

- Top performer model with few layers
- Clear roles and responsibilities
- Significant delegation
- Better relations with other groups
- Low overtime
- Less rush work and
- Higher productivity

Performance measure	before change	2 years after change
% rush jobs	80	15
% jobs scheduled for next day	10	80
Backlog I/E	None	3 weeks
Backlog Mech	1 week	6 week
Overtime	15%	5%
Number of supervisors	8 (say)	2
Number of workers	20	16

27.8 Some Problems

We were very comfortable with the results, but there had been problems. As always, they tended to be people problems rather than technical; and as always, effective change management was critical to success. In summary, these were the problems:

- Some technicians only wanted routine tasks, and didn't want to get involved in managing interfaces or doing complex diagnosis—they were allocated preventive/condition monitoring tasks, which didn't present too much of a challenge.
- Some foremen didn't want to take on the "engineering" facets of the zone engineer. They chose to leave, while some others chose to down grade to technician.
- Some engineers couldn't handle front-line technician issues without a foreman. They chose to leave while others moved to design activities
- The union was sensitive to being sidelined.

27.9 Lessons

- Technicians could capture their own jobs, run a backlog of about 20 jobs, order them to be done just in time, and do them effectively.
- Most technicians liked this way of working (Theory Y). A few of them wanted only routine structured work with little challenge and little decision making (Theory X).
- Productivity increased significantly on implementation.
- The Zone Management computer system brought high visibility to all the activities and highlighted the performance differences between technicians. The poor were dragged up by the better.
- Delays reduced as authorization hurdles were removed.
- Backlog reduced so far that we had to reduce maintenance staffing levels.
- The average supervisor is a barrier to work progress rather than an effective facilitator.
- Modern technology and software tools enable new ways of working that empower and harness the skills of the entire workforce. Prototyping software permits us to create bespoke solutions quickly.

27.10 Principles

We get the productivity we deserve. All too often we build barriers to effectiveness and do not use the full skills of the workforce.

Visibility of their performance can embarrass the poor performers into action. We found it much more effective than threats or sanctions.

The fewer the layers we have, the faster the decision making becomes. Empowerment of the workforce enables workers to manage their own work and taps latent talents.

Using IT systems to bring transparency ensures we delegate work while retaining control. Visibility of performance embarrasses the poor performer into action.

Additional Reading

i. McGregor, D. 1960. *The Human Side of Enterprise*, New York: McGraw-Hill. ISBN-10: 0070450927

ii. McGregor, D. 1967. *The Professional Manager.* New York: McGraw-Hill. ISBN-10: 0070450935

iii. Heil, Gary, Warren Bennis, and Deborah C. Stephens. 2000. McGregor Revisited. New York: Wiley. ISBN-10: 0471314625

PART 6: EXECUTE

Chapter 28

Trip Testing

The superior man, when resting in safety, does not forget that danger may come. When in a state of security he does not forget the possibility of ruin. When all is orderly, he does not forget that disorder may come.
Thus his person is not endangered, and his States and all their clans are preserved.
..... Confucious, 551BC to 479BC.

Author: V. Narayan
Location: 2.1.4 A Large Petroleum Refinery

28.1 Background

On taking up my position as area engineer, one of the things I found at this site was quite intriguing. We were doing 'function-testing' as the sole method of checking whether large rotating machinery tripped on overspeed (or other trip signals).

When there is an unsafe condition such as excessive axial displacement, the detector sends signals to the logic device (black box). This processes the signals and initiates corrective action by the actuator. Function tests do not exercise the actuators. Therefore, they (the final element as instrument engineers call it) may give a false sense of security—we cannot verify whether the valve or other final element would actually move (open or shut) as required. On the other hand, they do help to avoid the production downtime caused by full checking of the complete trip systems.

28.2 Hidden Failures

Some failures will be evident to the operators during the course of their normal duties. Thus, if a compressor's delivery pressure or flow falls to unacceptable levels, the control room operator will be alerted by the readings of the flow or pressure instruments. In some cases, there will be alarms that will highlight such events. If a seal leaks, there will be a pool of process fluid on the floor, so the field operator can see it. Worn bearings will produce higher vibration or noise levels so that the operator can call in the condition-monitoring technician to verify their condition. Similarly, a fused light bulb will be obvious to the observer.

There is a class of failures that are called hidden failures; these are not evident to the operator. It is not possible to look at the item and know whether it is still working or whether it has failed, as there are no indications to guide the operator. For example, it is not possible to say whether a gas detector, circuit breaker, or pressure relief valve is working at any given time. Similarly, we cannot say for sure whether a fire pump, emergency generator, or standby process pump will start on demand. A drain or vent valve on a vessel or pipeline may or may not open; furthermore, even if the wheel turns, there is no guarantee that the valve itself will work as required. The drain or vent pipe may be plugged with debris, preventing fluid flow. Another example is that of instrument protective systems, such as pressure relief or shutdown systems, which may not perform as desired when there is a real demand.

28.3 Technical Integrity (TI)

TI is the absence of foreseeable risk of failure that could endanger the safety of personnel, environment, or asset value. Loss of TI will harm the long-term viability of the facility, sometimes destroying the business itself. For example, after the Bhopal disaster [i], Union Carbide had to close down their operations in India. In the UK, Rail Track [ii] closed down after the Hatfield disaster and Ansett Airlines [iii] in Australia suffered a similar fate. In all these cases, the company involved lost TI. It is, therefore, very important to manage TI effectively. Hidden failures can seriously impair TI, so one of the things we have to do is to detect hidden failures and ensure that faults are rectified promptly.

28.4 Testing for Hidden Failures

In order to detect hidden failures, we have to test the items in a real or simulated situation. Such tests are also called detective or failure-finding tasks. Testing the sensing elements and logic units can be done during operation by defeating the signal to the actuators. This is what we call function testing. In such a test, the final element, such as an actuator or valve, does not get a signal, so it does not move. The advantage of doing function tests is that production is not interrupted, and there is no loss associated with such tests.

The alternative is to do a full test where the equipment, system, or plant shuts down. Such complete tests can cause large production losses. As a result, there is considerable reluctance to do them. Such a decision is often based on a limited understanding of the TI issues involved and of the resulting risks to the facility

28.5 Pilot Trials

In the Utilities Department, which provided air, electric power, and potable and sea water to the refinery, there were two 5 MW steam-turbine driven alternators. Only one of these was required during an emergency. The Utilities Operations Manager agreed with my proposal to carry out a full over-speed trip test. In general, these tests require the load to be physically disconnected, e.g., by dropping the coupling, to enable the turbine to go over the rated speed.

In larger machines, the over-speed device often consists of a small cylinder sliding in a radial hole at the end of the shaft. This cylinder is positioned slightly off center and its radial movement due to centrifugal forces is restricted by a compression spring. At rated speed, the movement of this cylinder due to the centrifugal force is kept within set limits by the spring. Above this speed, the cylinder movement compresses the spring further, thus moving away from the center. This slight extra movement opens a hydraulic port, allowing a sharp fall in the hydraulic oil pressure. As a result, the trip action of the main steam valve of the turbine is initiated.

In this design, there was a central axial hole in the turbine shaft, about 6″ long. Hydraulic oil pressure was communicated through this hole, by a 1/4″ tube, connected at the other end to the steam trip valve cylinder.

After dropping the coupling, we started the turbine and raised its speed. The machine should have tripped at 108% of rated speed. This did not happen, and at 110%, we tripped the machine manually. On inspecting the trip mechanism, we found that due to the turbine shaft temperature, there was considerable gumming in the port, and at the tubing seal. These were cleaned up, and a new test carried out. The second test was successful.

We carried out an over-speed test on the second turbine. In this case, we could not get the machine to run above the rated speed. An inspection showed that the 3/16″ shaft of the electronic Woodward Governor®™ was slightly bent. After renewing this shaft, we were able to complete the test satisfactorily.

28.6 The Case for Change

We had achieved a 100% failure rate with two tests and two failures. With this ammunition, it was not difficult to convince Operations of the need to test the final executive element. However, we had to limit or avoid production loss. In cases where it was possible, we agreed to limit the movement of the final element by restricting its stroke with a mechanical stop. This would allow the executive element, namely the actuator or valve, to move slightly, by 1/16″ to 1/8″. This movement shows that had the mechanical stop not been installed, it would almost certainly have moved its full stroke. The small movement did not affect production in any way, but it would de-gum moving parts. While this test was better than a functional test, it still did not guarantee, for example, that a shutdown valve would close fully and be leak-tight. Such tests could only be carried out just prior to a scheduled shutdown if we had to avoid production losses. Operations agreed with the proposal.

28.7 Results

We tested a number of turbines, typically those driving boiler-feed pumps or forced-draft fans. Of the first five steam turbines we trip-tested with the revised method, three more failed to trip. This confirmed the experience with the emergency generator turbines, and established the need to change our trip testing strategy.

28.8 Lessons

1. The need to avoid production losses can blind us from doing a very important duty of maintenance—to ensure TI and hence plant safety.

2. We can improve TI and avoid losses if we can accept a slightly less-than-perfect solution.

3. The alternative is to accept production losses so as to be able to assure we meet TI requirements fully.

4. Trip testing is an important safety check, and must be performed diligently.

5. There are ways to eliminate or limit the associated loss of production, so there is no justification to avoid testing of trip devices.

28.9 Principles

Reliability and safety are closely linked. Protective devices ensure safe operation of equipment and maintenance has a key role to play in ensuring that these devices work whenever they are required. We can ensure we have an acceptable level of technical integrity by testing trip devices at a suitable frequency.

References

i. http://www.newscientist.com/article.ns?id=dn3140
ii. http://news.bbc.co.uk/1/hi/uk/4221465.stm
iii http://www.theage.com.au/news/tips/crash-alert/2005/09/02/1125302695092.html
(Australia, through its Civil Aviation Safety Authority grounded part of Ansett's fleet in 2001 just before its demise, because of concerns about procedures in the maintenance department).

Chapter **29**

Work the Plan

The best leaders provide vision and direction, and establish a common strategy with common subordinate goals for ensuring organizational alignment, thereby ensuring pride, enjoyment, and trust throughout.
Ron Moore, Author

Author: Mahen Das
Location: 2.4.1 Medium Sized Semi-Complex Petroleum Refinery

29.1 Background

As discussed in Chapter 5, the refinery had lost much of its expertise and appreciation for asset maintenance over the years. The management of maintenance and reliability had the following characteristics:

- There was no formally stated philosophy.
- Maintenance was perceived as a cost center, a necessary evil.
- Maintenance business processes were not defined.
- There was a general lack of leadership.

The shutdown (called Turnaround in North America) maintenance process was characterized by:

- Lack of formal setting of premise and objectives
- Work-scope compiled from past history and inspection and operations wish-lists
- Inadequate preparation time
- Critical Path Planning (CPP) prepared but not updated; therefore, used only as wall-decorations for the shutdown cabin
- Willy-nilly changes to work-scope during the planning phase as well as in the execution phase
- Ineffective leadership during all phases of the process
- Preparation as well as execution of work of all disciplines carried out within their departmental boundaries, with little inter-disciplinary communication
- Gross over-runs in duration and cost

On arrival, I could see that there was a competent team, but one which had had little or no guidance and direction in asset maintenance. For the first few weeks, I focused on this situation. With the support of the General Manager and members of the management team, I initiated some key changes. One of these was to define the shutdown process. This is illustrated in Chapter 17, Figure 17.1. The main aspects of this process are as follows:

- Well ahead in time, when the question "Why do we need this shut down?" has been answered, management installs a team leader and identifies future team members in all disciplines, with clearly-defined roles. The premise of the shutdown is clearly established from which the objectives are derived.
- Timely compilation of the work-list, including a review of process-related issues e.g., catalyst regeneration.
- Business risk-based challenge of all items in the work list carried out by a multidiscipline team. The revised work list is the scope of work, which is then frozen.
- Imposition of a tough business hurdle for any new work proposed after the scope of work has been frozen.
- Identification of contractors at this stage.
- The next step is to do a multidiscipline integrated planning, scheduling, and resource (people, equipment, cost, etc.) optimization of all work in the scope. The result is a single plan for all disciplines optimized for all resources. Contractors participate in this activity.
- At this stage, alternative solutions for expensive items of work, e.g.,large scaffoldings, are explored. This is carried out through brain storming to evaluate, e.g., scaffolding rationalization, in a cross-functional team including contractors.
- The actual shutdown execution is a seamless and integrated process from the time the feed is cut off until the time finished products start to flow to storage. During this entire period, the Team Leader is solely in charge. The Leader manages daily coordination meetings, daily safety meeting, completion of inspection before the half-way point, daily update of plan, and a tough business challenge to emergent work.
- Top management team members, including the GM, frequently visit the site and gather a first hand "feel."
- Soon after completion of shutdown, a post-implementation review is carried out. Lessons learnt from this review are used to improve the process for the future.

In Chapter 19, there is a detailed process analysis of a shutdown, including a formal framework and timetable for action. There is also a discussion about the elements which need to be done correctly to bring success. Please refer to that chapter as well for additional information.

At this juncture, we had already scheduled a major shutdown in six month's time. We decided to apply the newly-defined process to this shutdown. We explained to all concerned that a critical success factor for making the plan work would be strict adherence to the key aspects of the whole process as described above.

After the kick-off meeting during which we set the premise, nominated the leader and the team, and derived the objectives, it was time to let the process roll, steered by the team.

29.2 Premise Setting

I obtained agreement from the management team to define the purpose of this upcoming shutdown, i.e., its premise, clearly and unambiguously. Since the commencement date was fixed, it was already a late start. Without a clear statement of purpose, which all parties understood and subscribed to, we would not obtain alignment or focus. I suggested the following premise.

"The purpose of the shutdown is to secure the reliability and technical integrity of the process plants for the next 4 years. All work which is proposed to be carried out during the shutdown period should be for this purpose only, unless there is a very strong economic justification for it to be otherwise. Work foreseen by all disciplines would be pre-planned. Though the duration has already been suggested, it would be finalized after the critical path planning of the agreed work-list has been carried out. The refinery economics were such that downtime should be kept as short as possible. Work would be executed with due consideration to the safety of plant, personnel and the environment."

I explained that this premise had been drafted jointly with two other managers who were present in the meeting. After some debate and clarifications, the premise was accepted as being appropriate

29.3 Roles of Team Members

We defined the roles of the main team members as follows:

Team Leader:

- Has overall charge of the total shutdown project from kickoff to on-grade product rundown, and the responsibility to "make it happen"
- Monitors the progress of the project initially on the basis of a milestone chart and then by a CPP
- Ensures good communication among all participants
- Ensures a seamless and smooth flow of tasks from pre-shutdown preparation, through feed cut-off and hydrocarbon freeing, to startup and on-grade product run-down

Operations Member:

- Participates in preparing the integrated CPP, not only for the shutting down and starting up activities, but also for all other activities as well from the perspective of operations
- Ensures dovetailing of maintenance and operations activities
- Together with the Team Leader, ensures a seamless and smooth flow of tasks from pre-shutdown preparation, through feed cut-off and hydrocarbon freeing, to startup
- Helps the team leader to "make it happen"

Planning Engineer:

- Takes the initiative in preparing the milestone plan and the integrated CPP
- Monitors progress and updates the plans as required
- Responsible for raising alarms if a deadline is in danger of being passed
- Helps the team leader to "make it happen"

Project Engineer, Inspection Engineer, Process Technologist, Materials Representative

- Participate in the integrated planning with respect to tasks of their discipline
- Ensure proactively that their tasks are dovetailed with other associated tasks
- Help the team leader to "make it happen"

29.4 Establishing Work Scope

The objective was to make one integrated plan for all disciplines and then execute it on an integrated basis and seamlessly. We gathered wish lists from all the disciplines and scrutinized them objectively. We could determine whether all listed tasks were in line with the premise and objectives, and justified in the business context. Risk-based techniques for optimization of maintenance work were unknown in the process industry at that time. In Chapter 26, we discussed a decision matrix and its use. From that matrix, we designed a simple risk matrix for classifying the items in the shutdown wish list, shown in Figure 29.1.

A team comprising the planning engineer, the operations supervisor, and a representative of the discipline whose list was under discussion scrutinized every task in the wish list.

From this we produced a consolidated list of work, which we froze at this stage. We used this list to plan and prepare the shutdown. Any additional work proposed after the list was frozen had to cross a tough economic hurdle. We justified the use of this hurdle as an aid to prevent the costly interruption to the progress of the planning process.

Risk Assessment Matrix

Noticeable $1,000	N	N	N	L	L	L
Important $10,000	N	N	N	L	L	L
Serious $100,000	N	N	L	L	M	H
Very Serious $1,000,000	N	L	L	M	H	H
Disaster $10,000,000	L	L	M	H	E	E
Catastrophic $100,000,000	L	M	H	E	E	E
↑ Consequence / Probability →	0.0001 Unlikely	0.001 Rare	0.01 Likely	0.1 Very Likely	0.5 Toss-up	1.0 Certain

N = Negligible L = Low M = Medium H = High E = Extreme

Figure 29.1 Risk Assessment Matrix

29.5 Hurdle for Additional Work

We set the following criteria for approval of additional work proposed after this stage:

- For maintenance and operations work, if not carried out, the economic risk value must be more than the cost of execution including any associated consequential cost.
- For plant-change or project type of work, we set the following guideline, as shown in Figure 29.2.
- Using this hurdle, we ask that not only should the project have the specified pay back, it must also bring in a specified minimum income in the very first year after implementation.
- Any consequential cost associated with the execution of the project, e.g., loss of revenue due to extension of shutdown duration, must be added to the cost of project before payback calculation.

The following example illustrates the use of this simple tool:

Consider two projects. The unit of cost and benefit could be any convenient measure to suit the size of the projects, say, 1 unit=US$1000.

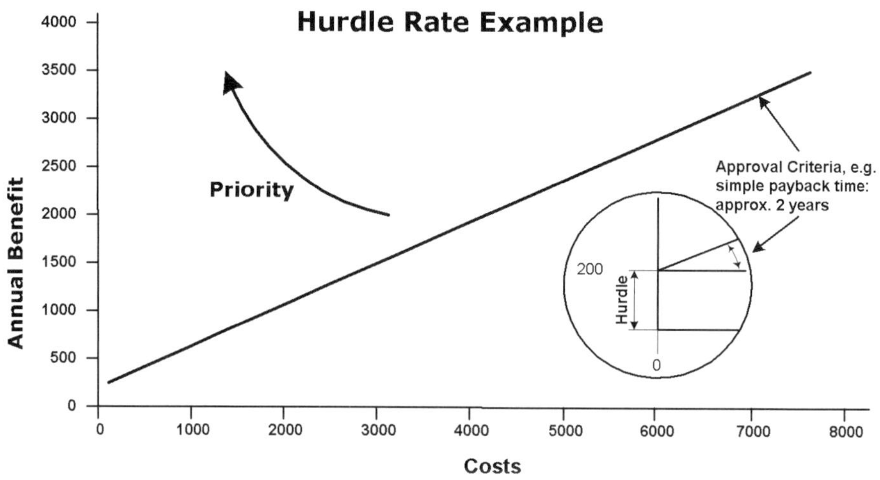

Fig. 29.2 Economic Hurdle for Emerging Work

Project A:

Total cost, including consequential cost, 2000 units
Annual benefit 1000 units
First year benefit 1000 units

Project B:

Total cost, including consequential cost, 100 units
Annual benefit 150 units
First year benefit 150 units

Project A, with its simple payback of 2 years and first year return of > 200 units, passes the hurdle.

Project B fails; although it has a payback of <1 year, which is far better than required, it earns less than the required 200 units in the first year.

With these criteria enforced, there was only one plant change, which made the additional jobs list.

29.6 Planning, Scheduling, and Resource Optimization

Integrated planning meant that there was one plan and one consolidated schedule for all activities starting with cutting out the feed before shutting down and finishing with on-grade products to storage after start-up.

To achieve this, a multidiscipline team was set up. The planning engineer and the operations supervisor were full time members. Representatives from various other disciplines joined in as part time members as and when their work was planned.

Contractors, who were already identified on the basis of unit rates, were invited to participate when work allocated to them was being planned. Other contractors, when identified, were invited to comment constructively on the plan and schedule.

The integrated plan was optimized for all resources. It was a totally transparent plan from which anyone could tell what was going on anywhere at any time.

29.7 Execution

About a week before the plant feed was cut out, the planning team moved to the shutdown cabin in the field. They would help the shutdown leader realize the plan. All execution supervisors had been involved in the preparation and planning of their respective work. They had in turn made their craftsmen fully aware of the schedules. Every one involved was ready to go when the time came.

The leader held a communication meeting every morning, about one hour after start of work. All supervisors were expected to attend. I attended daily as an observer. I encouraged the GM and other members of the management team to attend as observers whenever they could. This gave the meeting the high profile it deserved. The meeting lasted no more than 30 minutes. The planning engineer presented the updated critical path after the progress reported by all supervisors the previous evening had been worked-in overnight. Every one reported their expectations for the day, including any foreseen bottlenecks. At the end of the meeting, every one knew the current status in relation to the plan and the expectations for the day.

At the end of the shift, every supervisor reported the progress of his work to the planner for updating the plan before the next morning.

The leader spent most of his time on site. Other than the formal communication meeting in the morning, he went around walking and talking to the technicians at their places of work. In this way, he opened a direct line of communication with the front line. He often escorted the GM around the site. This radiated the message to every one that the GM gave due importance to this project.

Any emergent work during inspection was subjected to the same scrutiny as new work during the planning phase.

During the last week of the shutdown, I was able to report in the management meeting that the shutdown would finish one full day ahead of schedule and well within the budgeted cost.

29.8 Post Execution Review

We carried this out in the third week after the shutdown. Participants made several suggestions to improve the process. One such suggestion was that technicians be given refresher practical training in techniques which they are required to use only during the shutdown, i.e., once in so many years. Examples are: use of bolt tensioning tools, use of furnace tube expansion tools, etc.

The planner put forth some statistics, as follows:

- There was no lost time injury.
- Actual duration was 30 hours less than planned.
- Unexpected (emergent) work carried out was 6% of planned work.
- Total expenditure was 11% less than budget.

29.10 Lessons

1. A key factor for successful implementation of a plan is strict adherence to it. Some flexibility within an aspect of the plan may be acceptable as long as it does not jeopardize the premise and objectives. Deviations must, therefore, be carefully scrutinized before acceptance. Additional work may be accepted after the work-scope freeze without the first year cash-in criteria if it does not affect the total resource requirement and has a very high payback.

2. Keeping the same team for the total duration of a project, from planning to completion, creates a sense of ownershipand helps turn the plan smoothly into practice. A team working on a project develops a sense of joint ownership for the project. If members are changed along the way, two things happen. First, new members have to be brought "on board," which can take considerable effort. Second, the leaving members take away with them the characters which they had until then been imparting to their roles. Both can have a negative impact on the outcome of the project.

3. Communication and coordination are vital for good execution of multi-discipline projects. As every one in the armed forces knows, good communication is key to the success of any campaign.
4. Visible top management interest raises the profile of a project and general dedication to its success. If people know that top management is interested in the project, the inherent desire to "please the boss" will work for the success of the project.

5. Appropriate business hurdles ensure scarce resources are spent on actions which bring maximum profit to business. These effectively curb people pursuing pet projects which don't stand up to economic scrutiny.

29.11 Principles

A plan is prepared after much careful thought, in order to meet the objectives set for the project. These are likely to be missed unless the plan is strictly followed. Clear direction, well-understood objectives, a good system to manage work volume and flow, management visibility, and good communication are all vital to the success of any project.

Chapter **30**

Keeping to Schedule

Creativity can solve almost any problem. The creative act, the defeat of habit by originality, overcomes everything.
George Lois - Author, Advertising Legend.

Author: Mahen Das

Location: 2.1.3 Petroleum Refinery

30.1 Background

The refinery had a 3000 tons/day fluidized catalytic cracking unit (FCCU). The performance of the FCCU Regenerator cyclones, which recovered catalyst from the hot gases leaving the vessel, had deteriorated significantly over the years. A new set of cyclones had been ordered from a European vendor. There were five primary and five secondary cyclones, lined internally with an abrasion resistant refractory.

These would replace the existing five pairs of cyclones during the next FCCU Turnaround (or shutdown as it is called in Europe). The new cyclones were larger than the existing ones, and Figure 30.1 shows one such unit. Each cyclone weighed about 6 tons, and all ten were of the same size and geometry.

30.2 Cyclone Replacement Procedure

In most other locations, where availability of hoisting machinery is no constraint, the execution would have been as follows (see Figure 30.2):

- Order the new cyclones ready for mounting on the dome.
- Erect a derrick for removal and refitting the dome.
- Cut the dome of the regenerator beyond the mounting periphery of the cyclones.
- Remove the dome together with the 10 old cyclones; use a large capacity crane to lower the dome along with the cyclones to ground level.
- Place the dome on a steel structure at ground level, high enough to enable working under the dome.

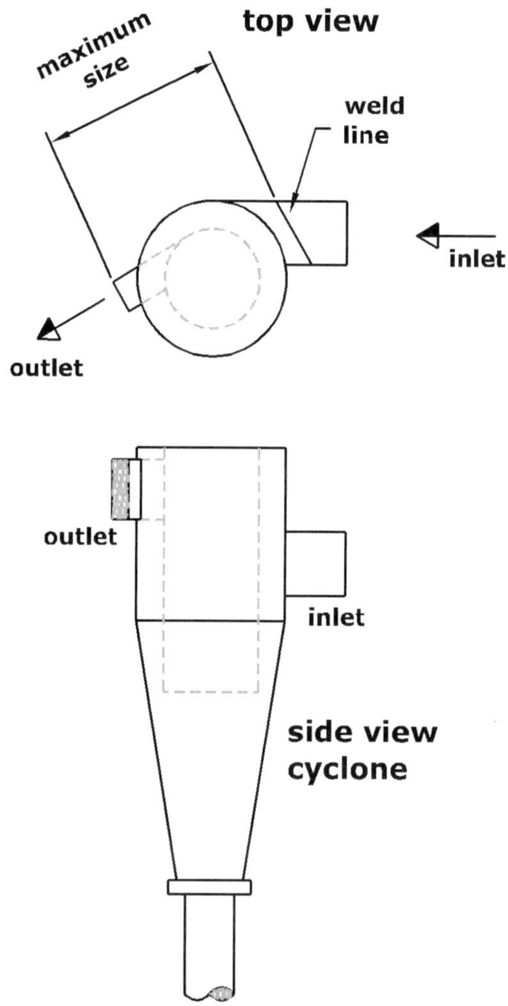

Figure 30.1 Cyclone

- Remove the old cyclones from the dome and attach the new ones.
- Lift the dome and place it back on top of the regenerator shell, using track crane and derrick.
- Weld the dome to the shell.
- Repair internal insulation.

At this location, it was not feasible to apply the conventional procedure, as it needed a long boom 200–250 ton capacity crane and a large derrick. At the time of these events, such a large crane was not available, and space constraints were such that a derrick could not be installed. So we knew we had a major problem on our hands.

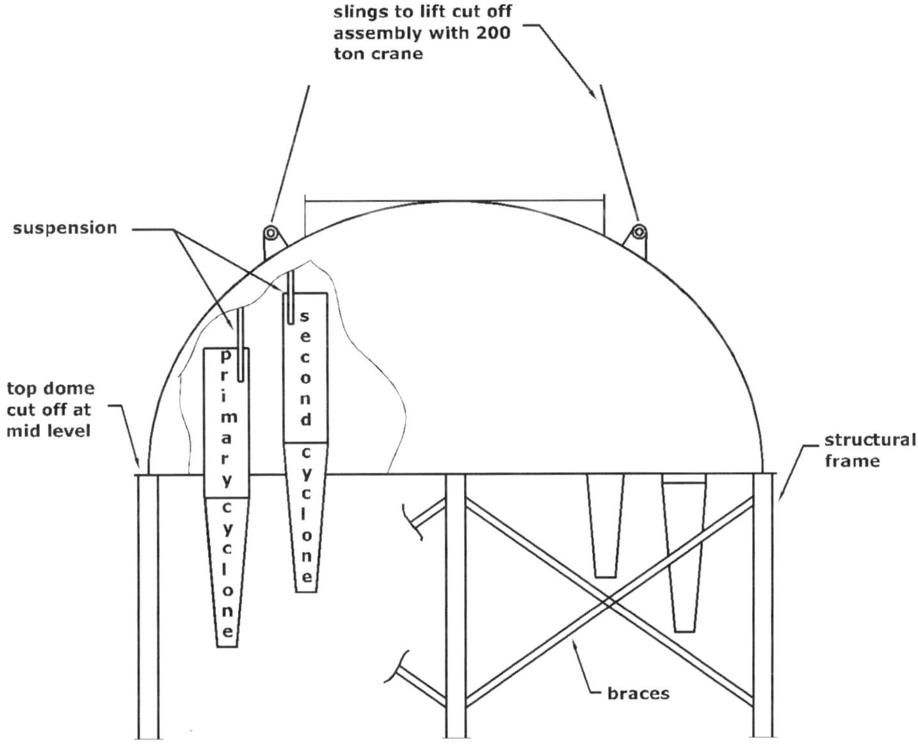

Figure 30.2 Cyclone Replacement Process—Conventional Method

30.3 The Cards We had been Dealt

The cyclones arrived about six months before the planned start of the shutdown. The new cyclones came in three parts; the largest of these parts was too big to go through the existing 60″ man-way of the Regenerator. In ordering the cyclones, this detail had not been specified, and the vendor made them in the normal manner. As planning engineer, I was also responsible for the execution of the plans. We had to find a way to execute the work safely without a large crane. A further challenge was to achieve this within the historical duration of the shutdown which was 30 days from feed cut-off to product rundown. Every extra day would mean a large business penalty.

The region had world class rigging skills, but was devoid of heavy lift machinery. A suitable derrick to carry out work as described above was out of the question. The refinery owned a track-crane of suitable capacity to hoist the cyclones, but with grossly inadequate boom-length. This crane was a legacy from the construction period, nearly 20 years earlier. The boom could just reach the 60″ man-way about half way up the structure. We had a diesel engine-driven winch which could handle the weight of individual cyclones, another legacy from the construction period.

We did not have a suitable communication system such as walkie-talkies or dedicated wavelength radios suitable for refinery use. Import controls were in place, and licenses for such items would not be issued by the Government.

30.4 Exploring Potential Solutions

I had a discussion with Vee Narayan, who was then the field maintenance engineer at the same location. Some ideas came up and we selected one that seemed distinctly attractive. We gathered a team which included riggers and crane-operators to brain-storm this idea. By the end of this session, we had evaluated the risks and found actions to mitigate them. We were satisfied that we had a way to accomplish the job safely.

The main steps were as follows:

1. Enlarge the 60″ man-way by cutting an opening large enough for the new cyclones to pass through it. Save the cut-out for re-fitting after the last cyclone entry.
2. Manually assemble a pulley support on top of the regenerator.
3. Place the track crane and the diesel winch, as shown in Figure 30.4.
4. Secure a pulley to the neighboring catalyst hopper vessel at suitable height.
5. Thread a wire rope and hook system from the diesel winch, through the catalyst-hopper pulley, to reach inside the regenerator via the top pulley and top opening.
6. Station the lead rigger and three other riggers, as shown in Figure 30.4.
7. With the help of two suitably hung chain-blocks inside the regenerator, transfer a cyclone from its suspension rods to the hook of the diesel winch.
8. Lower the released cyclone to man-way level.
9. With the help of the track crane hook and a suitably hung chain-block, remove the cyclone out of the regenerator, transfer it from the diesel-winch hook to the crane hook, and lower to ground level.
10. Repeat with all ten cyclones.
11. Install the new cyclones by reversing this process.

Item 1 in this procedure posed a potential problem. The nozzle of the 60″ man-way provided structural reinforcement to the shell of the regenerator. By removing the nozzle and flange, we would weaken the shell. This could cause the opening to close slightly in the vertical direction, as a result of the self-weight of the Regenerator. It had the potential to crease the shell at the horizontal diameter of the man-way. There was no guarantee that when we tried to refit the nozzle cut-out made earlier, we would be able to reinsert it, as the hole could now be slightly oval. We could not accept this risk. So we designed a reinforcement girder to strengthen the shell and compensate the temporary loss of the 60″ man-way nozzle and flange. This is shown in Figure 30.3. The procedure described above is illustrated in Figure 30.4.

30.5 Scale Model

We made a two-dimensional scale model to verify the feasibility and validity of these steps. After playing with the model, we knew we had to make a few minor changes, but all the main steps seemed to be in order.

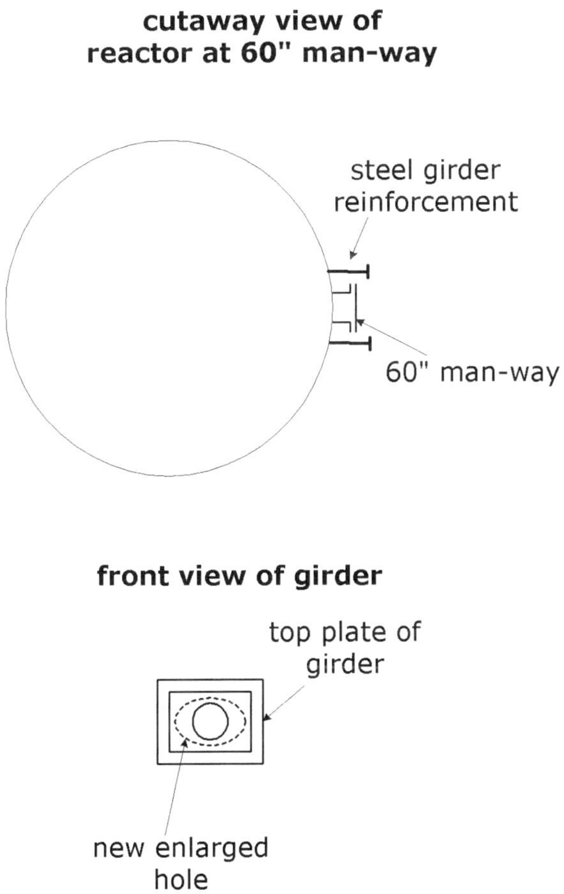

Figure 30.3 Reinforcement of Enlarged 60″ Man-Way

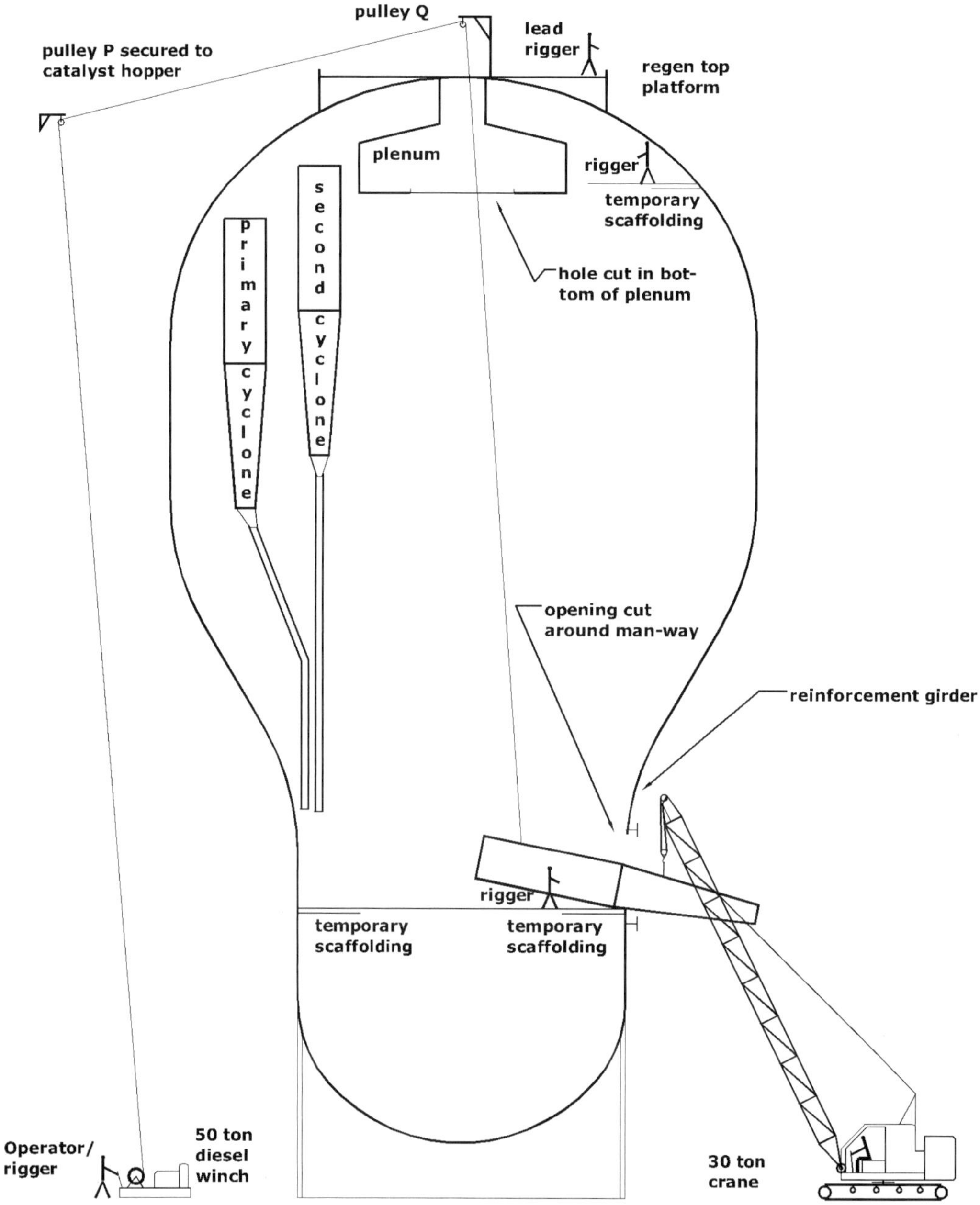

Figure 30.4 Schematic Rigging Arrangement

30.6 Safety Aspects

The operation required good rigging skills and excellent communication. We broke down the tasks in great detail and allocated each sub-task to an individual by name. As the work was to be carried out round the clock in two 12-hour shifts, there were two teams involved. A week before the start of the shutdown, we conducted a mock drill to enact one removal and one installation.

30.7 Results

During the actual exercise, we felt we had been doing this operation all our life. It went smoothly without a hitch. The shutdown was completed in 29 days, a whole day earlier than the historical duration. The exercise required innovation, teamwork, communication, and an enterprising spirit. But the most important ingredient for its successful completion was communication, both during planning as well as during execution of the exercise.

30.8 Lessons Learned

1. Brain storming in a cross-functional team delivers amazing results.

2. The vital ingredient for the success of a plan is extensive communication with all participants at all stages—i.e., concept, development of plan, scheduling—to make sure all participants fully understand their own roles as well as those of others.

3. For critical and complex jobs, it pays to carry out a "mock drill."

30.9 Principles

A maintainer's life can never be boring. There are genuine opportunities to demonstrate leadership and creativity. (Sometimes we may find ourselves parachuted into situations that we had no role in creating. Playing with the cards we have been dealt can be very challenging. Maintainers face such situations regularly, perhaps more often than some others.)

Chapter 31

Operators as a Maintenance Resource

The manager accepts the status quo; the leader challenges it.
Warren G. Bennis – Management Guru.

Author: Mahen Das

Location: 2.3.3 Corporate Technical Headquarters

31.1 Background

I was one of a team of experienced maintenance practitioners in the corporate headquarters, who provided technical support to the refineries. We focused on improving their reliability and maintenance performance

We developed a maintenance and reliability performance appraisal program with which we could identify improvement actions. The client refinery agreed to execute the recommended actions within a time frame and with a clear implementation plan.

The program was carried out during a visit to the client refinery, jointly with selected members of the client's own staff. This collaborative effort tapped their knowledge and experience while giving them a sense of ownership of the project.

31.2 Execution of Maintenance Work

We selected efficiency of execution of maintenance work as one of the target areas for performance improvement. The key success factors for this are:

1. Good planning, scheduling, and resource optimization
2. Full and effective utilization of resources

Aspects of good planning, scheduling, and resource optimization of daily work are described in Chapter 24, Long Look-Ahead Plan. The current chapter deals with full and effective utilization of a resource which is traditionally underutilized.

31.3 Operator Workload

When we looked for underutilized resources, plant operators stood out like sore thumbs. An internationally respected firm specializing in refinery performance benchmarking confirmed this view. In their studies, they found that at least 25% of plant operators' time is unstructured and, therefore, can be utilized productively for carrying out certain types of maintenance work.

Refinery managements are generally reluctant to accept this observation. The subject of operator numbers is traditionally closely protected, influenced by a perceived concern for operational safety.

During one of my visits to a client refinery, this topic came up for discussion in the course of the performance review. There were four participants from the client's side: the operations manager, the maintenance and engineering manager, the projects manager, and the chief inspector. I could not convince them of the validity of the observation. Not unexpectedly, the operations manager held the strongest objections.

I challenged him to call the front line operations supervisor to the forum and carry out a tally of operators' defined duties and time spent on these duties. He accepted the challenge. To the best of our knowledge, this was the first time that such an exercise was carried out in this manner. None of us foresaw, or could have foretold the results.

31.4 The Challenge

The operations manager invited the shift supervisor for this exercise. When he learned what he had to do, the shift supervisor requested that his senior panel operator also be allowed to come and help him. We scheduled the exercise to take place after the shift change when both these people would be free from their duties.

The two men came at the appointed time. Both were a bit nervous to face the forum in which their big boss, the operations manager, was also present. On my suggestion, the operations manager himself took the lead in developing the tally.

They generated the following Table 31.1,(shown on the following page).

The participants themselves found the result of the exercise incredible. Indeed, they could account for less than 75% of the time as shown in the table, and this included some maintenance work already being done by operators. This convinced them to accept additional alternative work for operators.

31.5 Type of Maintenance Work for Operators

Operators' unstructured time should be utilized only for alternative work, which we will call front line maintenance, as illustrated in Figure 31.1.

The following is a description of these tasks:

- Derive the tasks from RCM/RBI analysis results. These will be preventive maintenance tasks identified by a review in which operators have themselves participated. Hence they are more likely to accepted. Leave the specialist tasks for specialists.

Activity of outside operators	Hours/12-hour shift/operator	
Duration of shift		**12.00**
Shift hand-over and discussion	0.50	
Structured walk-abouts	3.75	
Infrequent operational work	1.00	
Minor Maintenance work (already done by operators)	0.90	
Other planned work (one year plan)	0.10	
Job preparation for maintenance work	0.50	
Safety standby for hot and dangerous work	0.10	
Training incl. manual review	0.50	
Breaks (lunch and tea-breaks)	1.50	
Total time for operator activity	**8.85** ⟶	**8.85**
Spare time per operator		**3.15**

Table 31.1 Operator Workload

- The order of preference is preventive maintenance tasks as described above, then condition monitoring, and last of all corrective maintenance.
- The primary task of the operator is to operate the plant safely. Therefore, only interruptible maintenance tasks are suitable as front-line work. This policy will enable release of operators should operators be required urgently to handle any emerging operational situation.
- The available operator time is treated as a maintenance resource. It is planned, scheduled, and accounted for, as for the regular maintainers.

Before operators can start doing some of these tasks, we may need to give some focused training.

31.6 Result

The refinery now uses nearly 25% of their operator time in front-line maintenance activities.

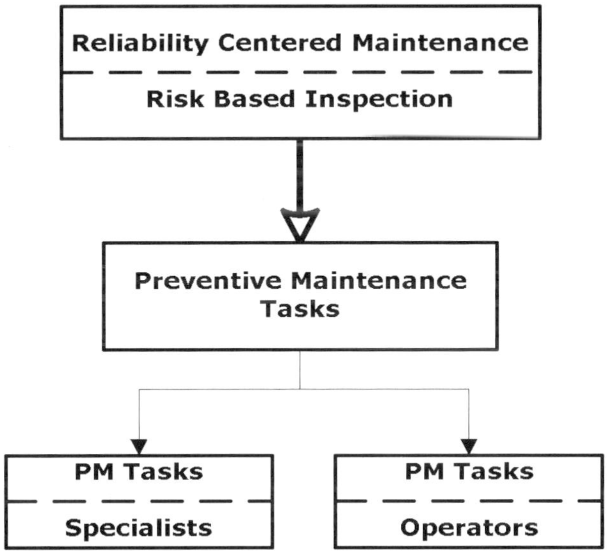

Priorities

1. Preventive tasks from RCM/RBI
2. Condition Monitoring
3. Minor corrective tasks

The primary task of the operator is "operating the plant safely; " therefore, all Front Line Maintenance tasks should be interruptible for prompt action in urgent situations.

Figure 31.1 Front-line Maintenance Tasks

31.7 Lessons

1. Use of resources can be critically scrutinized to reveal under-utilization.

2. To be useful, this should be done with an open mind with the full participation of the relevant parties.

3. Operators can carry out useful maintenance work after minimal training, without jeopardizing their primary duty of operating the plant safely.

31.8 Principles

We will always find many defenders of the status-quo. Challenging these and understanding the factual situation can help demolish such citadels.

Chapter 32

Overtime Control

If you don't measure it, you don't manage it
Joseph Duran, Quality Management Guru

Author: Jim Wardhaugh
Location: 2.2.2 Large Complex Refinery in Asia

32.1 Background

Controlling overtime is always a problem. The problem increases in importance when you are being benchmarked. You realize for the first time that the count is not just a headcount, but a count of all the maintenance man-hours being used. The benchmarking showed that maintenance was using too many man-hours and a part of the problem was poor overtime control.

We had tried the usual ploys of setting targets and banning overtime unless it was authorized by very senior managers. However the story telling skills of the supervisors did not find this hurdle much of a challenge and the overtime came down only marginally. See the first four years trend in Figure 32.1.

To say we were vexed would be to understate our feelings. We were supposed to be managing the plant and we could not even get a grip on overtime. It was time for serious action.

We had had significant success in the past by hitting relatively intractable problems with computer systems. The recipe had become almost standard:

- Computerize the business process.
- Collect facts.
- Make what was going on visible.
- Fine tune the system to encourage a better way of doing business.
- Embarrass into submission the people who were causing the problems.
- Gain improvements.
- Consolidate and institutionalize.

So this is what we did. As usual the prophets of doom in maintenance told us we were wasting our time. Production was the king and they had insatiable demands for overtime driven by their over-cautious approach

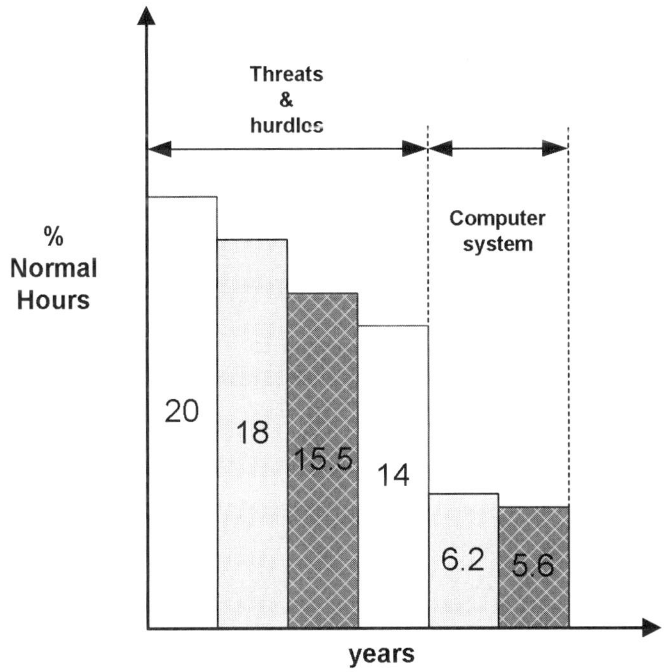

Figure 32.1 Overtime Reduction Profile

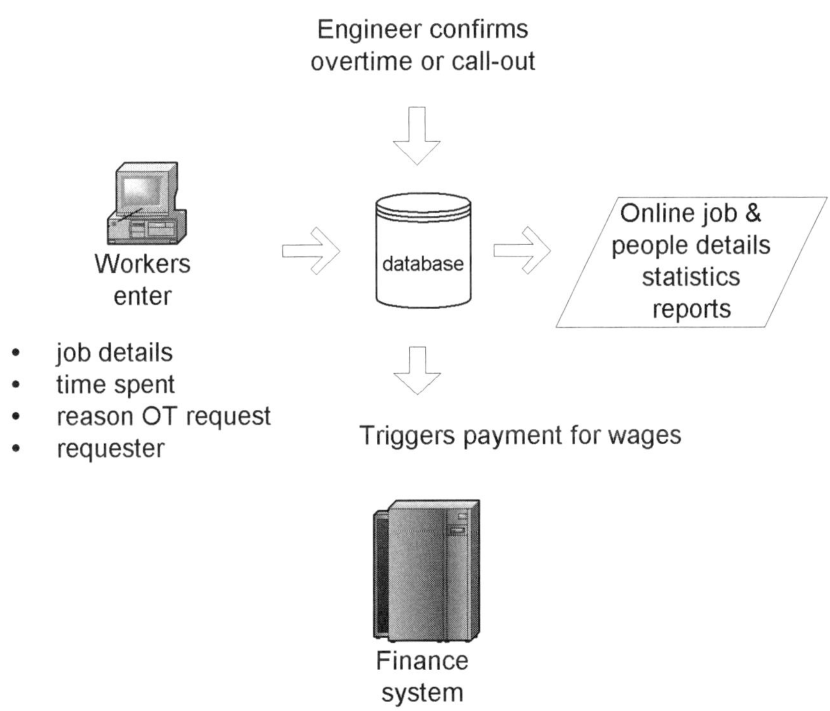

Figure 32.2 Overview of Overtime and Callout System

to risks. In essence, they didn't take any. Maintenance had tried their hardest but they failed because everyone from the Plant Manager to the lowliest operator was against them. We listened, we heard, we sympathized, but we carried on.

32.2 System Design

We devised a system shown in Figure 32.2. As always, we started simple and added complexity only when there was a real need. We never put things in just in case. For simplicity we decided that the system would initially be stand-alone and only linked to the finance system. But we'll hear more of that later. The steps in the process were as follows:

- We asked workers who were called out or asked to work overtime to enter details of the job, reasons for work request, who requested it, time spent on the job, technician skills involved, as found condition of the equipment, etc.
- The engineer for the area in which the work was done would vet the details input about the activity. He would ask whatever questions were needed to ensure that the story looked and sounded reasonable.
- All these details were visible almost immediately via on site PCs.
- The next working day the Chief Engineers and the Engineering Manager queried all out of hours work to confirm that it was justified.
- We consolidated this data and produced statistics showing on-line, out of hours worked and the requester of the work. We could scrutinize details of each job if necessary with a built-in drill-down capability
- We fed this data about monthly hours worked by technicians into the finance system to generate payments for the work done.

32.3 Some Features of the System

- On line, to make every detail of each callout or overtime activity highly visible to any interested parties
- Formalized request, authorization, and vetting procedures to ensure a clear structure
- Provide audit trails
- Direct payroll feed via flexible payment algorithms
- System-generated weekly, monthly and year-to-date statistics for comparison with annual targets

Versions were eventually made available for shift workers as well as day workers. We built the system using the prototyping approach and a fourth generation language; this required about 480 man-days of IT effort.

32.4 Some Subtle Features of the Design

- Technicians only got paid if they put the job into the system. It was a case of no job details, no pay.
- Each month we created a "Top Ten Job Requesters of the Month" list. In line with our approach of embarrassing people into submission, this was given a huge amount of publicity. People on this list were not amused; they moved heaven and earth to avoid being on the list

I can't stop here without a little story. One night the Plant Manager's electric shower wouldn't work. His wife then complained to the effect that he was the manager, she needed a shower, and he should do something about it. Suitably chastened, he rang the duty electrical technician and asked if he could come out and fix the problem; only one condition he wasn't to put the call-out in the system. The technician explained that meant he wouldn't get paid.

She didn't get her shower, not a hot one anyway, and certainly not that night.

32.5 Results

- We created a consistent site-wide terminology on what was overtime. This was consistent with the benchmark man-hours approach. It sounds easy, but it wasn't.
- We eliminated the clerical effort in administering overtime, and it

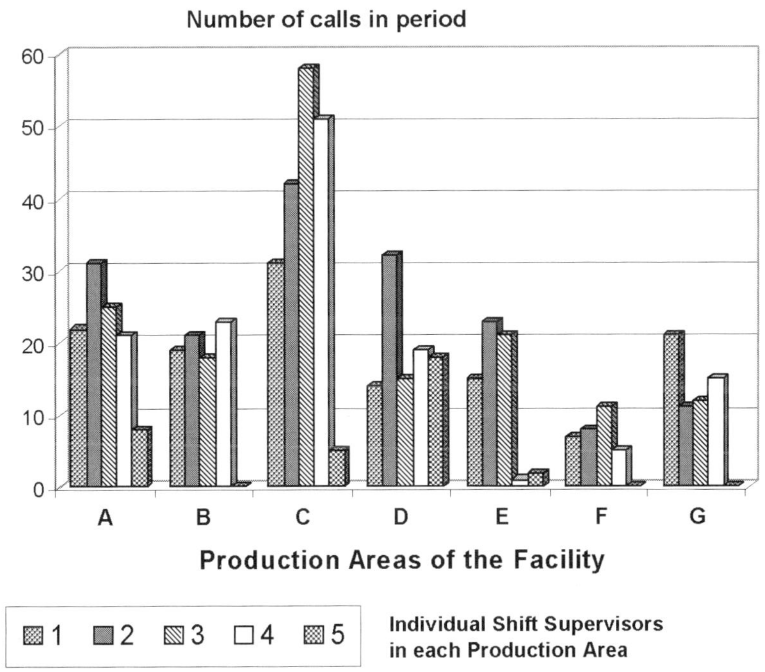

Figure 32.3 Analysis of Requesters of Callouts in Production Group

was quite a large amount. The system became almost error free, a new experience for us. Overtime activities became highly visible.

- We brought a universal understanding of when it was reasonable to call in. This was an instinctive response and not driven by any matrix type mechanism.
- We found that most requests for overtime were from Engineering—either Maintenance or Construction. This explains why we had had so little success with our previous approach. Because we lacked facts we were targeting the wrong people, the operators. Overtime was used largely as a reward to the worker for some sort of good performance or pure supervisory convenience.
- Inexperienced shift supervisors, either new to the job or on a new plant, also featured in the top ten lists. More experienced operators were better at managing risk and did not call out people unless absolutely necessary. Figure 32.3 shows the sort of statistics that could be produced.
- We halved overtime in all its aspects within a year—see Figure 32.1. The slow run down over the first four years came as a result of threats and continually raised authority hurdles. The last two years' results came because we implemented a new computer system.

32.6 Lessons

1. Attacking the wrong problem does not bring good solutions, so it is essential to work on facts. Logically-trained people can easily align to these. A characteristic universally found in poor performers is the habit of jumping to conclusions without proper analysis. Validate your assumptions and check your facts.
2. If you want people to enter data, make sure it brings them some benefit.
3. Validate your assumptions and check your facts.
4. Replace emotion in decision making with one based on facts.

Too often the design of computer systems gives one group all the work and another group all the benefits. A modern system development methodology (Soft Systems Methodology) provides checks and balances against this. However achieved, if you want people to enter data, make sure it brings them some benefit. Certainly make sure they are not going to be hurt.

32.7 Principles

It is important to understand what is happening before acting. Making assumptions is often fatal.

Performance indicators must focus on the key business processes. They should be designed to encourage good performers and embarrass poor performers.

Chapter 33

Manage Contractors

The leader must know that he knows, and must make it abundantly clear to those about him that he knows....
Clarence B. Randal, Author, Management Expert.

Author: Jim Wardhaugh

Location: 2.2.2 Large Complex Refinery in Asia

33.1 Background

At the time the contracting culture in Asia accepted as a norm low wages, low productivity, significant over-staffing, low quality, and the inevitability of accidents. Contractors tended to be small family businesses run by entrepreneurs as individual fiefdoms. The managers had poor managerial skills. Harvard MBAs were particularly thin on the ground.

In the refinery, our contracts were largely fixed-price lump-sum. These were attractive in some ways because they minimized the administrative effort involved in managing the contracts. However, because the price was fixed, our site supervision held a view that productivity was of no interest to us. So our supervisors took no interest in the quality of contract workers or in their numbers. That was his business not ours. To my western eye, this hands-off approach brought a number of problems:

- Larger numbers of contractors than necessary
- Many accidents
- Incompetent workmanship
- Low productivity and low mechanization of work methods
- High ongoing effort on safety and induction training
- Extensive infrastructure to support the large workforce (transport, food, toilets, washing, etc.)

There was little factual information available on the contractors—whether management, supervision, or workforce. All we knew was that each day:

- A mass of unidentifiable workers of questionable skill came to the plant.
- They worked on unidentifiable jobs.
- They worked for unidentifiable contractors or sub-contractors.
- They worked at unidentifiable cost effectiveness.
- They often worked unsafely.

We were putting in a considerable amount of training to raise the level of competence and safety awareness, but with a huge turnover of contract employees this was ineffectual. The situation was unacceptable in a cost and safety-conscious world. We knew we must drive down the numbers and attain a fairly small and stable workforce that could be trained to a high standard. We could not afford the hands-off approach any longer. Thus we started our drive to manage the contractors actively.

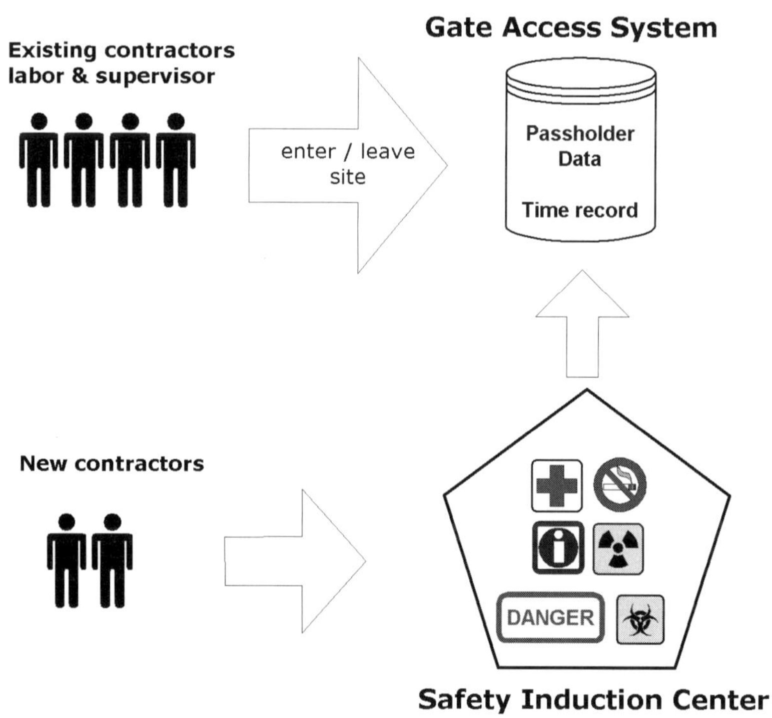

Figure 33.1 Gate Access System Overview

33.2 Gate Access System

We started our journey with the installation of a cheap gate access system with which we could control access strictly. Our objective was to make every worker immediately identifiable along with information on:

- Skill
- Work experience
- Sub-contractor, direct/indirect worker
- Safety training record
- Time of arrival and departure, overtime
- Absenteeism

We chose a proprietary swipe card security system to run on a stand-alone personal computer. Swipe card readers activated turnstiles for authorized personnel and captured their entry and exit times. Figure 33.1 gives an overview of the gate access system.

After some tough negotiations, we agreed on initial workforce numbers with each site contracting company. For all workers nominated by their company, we entered relevant details into the security system. Almost as soon as the system went live, we started to get factual information:

- About 30% of the people that the contractors had sworn to be key members of their site workforces hardly ever came to site.
- Almost 40% of the contractor workforce came late in the morning and left early in the evening.

This was reasonably consistent across contract companies.

Armed with these facts, we entered into discussions with the contracting company managements. We insisted on a rapid reduction in their core on-site workforces as, for once, we had access to facts and they had only a sketchy idea. There was little opposition to the next step as well, which was to improve timekeeping. We achieved our objectives within a few months (see Figure 33.2) and we saw:

- Significant reduction in numbers
- Improved timekeeping
- Reduction in accidents
- Somewhat better productivity and work quality

We did have some amusing attempts from the contractors to defeat the system. One which sticks in my mind was when our security people caught one of the contractor supervisors jumping backwards and forwards over the turnstiles with a fistful of swipe cards and swiping a different card for one of his company's absentees.

The result was a significant and rapid reduction in contractor costs.

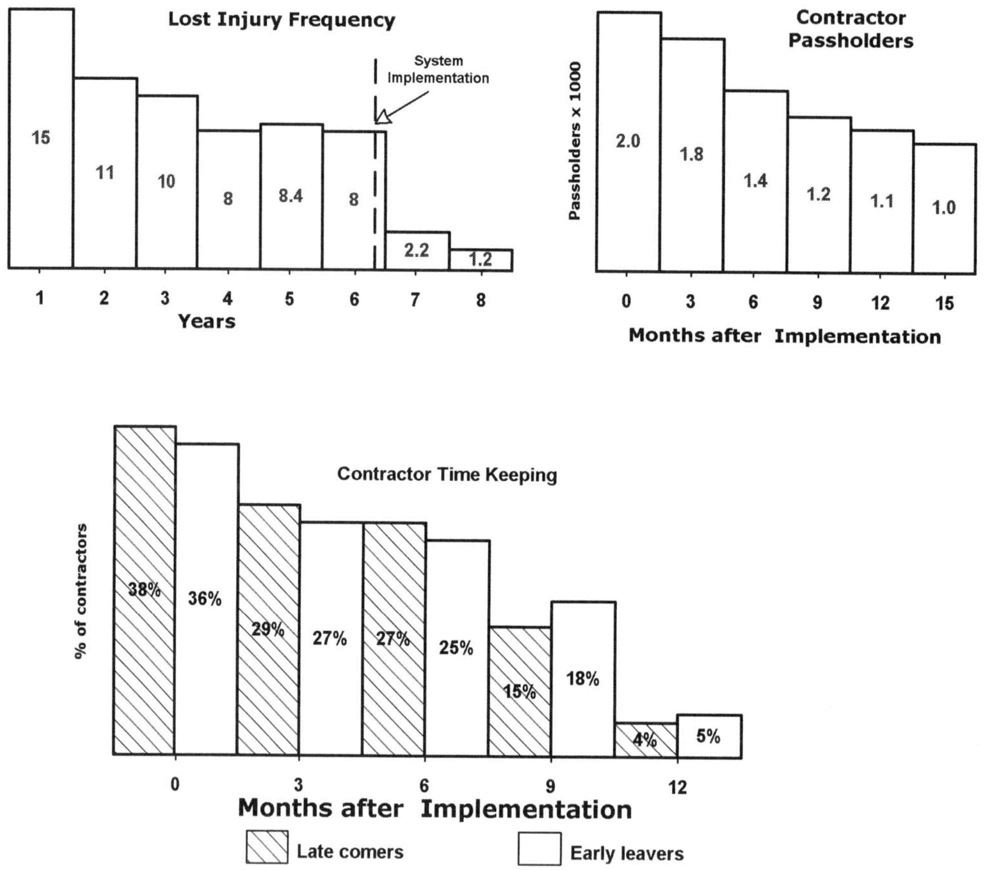

Figure 33.2 Trend in numbers, time keeping, and accidents

33.3 Contractor Management System

We had made significant reductions in contractor numbers and our focus on a smaller core of contractors was bringing benefits. We felt, however, that we could do much more. Productivity was still an issue. There was little mechanization, and work organization left a lot to be desired. As usual, we lacked facts on which we could make any decisions. It was time to move into the next phase of the game and get the facts necessary to actively manage the contractors. In this way, we aimed to get better value for money. We needed a new computer system to achieve that, but first the objectives:

- Every job must be vetted so that we could see:
- Staffing levels, foreman and supervisor allocation, hours worked progress, and profitability
- Every contractor organization must be assessable in terms of :

- Quality and sufficiency of labor and supervision, workload, time discipline, safety and training record, profitability, and competitiveness.

The overall system we put together is shown as Figure 33.3 and the daily input effort by various actors in the game is shown in Figure 33.4.

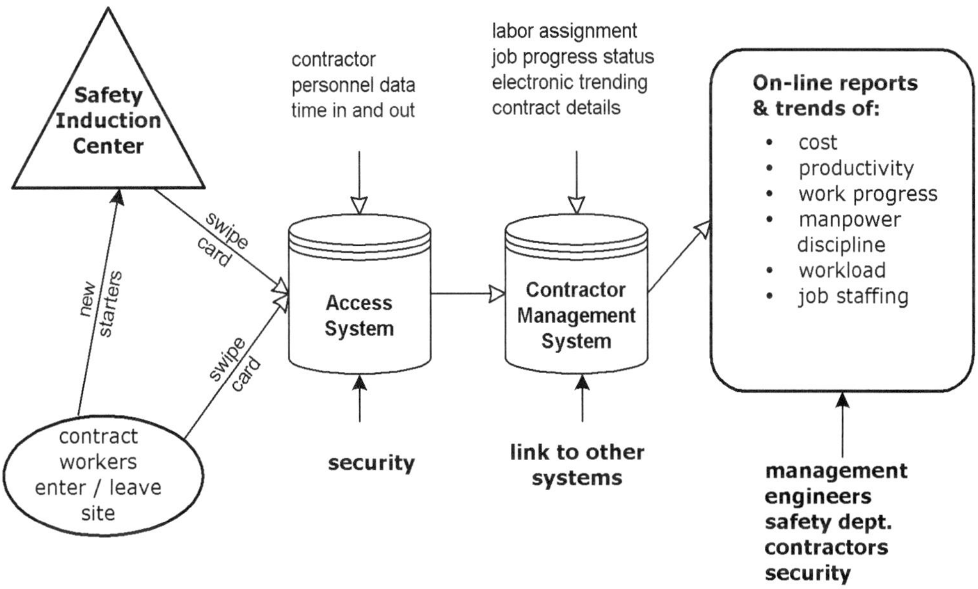

Figure 33.3 Contractor Management System Overview

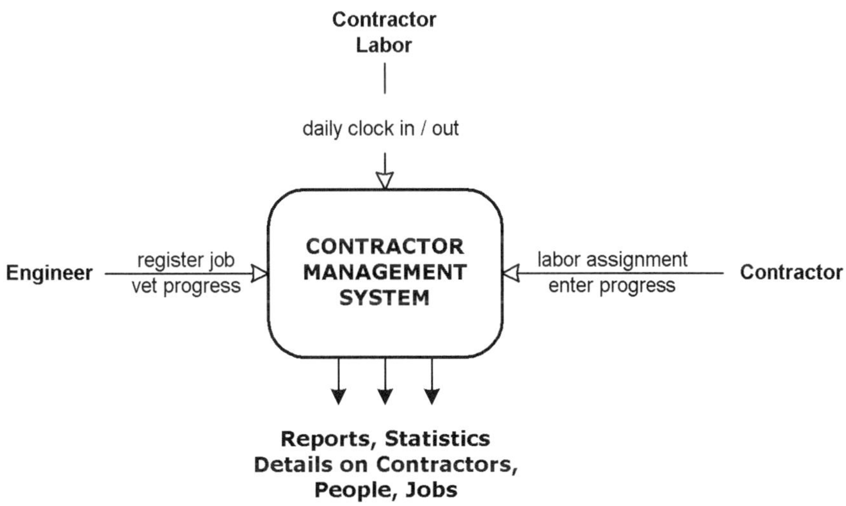

Figure 33.4 Summary Input Efforts

33.4 How Did It Work?

Security, Safety Department, and Administration maintained the personnel files, training in safety and competence, accident records, and contract details. Security Department issued the swipe cards and managed the issue of replacements as needed.

Contractors' administration clerks and supervisors had significant duties to:

- Confirm the daily labor assignment for each individual (i.e., what job each person would be working on). This identified where workers could be found and, via the reporting system, the staffing level and man-hours for each job.
- Plan the next day's work and labor assignments. This promoted a planning mentality.
- Update the progress status of all jobs weekly.
- Enter labor, material, and equipment price breakdowns via an electronic tendering module.

Company engineers had duties to:

- Review daily labor assignments and audit to ensure sufficiency and encourage accurate reporting.
- Vet weekly progress status of jobs for accuracy.
- Invite tenders for new work or repairs and advise the contractors of the tender results.
- Prepare robust and detailed labor, material, and equipment counter estimates for these tenders. We insisted that these should be on the basis of what the job should cost rather than what they felt the contractors might tender, even though it did result in some tedious debates with the tender board.
- Clarify with the contractors tenders that were significantly out of line with the counter estimates.

This might sound like a lot, but in reality the input efforts were relatively modest. The capture of this key data with analytical and reporting facilities made the monitoring of contractor activities easy and flexible. We could access information at virtually any level of detail and pinpoint problem areas as they occurred while avoiding reams of indiscriminate computer printouts.

33.5 Some Specific Aspects

We found that the breakdown of tenders into cost of man-hours, materials, and equipment gave us very powerful insights into the thinking of the contractors. Initially for the weaker contractors these numbers were mere guesses. However encouragement, pressure, and constant requests for clarification quite quickly brought a more professional approach to their tenders.

The breakdown also brought insights into productivity issues. We wanted to encourage them to move to a high productivity regime by better organization and mechanization. We showed a preference for low manpower tenders. The company tender board accepted our justification for accepting tenders which were not necessarily the lowest cost. Of course we applied strict criteria in making these exceptions.

We established a concept of man-day recovery rates which we defined as the labor cost element of the tender divided by the actual man-days used in executing the work as defined in the tender. Input of labor data (i.e. manpower on site as captured by the gate access system and allocated to that specific job by the contractor) gave us the ability to monitor actual use of labor against those defined in the tender breakdown. As always with contractors, we found it necessary to do sufficient audits of manpower actually at the job site to verify that allocations were reasonably accurate. We could now see the $ per man-day that each contractor was recovering in our jobs. We did some research via a tame contractor and soon had a good handle on expected returns. This gave us a good view inside the head of a local contractor. With this information, we drove down the man-day recovery rate significantly by the use of our tender clarification process—see Figure 33.5.

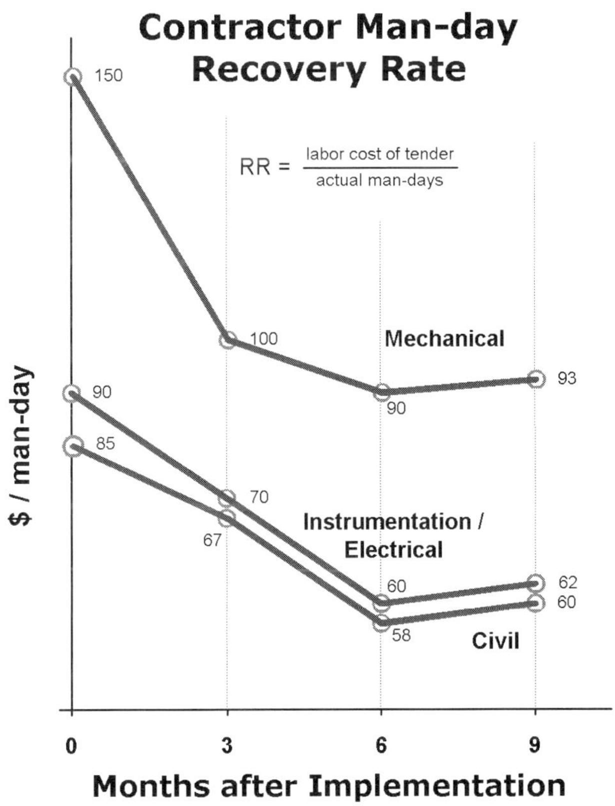

Figure 33.5 Contractor Man-Day Recovery Rate

We had missed one important aspect—the impact of our company supervisor on the recovery rate. In an earlier review of the productivity of the maintenance workforce, we found that the average worker spent only about 25% of the possible time doing work at the job site. This seemed to be at the same level both in normal day-to-day maintenance work and during shutdowns. Almost all of this could be laid at the door of less than competent supervision. When we analyzed these results by supervisor, we found that these averaged results hid a lot; good supervisors got a productivity of about 40%, while the poorer supervisors achieved about 20%.

It was rather frightening to discover that similar differences appeared when our supervisors dealt with contractors. Depending on the supervisor in charge, contractor productivity also varied. We discovered that the contractors factored these variations into their tenders. Good supervisors who could facilitate the job well brought in lower prices than poor supervisors. The cost differential could be as much as 30%.

33.6 Summarized Benefits

Against a background of high contractor activity we achieved:

- A better disciplined workforce
- Significant reduction in contractor numbers
- Significant reduction in size of the contractor infrastructure and effort in managing it
- Reduction in the number of accidents
- Contribution to savings in contract payments of about US $5m a year

33.7 Lessons

1. It is important to become knowledgeable about contractors, how they think, and what drives them. A hands-off approach is seductive, but a hands-on active management produces much better results. Do not, however, fall into the trap of micro-management.

2. A few simple facts, analyzed and presented appropriately, can make a lot happen.

3. A small, disciplined, contractor force with the necessary core competencies can outperform an undisciplined horde every day. All research confirms this simple thesis.

4. Poor supervision and organization cause many contractor performance issues. The root of many of the problems can be our own supervision with poor organization or people management skills. The best and worst supervisors have been shown to deliver significantly different productivity outcomes.

33.8 Principles

An important role for leaders is to shake people away from their comfort zones. Visible costs do not tell the whole story, and may only be the tip of the iceberg. Other direct and indirect costs also matter, and there may be a large intangible element hidden away. Lowest cost often means lowest visible cost, and that may not be the best way to judge a contract. Value for money is a better selection criteria than price in awarding contracts. Active management of contractors using factual information is essential for good performance.

LEADERSHIP

PART 7: ANALYZE

Chapter 34

Reliability Engineering in New Projects

He knew how to take what could be, and make it what is.
Wynton Marsalis, on Louis Armstong's musical improvisation ability.
Ken Burn's Jazz on PBS.

Author: V. Narayan

Location: 2.3.3 Corporate Technical Headquarters

34.1Background

The project team in the corporate headquarters was designing a new refinery to be located in the Far-East. They were working with a European design contractor in the latter's offices.

The refinery required steam and electricity for its operations. The unreliability of the local electricity company in the host country meant that the designers had to install a captive power plant 70 MW in size, along with a receiving station. The designers envisaged a combined cycle power plant to meet these demands efficiently. The facilities were estimated to cost about US$100m. As part of our maintenance and reliability team remit, we sought to influence project designers to build reliable and lean plants. It was best to do this at the front end of the project, and we convinced the project manager to invest US$30000 in mathematical modeling. We expected that by optimizing the configuration, we could reduce capital costs, as well as operating and maintenance costs over the life of the plant.

34.2 Functional Requirements

The power plant was required to produce

1. Electrical power 70 MW

2. High Pressure (HP) steam at 125 barg., Medium Pressure (MP) steam at 42 barg., and Low Pressure (LP) steam at 10 barg.

3. Instrument air at 10 barg.

34.3 Original Configuration

The designers planned to install three gas turbines (LM 2500 units, 19 MW). The hot exhaust gases from the turbines would be heated further in three heat recovery steam generators (HRSG) and used for raising high-pressure steam at 125 barg. Each gas turbine (GT) would be integrated with its dedicated HRSG. While the bulk of the HP steam would be used in the process, a part of it would be let down to the medium pressure (MP) steam header at 42 barg., using a back pressure steam turbine (ST 1). This would produce 16 MW. Some of the MP steam would be used to drive a condensing steam turbine (ST 2), which had two stages. At the end of the first stage the steam pressure is at 10 barg, some of which would be extracted and supplied to the low-pressure (LP) steam header. The remainder would drive the second (condensing) stage of the turbine ST 2, generating 17 MW of additional power.

There are four boiler feed water (BFW) pumps to supply water at high-pressure to the HRSGs. Two of these are motor-driven and two are steam turbine-driven. These pumps are large complex equipment, rated at 800 kW, provided with minimum flow controls and other protective devices. Figure 34.1 shows a schematic drawing of the original design.

34.4 Defining the Objective Function for the Model

As long as any four turbines are in operation, the power output requirements would be satisfied. The plant could continue to operate with just three machines available, by shedding non-essential loads. For normal operations however, the requirement would be 4 out of 5 machines, or a (4oo5) configuration, or 70MW of power.

The output from any one HRSG would be enough to meet the process steam requirements. Even at reduced power generation, at least one GT (with its HRSG) had to be in operation. Therefore, the steam requirements would always be met as long as at least 52 MW (19+16+17 MW) of power was available.

Instrument air would be extracted from the air compressors of the GTs. As in the case of steam, one GT could meet the full instrument air requirements. Thus, instrument air would always be available as long as at least 52 MW of power was produced.

Hence, all the functional requirements could be defined by a single objective function, namely, production of 70 MW of power.

34.5 Design Basis for Planned Shutdowns

1. GTs would be taken out of service for inspection/overhaul for 25-26 days a year.
2. HRSGs would be shut down for 14 days every 3 years for statutory inspection.
3. BFW pumps: overhaul—240 hours every four years; minimum flow valve—48 hours every six months; seal change—72 hours every two years.

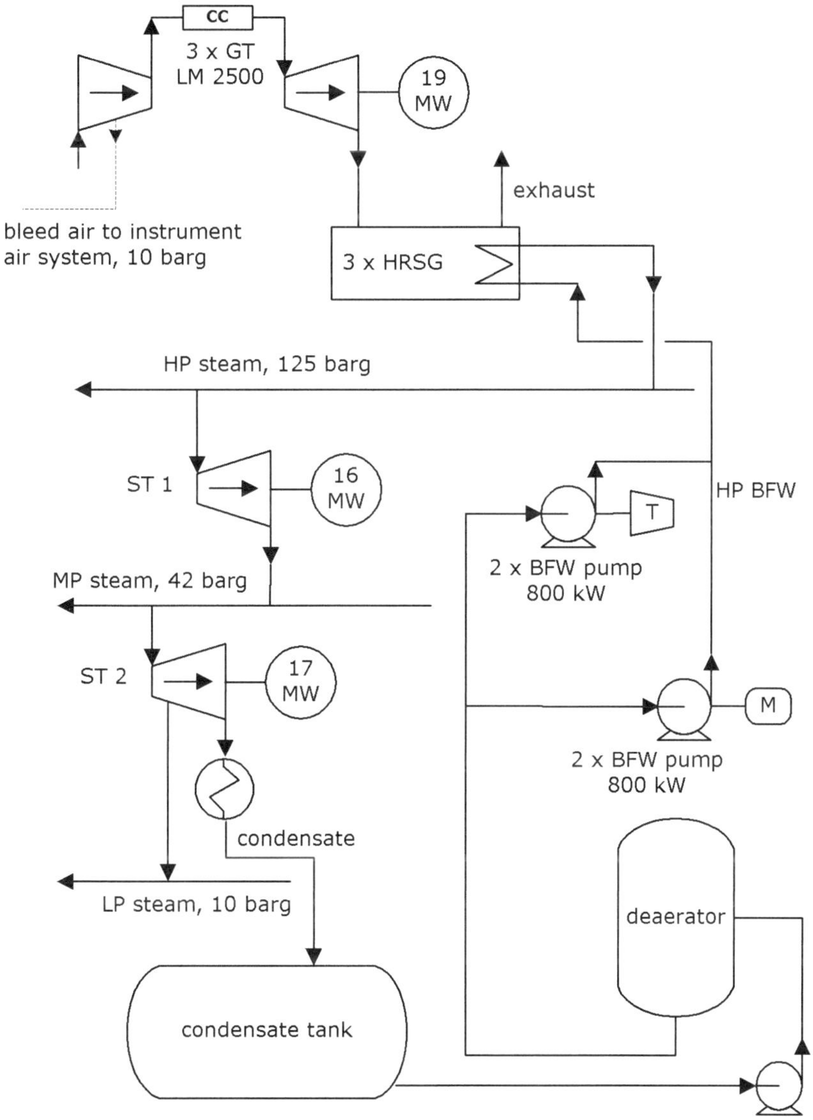

LEGEND	
GT - gas turbine CC - combustion chamber HRSG - heat recovery steam generator BFW - boiler feed water	ST - steam turbine T - turbine M - electric motor

Figure 34.1 Schematic Drawing of Power Plant

The HRSGs would be shut down at the same time as the corresponding GT. Hence, the HRSG down time need not be considered.

34.6 Analysis Technique

We appointed a well-known reliability consultancy firm to carry out the analysis. They planned to use Fault Tree Analysis (FTA) to model the system. While building the model, they worked closely with the project team members to ensure that the model represented the process flow scheme and that limitations in different scenarios were correctly represented. They used data from published generic sources such as IEEE Std. 500[i], OREDA II[ii], OREDA III[iii] and NPRD[iv] and applied error factors to the data to give 90% confidence limits. The software package had a built in Monte-Carlo simulator to enable probabilistic predictions. Appendix 34-A has more details about FTA.

Once the model was built, we could play tunes on it. The sensitivity of the final output to various changes could be checked. These included the addition or removal of individual items of equipment or in their configuration, changes in their reliability parameters, or shut down durations. This permitted us to find the most cost-effective configuration.

34.7 Outputs from the Model

Based on the model, 70 MW of power would be unavailable for 0.46% or 40.5 hours p.a. On average, this situation would occur on 2.6 occasions p.a. The main items that had scope to reduce this downtime (or improvement potential) were as follows:

1. Single HRSG 8 hours p.a; all 3 HRSGs 16.2 hours p.a.
2. Single GT 8.3 hours p.a, all 3 GTs 17.2 hours p.a.
3. Single Alternator 1 hour p.a., all 5 Alternators 4.2 hours p.a.
4. Motor driven BFW pump 0.52 hours p.a.
5. Turbine driven BFW pump 0.45 hours p.a.

Note that once the first item in a set, such as GT was improved, there are diminishing returns with the second and third GTs.

34.8 Recommendations Made to Project Team

Sensitivity studies showed that the fourth BFW pump contributed only 0.005% to the system availability. It was also clear that a second steam turbine driven alternator would have been considerably cheaper than having the third GT-HRSG combination. This would also have improved the steam balance. However, the GTs had already been ordered in advance, due to the long procurement lead time. Canceling the order would be very costly. At this stage, the lead time for procuring a condensing steam turbine was unacceptably high. However, we discovered that the project team was planning to order a complete LM 2500 GT as an uninstalled insurance spare, in addition to the spare rotor which was already on order.

We recommended that they cancel the order for the fourth turbine driven BFW pump and not proceed with the order for a spare LM 2500 GT. These two actions would save the project US$6 million. In addition, operating and maintenance costs for the fourth BFW pump would be eliminated. These savings made the project viable. The benefit to cost ratio was 20:1 based on just the capital cost reductions. The real benefit was, however, the knowledge that the project would confidently meet its functional requirements in spite of the reduction in investment.

34.9 Results

1. The optimized configuration resulted in a reduction of US$6m in capital costs.

2. The model gave us confidence that the power plant would meet its performance targets.

3. After the refinery was commissioned, the actual load proved to be just 55 MW, not 70 MW as projected originally. One way to manage the excess capacity was to sell power to other local consumers. The refinery did this very well, laying feeders directly to purchasers. They managed to sell about 20 MW, creating a new revenue stream.

34.10 Lessons

Mathematical modeling of new projects improves their Return on Investment.

It helps eliminate surplus equipment and associated maintenance costs over the plant life.

It allows what-if scenario development at relatively low cost.

We should have done the modeling work much earlier, at the conceptual stage of the project. This would have identified the advantage of having two gas turbine and three steam turbine generators at a stage where commitments for the GTs had not already been made. A further capital cost saving of up to US $10 million was thus not realized.

It turned out that the power requirement was grossly overestimated at the conceptual stage of the project. With a better estimate, we could have eliminated one HRSG-GT combination without having to order a third steam turbine generator, with further capital and operating cost reductions.

Modeling can be used in the operating phase as well, especially when operating contexts change during the life of the plant. Sensitivity analysis (what-if scenarios) offers a powerful way to estimate the outcome of changes in maintenance policy, resource levels, shutdown intervals or durations, and equipment replacement decisions.

34.11 Principles

The financial viability of major projects is greatly dependent on their capital costs. Project designers often concentrate on technical aspects of the design, but are less comfortable when dealing with economic aspects. Modeling tools offer cost-effective solutions and improve confidence in the design, by providing quantitative performance estimates. They permit designers to evaluate alternative configurations and operating philosophies to find optimal solutions. Marginal projects may cross the economic hurdles by eliminating non-critical items and thus trimming costs.

This is where reliability engineers can assist designers, but the communication between the two groups has room for improvement.

References

i. The Institute of Electrical and Electronics Engineers. 1984. IEEE Std 500-1984. Corrected Edition 2nd Printing, November 1993. New York: IEEE. Library of Congress Catalog Number 83-082816

ii. Det Norske Veritas. 1992. OREDA-92 Reliability Data Handbook. Hovik, Norway: (Offshore Reliability Data) DNV Industry.

iii. Det Norske Veritas. 1998. OREDA Handbook with data from Phase III. 2nd ed. Hovik, Norway: OREDA, Det Norske Veritas

iv. RAC. 1991. RAC NPRD Nonelectronic Parts Reliability Data. http://www.lricks.com/rac.htm

Appendix 34-A

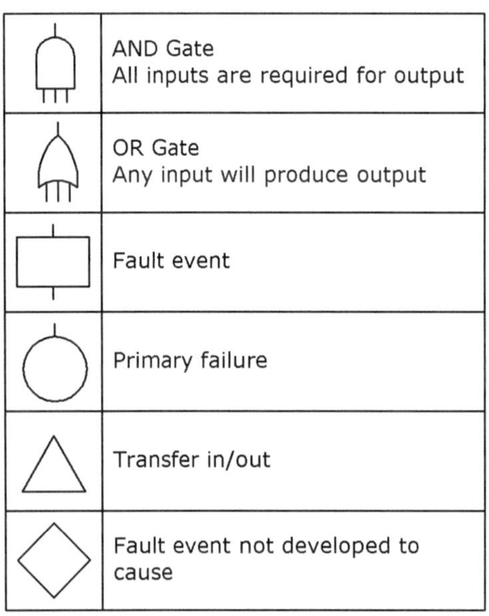

Figure 34-A.1
Explanation of Symbols

A Description of Fault Tree Analysis (FTA)

An FTA is a graphical representation of the relationship between the causes of failure and the system failure. From the 1960s when it was first introduced, FTA has been a popular method, especially with designers of safety systems.

An FTA helps us understand how a failure event could have occurred. The tree structure helps us derive the logical sequence of events leading to the undesirable consequence. Usually, the failure event we analyze has a high consequence, justifying the effort required for an FTA. We can incorporate the effect of human errors in FTA, so it is quite a powerful tool. You can see a simple example of an FTA in Figure 34-A.2. The symbols used in FTA are given in

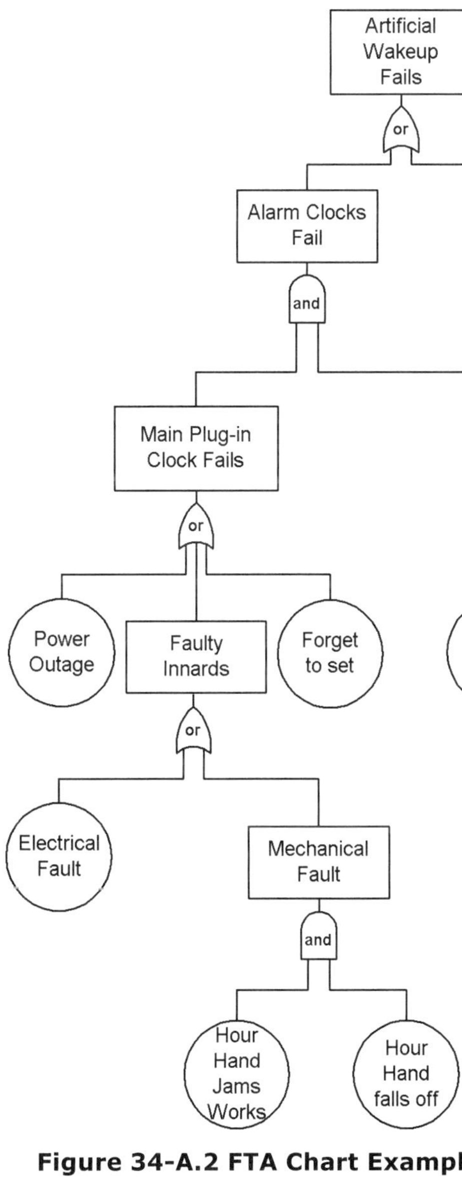

Figure 34-A.2 FTA Chart Example of Failure to Wake Up

Figure 34-A.1.

Failure events for which an FTA can be used include, e.g.,

- Wheels-up aircraft landing
- Train derailment
- Major oil spill
- Irretrievable loss of test data
- Loss of nuclear reactor cooling

FTA can be used as a reliability modeling tool. In our case, the top event is the failure to produce 70 MW of power. Readers can find additional information about FTA and its uses in the reference books listed below.

Additional Reading

Hoyland, A., and M.Rausand. 1994. System Reliability Theory. New York: John Wiley and Sons, Inc. ISBN: 0470593974.

Davidson, J. 1994. The Reliability of Mechanical Systems. Chapter 13. Mechanical Engineering Publications, Ltd. ISBN 0852988818.

Narayan, V. 2004. Effective Maintenance Management: Risk and Reliability Strategies for Optimizing Performance. Chapter 10. New York: Industrial Press, Inc. ISBN 0-8311-3178-0.

Chapter 35

Computing Reliability Data

Data is not information, Information is not knowledge, Knowledge is not understanding, Understanding is not wisdom.
Gary Schubert, Professor of Art & Computer Science, Alderson-Broaddus College, West Virginia.

Author: V. Narayan

Location: 2.3.2 Large Oil & Gas Production Company

35.1 Background

We need good reliability data to take decisions that will help improve safety, environmental, and cost performance. Such data is not readily available, so we tend to use generic data sources. Some of these data sources have well-defined taxonomies and good control of data quality. Even in these cases, there can be a large spread between the maximum and minimum values. While the operating context and maintenance effectiveness of the items in the data sets will be broadly similar, there will be differences that cause this large spread. In most cases, we end up using mean values, which may be significantly different from those applicable in our own operating context. The solution is to use our own data to develop a reliability database.

We had well-defined maintenance strategies and procedures in the company. With the help of a Computerized Maintenance Management System (CMMS), we issued Preventive maintenance (PM), testing, calibration and condition monitoring (CM) work orders, along with relevant procedures. Technicians recorded repair history in the CMMS and test results in calibration sheets.

We removed Pressure Relief Valves (PRVs) periodically for re-certification, and tested them before and after repairs or adjustments. The pre-overhaul data was recorded in relief valve data sheets. Wellhead valve and sub-surface safety valve test records were kept separately. Similarly, rotating machinery trip and bump-test results were kept in the equipment files. Test records from emergency shut-down (ESD) valves; fire, gas, and smoke detectors; deluge valves; and sprinkler-head tests were kept in calibration sheets. All this amounted to a veritable data mountain, available for inspection by any interested party, including the Regulator.

These records provided an audit trail and helped prove that the company was behaving responsibly and openly. The records were proof of its diligence. Nobody had considered the possibility of milking this data to improve safety and cost performance.

35.2 The Opportunity

This large volume of data presented an opportunity to compute reliability parameters for equipment using our own data. The numbers, based on our operating context and maintenance history, would be far more relevant and applicable in managing risks that we faced in our operations. Hitherto, reliability data from generic sources were used for such efforts as quantitative risk analysis and Reliability Centered Maintenance (RCM) studies.

35.3 Reliability Parameters

Many of the items under consideration were subject to hidden failures, i.e., failures that would not be known to the operator under normal conditions. For example, failures of PRVs, ESD valves, firewater deluge valves, gas detectors, etc., would not be known to the operator. If they failed during a test, they would be replaced with a new item or previously shop-repaired item (to as good as new or AGAN) immediately. An applicable reliability parameter to use in these cases is the mean time to failure (MTTF).

The operator can know about evident failures, because they will result in some local effect such as low flow or pressure, a pool of product on the floor, high current, etc. Once the operator knows of the defect, corrective action can be initiated, and the item would be repaired. Such items are termed repairable, while items subject to hidden failures are termed non-repairable. When items are repaired, it could be to AGAN standards as before and in such a case the parameter to use is MTTF. In most cases, however, the repair is to a lower standard than that of a new item. There are many reasons for not being able to achieve AGAN standards in the field. These include, e.g., unavailability of special tools, measuring instruments, skills, controlled environment, etc. The reliability of the repaired item is not 100%, as is the case with AGAN repairs. The applicable parameter for such repairable items is mean time between failures (MTBF).

The formula for calculating MTTF and MTBF is identical. In both cases, we divide the cumulative time in operation by the number of failures in that period. In this chapter, we will use the terms interchangeably, knowing that they are not the same and with apologies to the purists.

Cumulative time in operation = Sum of operating time of all items in set.

Number of failures is the sum of all the failures in the time period under review.

35.4 Pressure Relief Valves (PRV) Data

PRVs are bench-tested before cleaning and inspection. Data recorded in the calibration sheets include

a. Leakage, if any, below 90% of cold set pressure
b. Pressure at which valve lifted
c. Simmering or chattering between 90% and 100% of set pressure
d. General internal condition, including e.g., fouling
e. Date of installation
f. Date of removal from service

We collected all the PRV data sheets (archived over the years) from storage cartons, and entered the data into a spreadsheet. The first item (a), namely leakage below 90%, is useful for calculating the mean time between failures (MTBF) for the failure mode—internal leakage. In cases where item (b) is more than 110% of set pressure, these are marked as 'failed to lift on demand.' A count of these failures allows us to compute the MTBF for the failure mode—failed to lift at set pressure. The number of (calendar) days in service is computed as the difference between items (f) and (e).

As far as the operating conditions are concerned, we divided the PRVs into broad groups, such as oil, gas, produced water, and air. We divided them further in pressure ranges e.g., 0–100 psig, 100–500 psig, 500–1000 psig, 1000–2000 psig, over 2000 psig. In selecting the mechanical design features, instead of using the make, model, and size, we used the valve type, e.g., conventional spring-loaded, balanced-bellows, or pilot-operated. While we recorded the manufacturer's name, model number, and size, we did not use this for sorting the data.

For each set of PRVs, sorted by service, pressure rating, and valve type, we computed the cumulative days in service by adding up the days in service for each valve in the set. We added up the number of failure events in two separate lots, namely those that leak below 90% and those that do not lift above 110%.

(Internal leak) MTBF = Cumulative days/No. of valves leaking below 90%

(Fail to lift on demand) MTBF = Cumulative days/No.
not lifting above 110%

We divided these MTBF values by 365 to give their value in years.

35.5 Potential Sources of Error

In computing reliability parameters such as MTBF, two conditions have to be fulfilled. The first is that the items must be identical in design and operating context. The second is that the performance of any item in a set should not influence that of another item in the set. These are the so called 'identical and independent' conditions.

When we do statistical analysis, we need large data sets. If items are pooled together to obtain larger data sets, all items in the set must have the same operating context and be of the same make, model, and size. Further, no item should influence the performance of another item in the set. In many cases, it is not practical to find many of the same make, model, size, and operating in identical conditions. So we make approximations of the type described in the case of PRVs above.

Let us look at a simple situation in a company operating a fleet of automobiles. Let us say we want to compute the MTBF of tires. If there are 100 cars, our data set may appear at first sight to be 400 tires. Is this true? Does the wear of the right side tires influence that of the left side tires? Does the wear on the front tires affect the rear tires? Clearly tires have an influence on each other, so they cannot be considered independent. Next, are the front tires performing the same function as the rear tires? They are used for steering, and they experience higher wheel loads than the rear tires for most of their life. Depending on whether the car has front-wheel, rear-wheel, or four-wheel drive, the traction effort and, hence, tire wear will also differ.

From this discussion, it will be clear that the operating context for each tire position is different. Though the physical construction and design of all the tires is identical, their operating contexts are different, so the second condition is also not fulfilled.

A new complication arises from the maintenance policy applied during servicing. If during pit-stops, the tires are rotated, i.e., installed in different positions, then each tire sees different operating conditions over its life. In practice, we often ignore these effects, thereby introducing errors. This discussion should make us aware that the magical MTBF figures produced by the analysts may not be as accurate as they seem.

35.6 Gas, Smoke, and Fire Detectors

There are many of these items installed in an offshore platform. Most of them will be in naturally ventilated process modules. A few would be located in closed acoustic enclosures of equipment such as gas turbines. Some would be in artificially ventilated areas such as the living quarters. Such differences can produce significant variations in MTBFs. Hence, the data sets are sorted using external environmental considerations, as well as by type of detector.

We count the number of installed items in each such set (P), Next, we note the number of recorded failures over a given period, say one year (for the failure mode—failed to detect, Q). This number is obtained from the periodic tests done every 2 or 3 months. The reliability parameters are computed thus:

MTBF (fail to detect) = Px1 year/Q years between failures

Failure Rate (fail to detect) = Q/(Px1 year) failures/year

These items may also fail by detecting unsafe conditions when in fact there are none. Such failures are termed spurious or nuisance events. These are also unacceptable, as they can cause production losses or divert operators' attention unnecessarily. Spurious failure rates can be computed using the same method as above. For this purpose we need the number of false detections during the year. This number has to be obtained from the operations log.

35.7 Emergency Shut Down (ESD)and Blow-Down Valves

These valves are meant to operate in an emergency, providing quick isolation (ESD) or quick depressurization (blow-down). Testing these valves is not an easy task because it would mean shutting down the system, platform, or the whole network of platforms. So these tests are done in two stages. By disconnecting the actuator of the valve mechanically from the valve, it is possible to test whether the valve would have worked when it received a command signal. These are called functional tests. The real proof that the valve works is when it closes (or opens in the case of blow-down), within the required time limit.

The result of these (full) tests gives us the data for computing the reliability parameters. ESD valves will range in size from say 2″ to 48″. Some will be in oil service, others in gas service. Operating pressures may range from 150 psig., to 6000 psig., or higher. So this data set needs to be sorted by service, size, and pressure range.

Once again there are two possible types of failure. The valve may fail to operate on demand or it may operate when there is no demand. During the periodic tests, we can detect failure to operate on demand. As before the MTBF is computed thus:

(Fail to operate on demand) MTBF = Valve-years in service/No. of failures

Spurious events will be recorded in the operating history. Using these numbers, we can compute the MTBF for spurious events in the same manner.

35.8 Fire Water Pumps

If a fire pump fails to start, it can have serious consequences. Laws or insurance regulations will often prescribe a test frequency. In many cases, these tests are done fortnightly, so each pump is started at least 26 times a year. Failure to start at first attempt is what matters. There will be plenty of data points, so computing failure to start MTBFs for these items should be fairly easy. However, this data for computing MTBFs is not usually available in the CMMS, and has to be obtained from the operating log.

35.9 Evident Failures—Weibull Parameters

So far, we have considered hidden failures. The bulk of the recorded data relates to evident failures, such as those of bearings, seals, and couplings. In our case, there were about 900,000 records to examine. If these had been on paper, the stack would have been nearly 300 feet tall! It would have taken a person 4 to 5 years to analyze such a large volume of data. There would be many errors, especially if we put in many analysts to speed up the work. The analysis would in any case be out of date by the time it was complete. We had to look for a more efficient solution. Since the data was available electronically, a software-based search appeared viable.

In order to understand the method used in searching for the applicable work orders, some of the vagaries involved in data entry have to be understood. Typically, we may find one or more of the following difficulties in searching in the history.

- incorrect spelling, such as `chocked' instead of `choked'
- homonyms, such as `ceased' when meaning `seized'
- typographical errors, e.g., `baering' instead of `bearing', or ';' instead of 'l'
- different words with similar meanings, such as `shaft' in place of `stem' or `sleeve'
- alternative spellings, such as `sea water', `seawater', or `sea-water'.
- importance of word order; oil seal and seal oil (or Venetian blind and blind Venetian) clearly have different meanings
- absence of any words in the work order by which we can identify the failure mode
- context sensitivity of words, e.g., bearing may mean something on a pump shaft or an angle in relation to navigation

We found data mining software that had a built-in lexicon and used context sensitive searches. It could identify word order, such as oil seal vs. seal oil or words which were spaced differently but meant the same thing such as 'bearing seized' and 'bearing examined and found seized'. It could recognize synonyms and homonyms. It had error-forgiving rules built in, so that it recognized P 1234, P-1234 and P:1234 as the same item. With all this sophistication, we could locate the work orders pertaining to a single item or of a group of items—such as pumps—rapidly and sort them out by failure modes. All this could be done in days rather than years, and repeat searches produced fairly consistent results. About 35,000 failure data points were identified, but less than 20% were useable. The rest were about failures that were trivial or otherwise uninteresting.

Reliability engineers have found that many failures are distributed according to a distribution called Weibull, named after a Swedish mathematician who introduced it. The distribution itself allows us to approximate many other common distributions by changing the values of some of the parameters in the Weibull equation. We found good Weibull software fairly easily. Once the failure data was entered, we could get nearly 1000 sets of Weibull parameters.

35.10 Results

These were published internally as an on-line database. This database also had all the other reliability data relating to e.g., PRVs, ESD valves, fire and gas detectors, deluge valves, etc. These were used for RCM studies, Quantitative Risk Analysis, etc.

35.11 Lessons

Maintainers are generally very good at collecting data. We use sophisticated CMMS to record the history, using failure and repair codes to speed up analysis. We also collect calibration data from tests on protective instrument loops, relief valve bench tests, trip tests, etc. Do we use this vast pile of data effectively?

We can compute important site-specific failure data that we can use to adjust our maintenance strategies to improve safety levels and cost performance. There are hurdles to cross, but these are within our reach.

Using our own data is better than using published data from others. Knowing the MTBFs, we can reset the test frequencies logically (see pages 163 to 172 in Reference [i]).

35.12 Principles

Degradation depends on the original design and build-quality as well as the way we operate and maintain the equipment. Hence the operational reliability of seemingly identical equipment varies from plant to plant. Maintenance strategies should be based on the physical degradation mechanism, the rate at which it occurs, and the consequence of failure. This policy will help minimize operational risks—to safety, environment, and profitability.

The rate of degradation can be defined accurately with reliability parameters. In order to be meaningful, one should as far as possible compute these parameters using one's own operating context. The use of generic databases is sometimes necessary, but we must be conscious of the possible sources of error.

References

i. Narayan, V. 2004. *Effective Maintenance Management: Risk and Reliability, Strategies for Optimizing Performance.* New York: Industrial Press. ISBN 0-8311-3178-0

Chapter 36

Turnaround Performance Improvements

A system cannot understand itself. The transformation requires a view from outside.
W. Edwards Deming, Quality Management Guru.

Author: V. Narayan

Location: 2.3.1 Large Complex Petroleum Refinery

36.1 Background

This refinery had a clear technical edge over its competitors. In benchmarking studies, however, it proved to be only an average performer. As there were many process units to maintain, they carefully spaced their turnarounds (called shutdowns outside North America—and in the rest of this chapter) to balance the workload, costs, and product availability. A central shutdown planning and execution team managed most of this work, in cooperation with staff from the relevant units.

This brought two benefits. Lessons learned from previous shutdowns could be applied immediately in the following ones. Using the same teams meant that they acquired specialized skills in planning and executing these shutdowns. Staff from the relevant units played an active role at all stages, so the danger of alienation was avoided. It was therefore surprising to find that they were poor performers in shutdowns as well, according to the benchmarking studies.

They were keen on rectifying this situation and requested the corporate technical headquarters (location 2.3.3) to assist them. For this purpose, I was an independent observer at a major multi-unit shutdown, lasting 8 weeks. The people involved were receptive and cooperative, so my presence was not seen as an intrusion. They, and many others in the refinery, were aware that despite their best efforts they were seen as poor performers and could not understand the reasons. It was not a blame game; once they knew why, they could commence their corrective actions.

36.2 Outline of Shutdown of Crude Distiller and Associated Units

The crude distillation unit, capacity 180,000 barrels per day, was on a four-year shutdown cycle. It would be off-stream for 55 days: 40 for engineering work, and 15 for operations to shut down and start up the unit. Most of the engineering work was executed on an eight-hour day-shift basis. Refractory repair work inside the heaters and some cleaning activities were being executed on a 24-hour 3-shift basis, as these were on the critical path.

As a swing refinery, with capacity to accept ad-hoc cargos, every day of plant availability was valuable, with an income of about US$ 300,000 per day, at the time of this shutdown. Major shutdowns of this kind were always done during the warm weather months. The refinery was located in an area where there were many other industrial plants, who wished to shut their plants down in the same period. The companies agreed to schedule their shutdowns to balance the demand on contract resources. This was one reason they worked on a day-shift basis on most of the work.

36.3 Review Process

The Terms of Reference for the review were to:

- observe and critically review the shutdown
- highlight deviations from the plan
- identify improvement areas and recommend action
- prepare a short report and present the findings

An examination of the shutdown plan showed that they had done it very thoroughly. The initiating departments carefully justified the work-lists. The company was a leader in the application of risk-based methods, and used a forerunner of Risk Based Inspection (RBI). Their inspection team had obtained approval from the Regulator to stretch the interval for furnace inspections from 2 to 4 years, vessel inspections up to 8 years, and when there were three or more identical heat exchangers, up to 12 years, by inspecting only one every 4 years. They collected relevant data and demonstrated to the Regulator that the risk levels with the new intervals were tolerable.

The review process included physical observations at site, attendance at the morning coordination meetings and discussions with key personnel. Audits of the issue of permits-to-work (PTW) highlighted some productivity issues. Similarly, work-sampling showed whether the workmen were able to get blocks of time long enough to work effectively. Analysis of the main contracts showed the distribution between lump sum and unit rate contracts. The latter need more administrative effort than lump sum contracts. Such paperwork diverts the shutdown team's attention from their primary roles.

36.4 Potential for Improvements—Planning

There were three separate plans: the first for operational activities, a second for project work, and the third being the main plan for maintenance work.

It was not possible to coordinate and manage changes that would occur from time to time in individual plans. Resource balancing was not global and thus not optimal. Equipment status charts were maintained separately by operations, inspection, and maintenance. This made it difficult to get an overview, resulting in avoidable delays.

Some items of work could have been reduced or eliminated during the work list challenging process. For example, in order to measure wall thickness of heater tubes ultrasonically, they erected scaffolding along the walls of the heater cells. Then they scraped and cleaned the tubes. Oil firing produced sulfur deposits on the tube surfaces which were quite acidic and posed a health hazard to people working inside the heater. They cleaned these using mechanical scrapers before allowing the main group of workers inside the furnace cells. Once the heaters were scaffolded, they ground spots on the tubes to bare metal, where the measurements were required. Records showed that not even a single tube had ever been rejected in the past as a result of these measurements.

The only tubes that needed replacement were those that were damaged by flame impingement. None of the 320 tubes (6 5/8″ dia., 44′ long, and 7/16″ wall thickness) showed excessive loss of metal during the current inspection. All readings were within 10% of nominal thickness, while the rejection limit was 50%. These readings did not help make replacement decisions, so it was unnecessary work. Convincing the Regulator, however, was a different ballgame.

Some weeks later, in a frank discussion with the Regulator, the refinery tabled all the findings over the years. With these, they convinced the Regulator to let them reduce the number of measurements significantly. Instead of measurements every half meter over the whole length of the tube, the Regulator permitted them to limit measurements to the most severely affected locations on the tubes. This reduced the ultrasonic measurement work by about 70%.

For future shutdowns, I suggested a switch over to natural gas firing for two days prior to shutdown. This change would help vaporize the acidic sulfur deposits and eliminate the need for extensive cleaning. Together, these changes would reduce heater work duration by 16 hours. Gas firing was expensive, and the difference between the cost of gas firing and reduction in maintenance costs was estimated at US$15,000. This increase was minor, but the reduction in health risks was quite considerable.

A total of 88 non-return valves of varying sizes and pressure ratings were opened for inspection. Of these, 19 had damaged internals and 6 others were fouled and needed cleaning. Of the 19 valves with damaged internals, at least 50% of the damage occurred while dismantling them. Additional pre-shutdown non-intrusive inspection could have identified many of the fouled or defective valves. The number of valves to be examined could have been reduced significantly.

The software package used for planning and scheduling was quite satisfactory as far as its main function was concerned. However, it did not produce Gantt charts or lists of pending items by zone. All these were done manually by the supervisors themselves. A different software package could eliminate such time-consuming manual work.

Evacuation drills were not part of the plan. In a shutdown of this size, with nearly 500 contract workers at site, it is not sufficient to do a classroom safety induction. Fire and gas alarm conventions differ in different companies and escape routes need familiarization. Evacuation drills are necessary to ensure that people can get away safely, should there be an incident. A common misconception is that such drills would result in a significant loss of productive time. If these are conducted just prior to a planned break, such losses can be minimized.

36.5 Potential for Improvements—Productivity

Issue of permits to work (PTW) can be a major problem affecting productivity in any plant. In this shutdown, operations had streamlined their work very well. Between 3 and 4 pm., they discussed the safety aspects of the following day's work with the relevant contract supervisors. The PTWs were ready in all respects except for the signatures of the two parties. Next morning, operations merely updated the contract supervisors with any changes in safety requirements and then signed it. They asked them to meet the field operators before commencing work. Meanwhile, they contacted the field operators so that they could ensure that the work commenced safely. Using this procedure, they were able to release, on average, three PTWs per minute.

On day 4 of the shutdown, an audit showed that by 8 am., only ten permits had been released. The distribution can be seen in Figure 36.1. In a discussion with the shutdown leader, a solution that emerged was to ask the contract supervisors to come in earlier, at 7.30 am, allowing them to have the permits in hand before their crew arrived. This plan was implemented and an audit on day 8 showed a slight improvement. An audit on day 15 showed that the problem had been substantially resolved. These results can be seen in Figures 36.2 and 36.3.

Figure 36.1 Issue of PTWs on Day 4

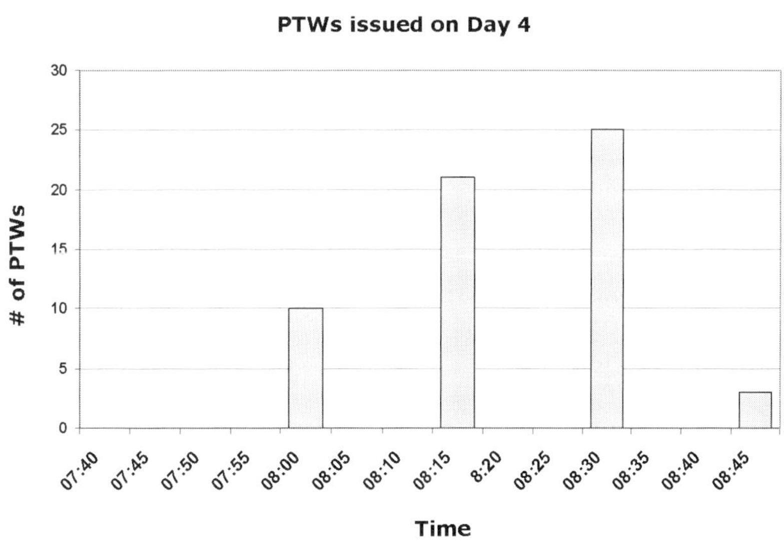

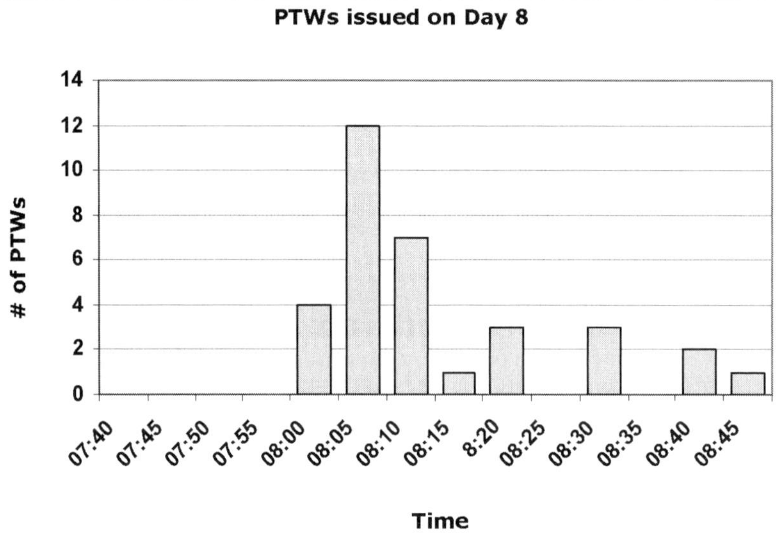

Figure 36.2 Issue of PTWs on Day 8

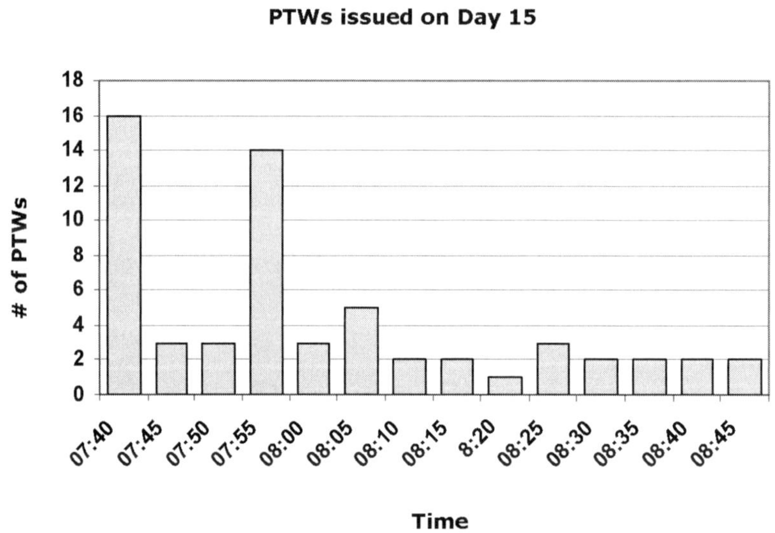

Figure 36.3 Issue of PTWs on Day 15

Another significant cause of loss of hand-on-tools time was the way working hours are organized. At this location, shutdown work for non-critical activities was done on an eight-hour day-shift basis. Contractors transported their workers from local bus stations, designated car parking lots, and the refinery gate to the central washroom area. Then the workers walked to the shutdown area. The process was repeated in reverse at the end of the day. Each of these walks took about 7–8 minutes.

Supervisors gave tool-box talks to their crew before they left for their work site. These talks were quite short, lasting 2–3 minutes. Since each supervisor had 3–5 crews, this process could take 15 minutes. It took 3–5 minutes to walk to the work site, sometimes longer if they needed to climb stairs or scaffolding. By the time the crews started work, it was invariably about 8:15 am.

Contractors were responsible for providing coffee and tea to their workers. They used vans for this purpose. The first of these arrived at 9:30 in the morning and 2:15 in the afternoon. Human nature being what it is, as soon as the first van was sighted, someone whistled, and everybody came down. This meant that the first work period could not exceed 45 minutes. After the morning break, the crews could work uninterrupted till 11:45. They were allowed 15 minutes to wash up before lunch at noon. The afternoon coffee break lasted 30 minutes instead of 15, for the same reasons as in the morning. Traditionally, the workers had been allowed 45 minutes at the end of the day, to allow time to wash up and walk to the gate. The central washroom was crowded and buses left promptly, so extra time was needed to ensure that the workers reached in time. This meant that after the afternoon coffee break they could work only 11/2 hours. Thus, the available time to work was only 51/2 hours. Of this, they worked perhaps for 31/2 hours. This type of situation must have been experienced by a number of readers in their own work sites.

The main players in the shutdown team discussed the results of the analysis, and evaluated possible solutions. I suggested we add 11/2 hours of planned overtime every day, making it a 91/2 hour day. They rejected it immediately. In continental Europe, there is strong social legislation. Overtime work needs governmental approval, and planned overtime needs prior approval. Team members felt that the authorities would turn down our request. Later, I discussed this with the person who dealt with the government agency concerned. He said it was perfectly feasible, exploding another myth. This was something we could explore at least for future shutdowns. The benefit of 91/2 hour vs. 8 hour days is illustrated in Figure 36.4. The main advantage is not the total additional time, but the fact that the blocks of time are longer and thus more effective.

This still left us with the problem of the 45 minutes lost at the end of the day. There was no easy solution to the washroom crowding problem. For various reasons, changing departure time of buses couldn't be discussed at that time. Traditions are hard to change, so we decided to accept this loss, at least for the present. After all, with the 91/2 hour day, we would gain 2 hours of available working time. More importantly, these were in longer and more meaningful blocks of about 11/2 hours. One could reasonably expect that the actual work output would rise correspondingly by 30–35%. This meant that the 40 days duration could in theory be 30 days. In practice, one might expect a smaller reduction, say 5 days, with a value of US$1.5m.

Benchmarking studies showed that in plants of comparable size in other countries, Operations needed much less time to shut down and start up the unit. The potential reduction was about 9 days. Environmental and safety considerations resulted in additional work, and Operations' resources were severely limited, in spite of working on a 2x12 hour basis. Another brainstorming session ensued, with some interesting ideas. In the short term, the idea

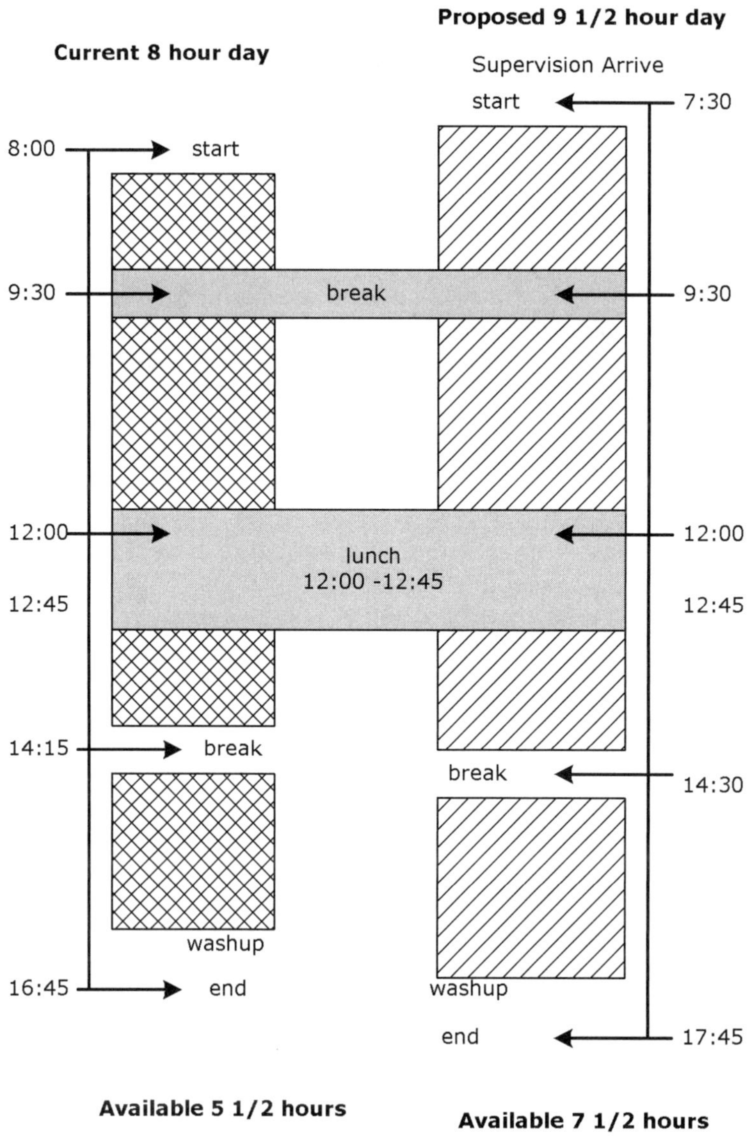

Figure 36.4 Effect of 91/2 Hour Day

was to hire skilled operators from contractors during the busy shut-down and start-up periods. For the long term, an idea that was floated was to train the mechanical crew to support operations in valve operation, shutting down complex machinery, etc. After all, during the shutting-down process, mechanics were largely on standby, doing preparatory work. This idea had many pitfalls, so we parked it for later action. Contract operators would cost about US$ 200,000. A modest reduction in duration of, say, 5 days was worth US$1.5m.

Similarly, one reason for the criticality of the refractory work was because they worked 3x8 hours. Refractory masons were in short supply, so the work had to be done in sequence. Had they been able to work in parallel, this work would not be on the critical path. Two shifts of 81/2 hours each would release one complete shift crew, enabling work in parallel. Applying the same approach as earlier with the 91/2 hour day work, we could get a much higher 'available time,' thereby increasing productivity. As a result of traditional thinking, it was assumed that critical work must be done on a 24-hour basis. That this is not always true is evident from the above discussion.

During shutdowns, normal weekly working hours were 5 days x 8 hours. As a result of all these discussions, the refinery decided to work 6 days x 91/2 hours. On critical path activities they worked 2x9 hour shifts for 5 days and one 9-hour shift on Saturdays. In a few exceptional cases, they used 2x12 hour shifts. Note that skilled labor availability and prevailing legislation were limiting factors. They used the weekend to catch up on delays by doing additional overtime work.

36.6 Analysis of Previous Contracts

The refinery depended on contractors to provide the bulk of the engineering workforce for their shutdowns. Clearly definable work was awarded as lump sum contracts, but a significant volume was awarded using unit rates. An analysis of the 68 shutdown contracts awarded over a three-year period showed the following results, divided into three groups marked A, B, and C.

A. 16% of the shutdowns accounted for 53% of the cumulative costs.
B. 22% of the shutdowns accounted for 26% of the cumulative costs.
C. 62% of the shutdowns accounted for 21% of the cumulative costs.

Shutdowns in the A group cost over US$ 2m each, those in the B group cost US$0.75 to 2m each, and those in the C group cost less than US$ 0.75m each.

The existing shutdown procedures applied the same controls on all three groups. Implementing the following changes would help improve management control.

1. Finalizing and freezing of main inspection and operations work lists 12 months in advance for groups A and B, 6 months in advance for group C.
2. Finalizing and freezing of main operations/maintenance lists 4 months in advance for group A, and 3 months in advance for groups B and C.
3. Finalizing and publishing work program 3 months in advance for the group A, 2 months in advance for groups B and C.
4. Awarding main contracts 2 months after freezing main work scopes.

36.7 Reducing Planning Effort and Time

The planners produced excellent estimates for each shutdown. They could easily reuse many of them. There is a strong case for producing a reference database, so that could be used readily. Related isometric drawings were available electronically, but were currently accessed each time from a different database. Storing these in a linked fashion to the estimates would help speed planning work. Obviously drawing revision numbers had to be verified each time to ensure they were current.

The refinery initiated an upgrade of their IT systems and computerizing of their archive to resolve these problems.

36.8 Contracts Administration

Prior to this shutdown, contracts worth about US$3m had been awarded. Of this, lump sum (LS) contracts accounted for 63% and unit rate (UR) for 37% of the total. The main UR contracts related to scaffolding, insulation, and cleaning work. A significant proportion of the scaffolding and insulation work could have been awarded as LS contracts. The advantage of LS contracts is that they need much less administration effort. For example, the UR insulation contracts required 220 documents to be filled in, vetted, and approved. These required a total of 375 calculation sheets, based on a similar number of quantity surveys. Had these been LS contracts, about 120–140 supervisor-hours would have been reduced. They could spend this time actually supervising work instead of shuffling paper.

The refinery now uses LS contracts wherever possible. They also combine contracts like scaffolding, insulation, painting, and NDT. One of the contractors is the co-ordinator of all these works.

36.9 Management Presence

A shutdown is a major project, with a daily expenditure of a comparable size. Yet the level of management involvement in the shutdown was relatively low. Seeing senior managers at site motivates staff and demonstrates management commitment.

The management team has taken on this recommendation quite enthusiastically.

36.10 Results

The refinery was receptive to suggestions, and incorporated some during the early stages of the shutdown. For example, the idea of calling in supervisors 30 minutes early was implemented within 3 days of the first PTW audit. There were open discussions about the improvement opportunities, some of which could affect cost and duration quite significantly. I discussed most of these ideas with the relevant people while the review was in progress, and some were implemented right away. The refinery adopted many of the key recommendations after an internal review.

Three recommendations were accepted completely, for all their future shutdown work. These were the use of 91/2-hour working days, getting supervisors to come in 30 minutes early, and provision of additional operating crews during shutting down and starting up of the units. In the case of the distillation unit, these three had the potential to reduce duration by 12–15 days from the current 55 days. In practice, one could expect a saving of 8–10 days, with a value of US$2.5–3 m.

The refinery considered the four points made in Section 36.6 so important that they included them in their Shutdown Handbook. They see them as key targets for shutdown planning and use them as Key Performance Indicators during shutdown preparation. They are convinced that there is a direct relationship between the quality of the preparation of a shutdown and execution effectiveness, impacting on safety, duration, quality, and emergent work.

A number of other recommendations were specific to the distiller shutdown, and they took these up as well.

36.11 Lessons

1. Good planning is essential; work scope definition, freezing of scope in time, etc., are all well-known principles, but are not applied rigorously.
2. Eliminating unnecessary work in the first place is better than doing it faster.
3. Challenging 'the way we always do things here' is the first step towards making a step change in performance.
4. Managing work permit issues and setting working hours sensibly can make a huge difference to productivity and, hence, duration and cost.
5. People often think that 'beating the drum faster' is the only way to get high productivity. Often, good planning and scheduling can be far more effective.
6. Workers must be allowed blocks of time that match the type of work. Once they have prepared the work, there must be enough time to complete a substantial chunk of work without interruption.
7. Good communication is essential for efficient execution; a single integrated plan helps communication between the main parties.
8. External examination of shutdowns can reveal some practices that are not economically ideal, even in well-managed companies.
9. On a personal note, I gained a lot from being an observer.

In Chapter 17, Mahen describes the principles involved in managing shutdowns efficiently. In Chapter 19, Jim has explained the shutdown business process, best practices, and related metrics. In the next chapter, I will describe a practical way of achieving improved shutdown performances.

36.12 Principles

Technical expertise is a necessary but not sufficient condition for business success. Engineers enjoy technology and are less comfortable with economics. This is one reason they have a smaller impact in the Board Room than their colleagues.

An external view can help regain focus on what really matters.

Chapter 37

Reducing Shutdown Duration

(A practical illustration)

There is nothing more difficult to take in hand, more perilous to conduct, or more uncertain in its success, than to take the lead in the introduction of a new order of things.

Machiavelli, in The Prince, 1532.

Author: V. Narayan

Location: 2.1.4 Large Petroleum Refinery

37.1 Background

I was late in taking up my new assignment in this refinery, as it took longer than expected to obtain the work permit from the host government authorities. On arrival, I was to execute a major shutdown (called turnaround in North America) that had already been planned. Due to this delay, my predecessor stayed back to execute it. Thus I did not have an executive role for a few weeks and was able to observe the shutdown independently.

The Thermal Cracking Unit (TCU) added about US$60,000 of value per day of operation. In this unit, the long-chain heavy residue molecules were broken down by a high temperature cracking process. This yielded short-chain products like naphtha, which were valuable as feedstock to petrochemical plants. Figure 37.1 shows a schematic flow diagram of the process. The cracking of the long-chain molecules liberated free carbon which was then deposited as coke in the heaters, columns, vessels, and pipelines. In due course, the coke deposits resulted in a sharp fall in operational efficiency. The unit had to be shut down every six months to remove the fouling by coke deposits. Past records showed that these shutdowns lasted 21 days or longer, even though the number of equipment items involved was quite small.

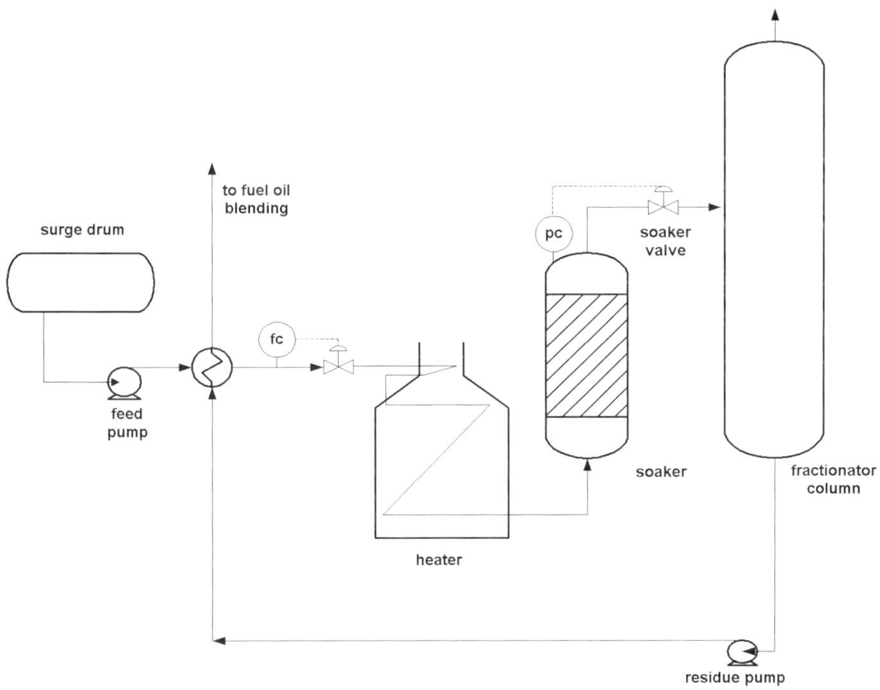

Figure 37.1 Process Flow Scheme

37.2 Observations

During the TCU shutdown, I observed the existing operational and maintenance activities. The shutdown planner assisted me in this process.

We examined all the major activities, to understand why they were being done, the sequence followed, and the results achieved. Since de-coking and cleaning were the primary activities, we kept a record of the volume of debris removed and the duration of each cleaning activity. If work was held up on a planned or unplanned basis, we recorded the reasons. An analysis of this data showed that there were some opportunities to reduce the work volume, duration, or both. We will discuss some of these in the following sections.

37.3 Soaker Vessel

This vessel provided a short residence time for the hot residue coming from the heater. The liquid pressure dropped as it entered the vessel, and this initiated cracking of the long chain hydrocarbon molecules. The Soaker is a vertical vessel about 10′ diameter and 60′ high, mounted on a cylindrical steel skirt about 10′ high. It had a 6″ thick insulation layer to minimize heat loss by radiation. There were 7 simple trays in it, each with a number of 6″ holes. As the hot liquid flowed up the vessel, the long-chain molecules were broken down into shorter chain molecules. During this process, free carbon was liberated. At the end of a run, the coke buildup was nearly 5″ thick on the trays and the holes were covered up so they were just 1–2″ in diameter. There was coke on the vessel walls and dished ends. This coke had to be removed by high pressure water jet cleaning (hydro-jetting). Prior to vessel entry, it was steamed out with low pressure (LP) steam for eight hours to remove hydrocarbon vapors. Then it was aerated for four hours to make it safe for entry. The arrangement is illustrated in Figure 37.2.

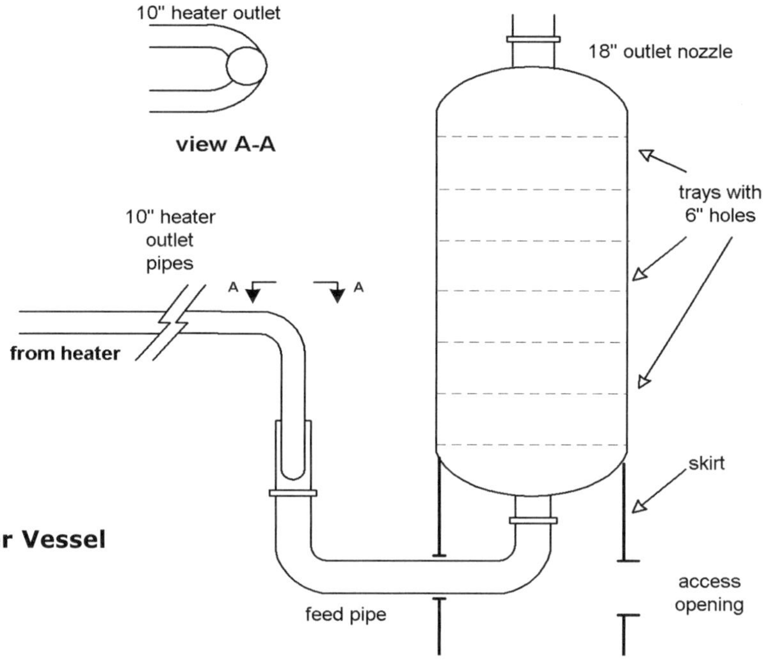

Figure 37.2 Soaker Vessel

The vessel remained hot for about three days after the steaming was completed. There was a 16″ nozzle at the bottom, through which the oil from the heater entered the vessel. This opening provided access into the vessel once the flange was opened, and the pipe removed. A small scaffold had to be erected in the vessel skirt, to reach and unbolt the flange. This could only be done about 2 days after operations handed over the unit, because of the intense heat inside the vessel skirt.

On entering this vessel, I observed that it was very oily and slippery. Workers had to be extra careful, as the reaction forces from the hydro-jet nozzle could make them slip and fall down. The coke was not brittle, so it had to be gouged out in chunks. This meant that just to clear the coke it took nearly twelve days, working 24 hours a day.

I had three concerns:

1) safety hazards for the workers in the oily environment,
2) delay in the opening the vessel for entry, and
3) the time it took to remove the coke.

After the shutdown, I discussed the steam-out procedure with the unit manager. LP steam can be slightly wet and we explored the use of high pressure (HP) steam, suitably throttled to low pressure so it would be very hot and well superheated. I asked that the steaming duration be extended to 36 hours to improve the dryness of the coke. This meant that operational activities would take longer and cost more, but there were clear benefits in terms of safety. I promised him that the additional steaming time would be recovered as overall duration would be reduced.

37.4 Fractionating Column

The vapors from the top of the Soaker entered the Fractionating Column through an 18″ transfer pipeline. The different components in the vapor have different boiling points, depending on their molecular weight. These components are separated in the column, which was about 8′ diameter, 100′ tall with 36 trays. The light molecules, such as butane and propane flow out of the top, while the heavy oils are pumped out from the bottom of the column. Quite a lot of coke was deposited in the lower sections of the column, especially on the bottom dish. There was a large alloy steel strainer inside the column to prevent coke pieces entering the pump suction. The strainer was conical, about 36″ diameter at the base and 24″ diameter at the top, and 48″ high. It had slotted holes and covered the 10″ outlet pipe at the center of the bottom dish completely. On inspection, we found that 60–70% of this strainer was plugged with coke.

The column had a 4″ thick insulation layer, so that heat loss by radiation was minimized. As a result, the column remained hot for a few days after it was ready for entry. Workers had to come out every 30 minutes to cool off and drink water. When working inside columns, they would enter through one of the six column man-ways, and work their way up through the trays, by opening the tray man-ways. Coming down frequently, meant that the actual work done in each trip was quite small. The work inside the column needed 13 days for execution.

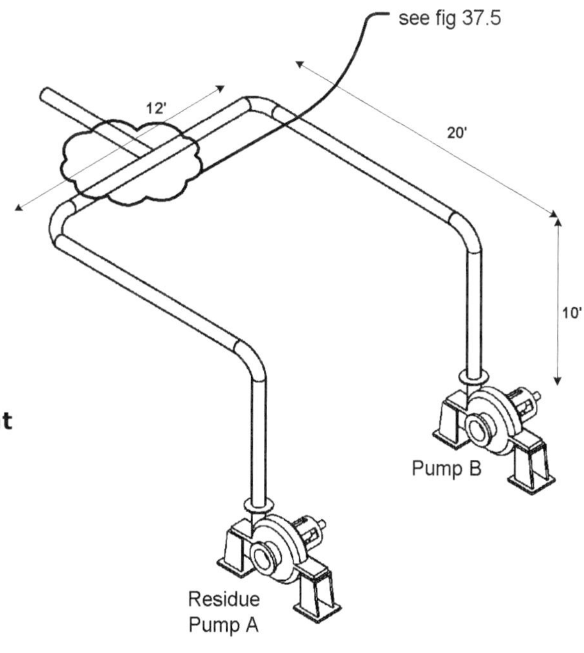

Figure 37.3 Piping Arrangement

37.4 Residue Pumps

The hot, heavy liquid from the bottom of the column was pumped out as a residue, to be used as a fuel oil blending component. As the liquid went through the pipeline, some carbon particles continued to deposit on the pipe walls. The geometry of the alloy steel piping was such that it became very difficult to clean them effectively. The piping layout can be seen in Figure 37.3.

Towards the end of a production run, the pump capacity dropped significantly, and cavitation noise was audible. Seal failures occurred more often at this stage.

37.5 Pipeline from Soaker to Fractionating Column

The 80′ long, 18″ transfer pipeline from the top of the Soaker to the Column was flanged at the two ends. At an early stage of the shutdown, this pipeline was lowered to the ground to enable its internal cleaning. This pipe was reinstalled after all other work was completed, adding eight hours to the shutdown duration. The lifting operation usually resulted in damage to its insulation. This pipe layout is shown in Figure 37.4.

The coke pieces that emerged during the hydro-jet cleaning of this pipe were only about 1/8″ thick, and the quantity was about 8 lbs.

37.6 Heater Outlet Pipes to Soaker

Fouling of the heater tubes determined the timing of the shutdown. The heater tubes themselves were cleaned by the operators, using a process called steam-air decoking. The coke was burnt off in a controlled manner by the use of air for combustion and steam to regulate its speed. Steam also

Figure 37.4 View of Soaker Overhead Line

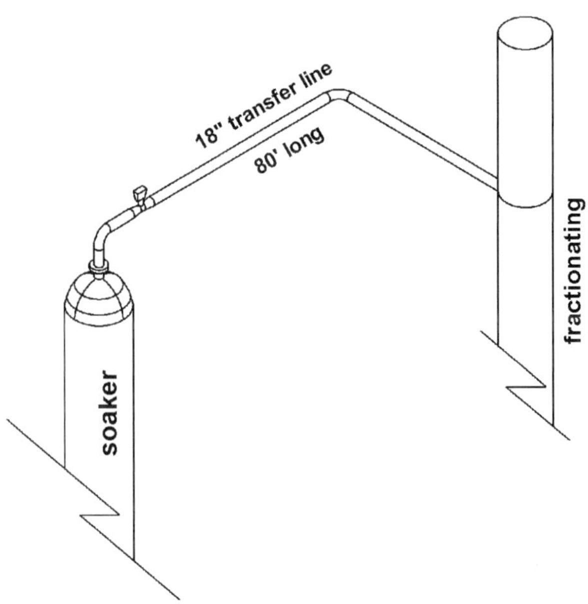

helped to keep the tubes from getting overheated and to blow away the unburnt coke pieces. The decoking process was fairly fast; it took about three days to decoke both the heater cells.

There were two 70′ long, 10″ outlet pipes, one from each cell, flanged at the heater end and on the 16″ soaker inlet pipe, as shown in Figure 37.2. The location of the flanges at the Soaker end made it difficult to clean the horizontal section of these pipes. Similarly, the location of the flange on the 16″ pipe leading to the Soaker made that section unwieldy to handle.

37.7 Improvement Opportunities

The planner and I were confident that the shutdown duration could be reduced significantly. We analyzed the main activities described above and formulated some ideas, as follows:

1. Soaker—Use of HP steam would improve the safety of the cleaning activities significantly. A longer steam-out period would speed up the cleaning process and help reduce overall duration.

There was no simple solution to minimize the access time for scaffolding inside the skirt of the Soaker. A few months later, the maintenance supervisor came up with an interesting idea. While this would not solve the access time problem, it was still worth pursuing; more about this later.

2. Fractionating Column—In another part of the refinery, nitrogen cooling was used in the reactors of the Hydro-Cracker Unit. Nitrogen was supplied from a portable liquid nitrogen plant. These units liquefy air and fractionate it to get liquid nitrogen. The contractor who supplied the unit confirmed it was possible to supply cold air, using the same unit. We decided to hire the unit for the next TCU shutdown to cool the column quickly, and to provide a comfortable working environment for the people. With ambient temperatures in the Middle East over 100°F for most of the year, this would be a welcome change. The concept was a somewhat revolutionary welfare measure, and nobody had so far considered air conditioning a column. I was thus at the receiving end of jokes.
3. Pump piping—We discussed the problem relating to the cleaning of the discharge piping of the residue pumps with the process technologist. One possibility was to introduce a 2″ nozzle at the tee junction, as illustrated in Figure 37.5. We evaluated the risks and initiated a Plant Change.
4. Fractionating Column strainer—It was evident that the coke build-up in the strainer was contributing to the pump failures. The loss of pump capacity was so severe that it was not feasible to operate the unit beyond six months. The process technologist and I came to the conclusion that the strainer design had to be modified to overcome this problem. He designed a strainer with a larger number of holes, so that the pumps had sufficient net positive suction head (NPSH). This was another Plant Change.

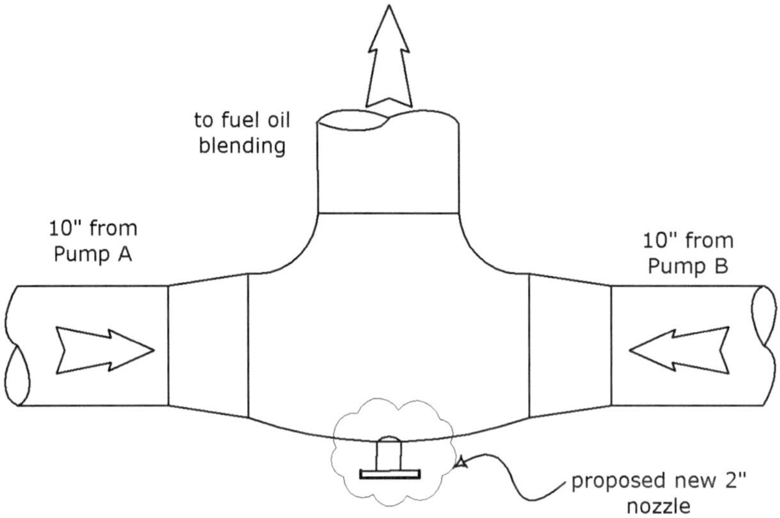

Figure 37.5 Cleaning Access Nozzle

5. Soaker overhead pipe—The unit manager and I discussed the logic of lowering the 18″ Soaker-to-Column transfer line for cleaning. After seeing the evidence, he agreed that this line need not be cleaned regularly in future. This item was deleted from the work list, clipping at least eight hours from the duration.
6. Heater outlet lines—Cleaning of the heater outlet lines would be much simpler if we could relocate two pairs of flanges. Similarly, relocating one pair of 16″ Soaker inlet pipe flanges would speed up access to the Soaker. The process technologist agreed to raise a Plant Change, illustrated in Figure 37.6.

37.8 Results

1. The use of HP steam removed the oil slick entirely, so the safety problems reduced significantly. The superheated steam dried out the coke and made it hard and brittle. As a result, cleaners were able to shatter the brittle coke rapidly, and the entire Soaker cleaning took only six days.

The maintenance supervisor's suggestion was to cut four horizontally-oriented holes in the skirt and install 3″ pipe sleeves. We could thread 2″ scaffolding pipes through these from outside the skirt, so the heat inside was of no concern. As soon as the vessel cooled down, we could lay planks on the scaffolding pipes and start the work of opening the 16″ flange. This would save 2–3 hours, and every little saving counted. As a new Plant Change request had to be raised, this action was completed only after three more shutdowns.

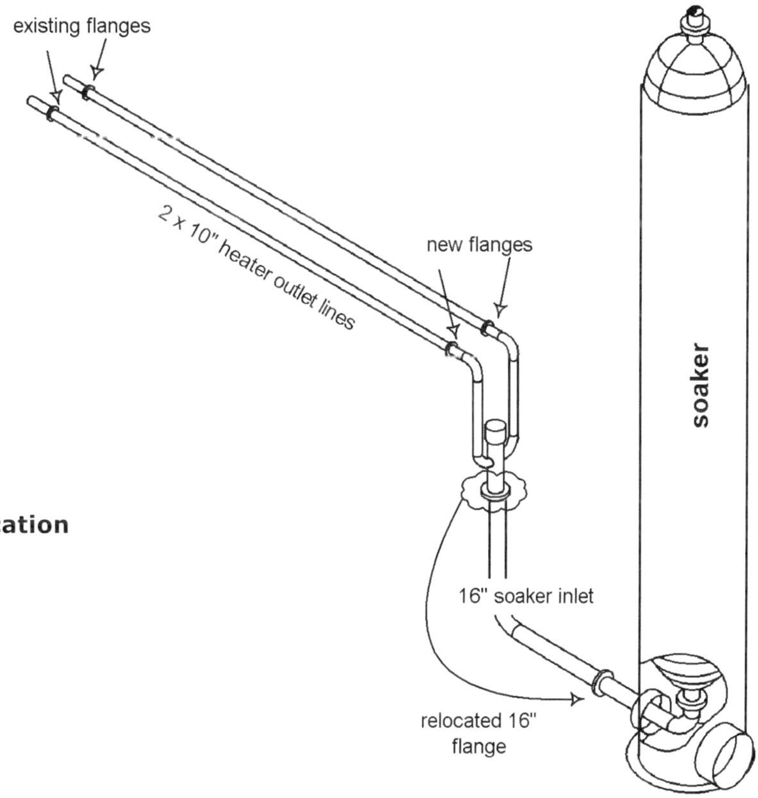

Figure 37.6 Relocation of Pipe Flanges

2. As in the case of the Soaker, HP steam cleaning worked very effectively in the Fractionating Column, removing all traces of oil. The use of the air-conditioning unit was a tremendous success. Work inside the column progressed rapidly, as the need for frequent breaks was eliminated; in fact, it was hard to get the workers out of the column! The cleaning, repair, and inspection work could now be completed in six days. Those who joked about my 'touchy-feely' approach started thinking anew.
3. The new 2" nozzle in the pump discharge piping was a boon to the cleaning crew. More importantly, the rundown pipeline was cleaner and improved product delivery flow.
4. Enlargement of the column strainer also proved successful. It was now possible to increase the shutdown interval to eight months, provided the rate of fouling in the heater could be reduced at the same time. Re call that the fouling rate of the heater tubes determined the shutdown timing. The technologist worked out the economics of operating the heater at a slightly lower severity. This would reduce the cracking efficiency slightly as well as the rate of coke build-up in the tubes. The overall economics were sound, and we could change the shutdown in terval to 8 months. The larger strainer increased the pumps' NPSH, improving their reliability and performance. It also helped make the longer shutdown intervals possible.

5. Eliminating the work of lowering the 18″ transfer line reduced the shut down duration by eight hours. Some cleaning effort and insulation repair costs were also eliminated (ca. US$ 10000 per shutdown).
6. On implementing the Plant Change for relocating heater outlet and Soaker feed pipe flanges, there were two positive outcomes. The first was that there was a significant improvement in the cleaning of the heater outlet lines. The second was a reduction of six hours in the work relating to the Soaker.

As a result of all these actions, we brought down the TCU shutdown duration in stages, to 14, 11, and finally 9 days. Once the strainer was enlarged, we extended the interval to eight months. We completed all these changes within three years. On an annual basis, the value added by these actions was US$ 1.7m.

37.9 Lessons

1. Accepting existing practices without a challenge can lead to stagnation, especially with regard to shutdowns.
2. Unbiased observations can yield large benefits, especially if they are made holistically, looking at the whole process, not just the maintenance part.
3. Business benefits must be the drivers for maintenance improvement, not merely technical excellence.
4. Team members, whether in operations, process technology, or maintenance, can make important contributions to the business. It does not matter whether they are technicians, supervisors, or managers; ideas come from unexpected sources. By drawing on the knowledge, experience, ingenuity, and support of all those involved, it was possible to make the improvements fairly quickly. This requires good team work and communication.
5. By computing and demonstrating the value added – in this case US$ 1.7m p.a., we raised the maintenance profile, and demolished the myth that maintenance was merely a cost.

37.10 Principles

Any business situation can offer improvement opportunities. But first we need clear goals so we can recognize these opportunities.

Looking after people makes good business sense.

Chapter **38**

A Small Matter of Cleaning

A single idea, if it is right, saves us the labor of an infinity of experiences.
...Jacques Maritain, French philosopher.

Author: V. Narayan

Location: 2.1.3 Petroleum Refinery

38.1 Background

In a Fluid Catalytic Cracking Unit (FCCU), the waxy long-chain hydrocarbon molecules are split into shorter-chain, high-value products such as gasoline. At the heart of the FCCU are two pressure vessels working in a loop configuration: the Reactor and Regenerator. The Regenerator is a large vessel operating at about 15 psig., and at temperatures over 1000°F. In the Reactor, the catalyst powder gets coated with carbon released in the cracking process. This carbon is burnt off the catalyst surface by combustion in the Regenerator. A large axial-flow compressor provides air for combustion. Air is distributed inside the Regenerator with a set of pipes called an air-spider. There is a large and continuous fire inside the vessel. The steel wall of this vessel is protected on the inside surface by two layers of thick refractory insulation, keeping the shell relatively cool.

In the FCCU we are discussing here, the lower part of the Regenerator was about 20′ diameter and 30′ high, while the upper part was about 35′ diameter and 60′ high. The two parts are joined together by a swaged section, making the vessel about 120′ tall. In Chapter 30, Mahen discussed cyclone replacement in this vessel. You can see a cross-sectional view with some internal details in Figure 30.4. The combustion inside the vessel heats it up significantly and it grows as a result. This causes flexing of the refractory lining, especially at the swaged section. Such flexing occurs mainly during the startup and shutdown phases. The lining damage due to such flexing could be quite large, requiring a fair amount of repair work.

During the planned shutdown (called turnaround in North America), we were to renew the cyclones fitted at the top of the vessel (see Chapter 30, Figures 30.2 and 30.4). Another job was the renewal of the air distribution spider, fitted at the bottom of the vessel.

38.2 The Cleaning Problem

We expected a large volume of lining repairs, a process that required 'gunning' of the insulation concrete (Verilite®™ and Tuffmix®™) onto the walls. Apart from the lining material removed for the renewal work and repairs, quite a lot of 'rebound' material from the gunning fell to the bottom of the vessel. The volume of such debris was expected to be in excess of 1000 cubic feet. The standard cleaning procedure was to lift the debris in 4-gallon buckets up to the 60" man-way about 22' above the vessel floor, using pulleys and tackle, and then drop it into half-drums. These half-drums were taken down the elevator and carted away in trolleys to the dump site.

The cleaning process added 2 days to the duration of normal shutdowns. During this shutdown, due to the larger volume of debris, this process would have added 5–6 days. Cleaning was always the last activity, completed after all the other work was done. Since people were working at different elevations, cleaners could not work at the bottom of the vessel till all other work ceased. As discussed in Chapter 30, the shutdown duration had to be restricted to 30 days, so an unconventional solution was required.

38.3 The Air-Spider Replacement

Large volumes of low pressure air are blown into the Regenerator in a balanced manner through this air-spider. The arrangement is illustrated in Figure 38.1.

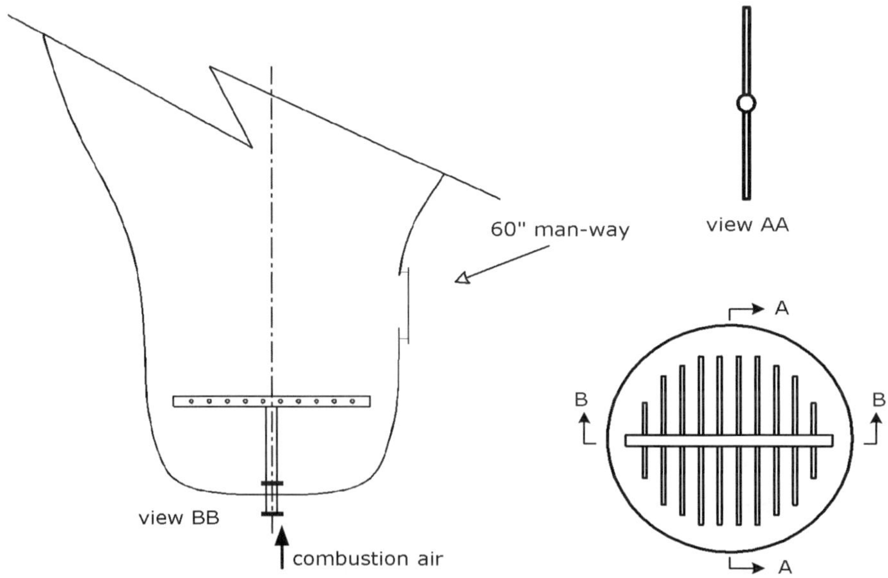

Figure 38.1 Air Spider

The air spider was made of (heat and oxidation resistant) alloy steel. The 18″ central pipe with the branch stubs could go through the 60″ man-way in one piece. We had to weld all the branch pipes in-situ, as the whole assembly was too large to go through the man-way opening. If heavy lifting operations were going on, e.g., in replacing the cyclones, we could not allow work on the spider. So we had to schedule this work carefully to minimize the hazards to the people working on the spider.

Meanwhile, debris from the upper levels would continuously rain on the bottom dish. Unless this was removed, the fitters and welders could not work on the spider. In a normal shutdown, most of the internal cleaning work would be done at the end. In this shutdown, however, we had to do cleaning continuously to enable work on the spider.

Once the old spider was removed and the new central spider pipe installed, we built a scaffolding platform at the level of the bottom of the 60″ man-way. A solid floor of wood planking covered with a thin sheet metal layer provided adequate protection from small falling objects and weld spatter. This platform prevented the bulk of the debris from reaching the bottom dish. Even so, the bottom dish had to be cleaned regularly. The platform arrangement is illustrated in Figure 38.2.

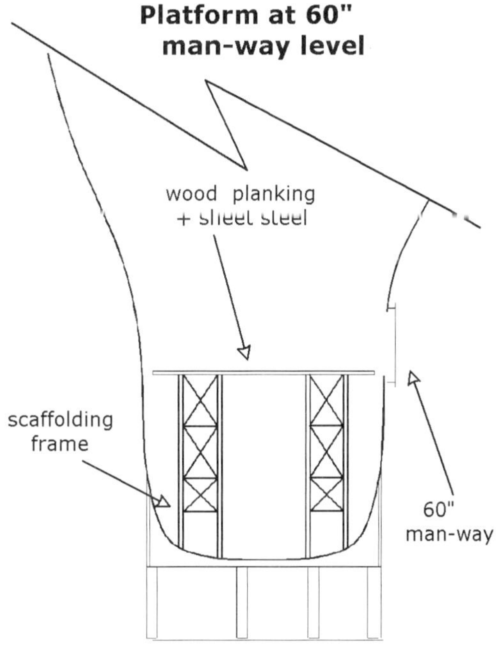

Figure 38.2 Platform over Lower Work Area

38.4 Analysis of Cleaning Activity

The civil engineering supervisor, the Planning Engineer (and co-author) Mahen Das, and I analyzed the cleaning process. The existing process was resource intensive and clumsy. The four-gallon buckets held fairly small quantities, and larger buckets were unwieldy. Transporting half-drums up and down in the elevator was a slow process as there was only one elevator. The removal of the filled half-drums was a logistical nightmare. The whole process needed to be re-engineered.

38.5 Questions

The civil engineering supervisor was toying with an idea, but realized he needed help. He put forth a set of questions, as follows:

- Can we use a dump truck rather than trolleys to cart away the debris?
- There is already one central 18″ nozzle for the air inlet to the spider. Can we introduce a new 24″ nozzle on the bottom dish of the vessel?
- Will we allow cleaners to work in the bottom dish along with the welders and fitters?

All of these were feasible, but we had to examine some details. We had to check the relevant codes before proceeding with the design of the new 24″ nozzle, and complete the risk assessment; these steps were required in any case to fulfill the change control procedures.

38.6 Solution

The supervisor's idea was innovative and practical. He suggested we drag and drop the debris on the bottom dish through the new 24″ nozzle to a temporary chute placed below this opening. The chute would slope down towards a loading point, where the debris could be dropped into the bed of a dump truck. We checked it for feasibility and engineering design code requirements.

Figure 38.3 Proposed Arrangement for Removal of Debris

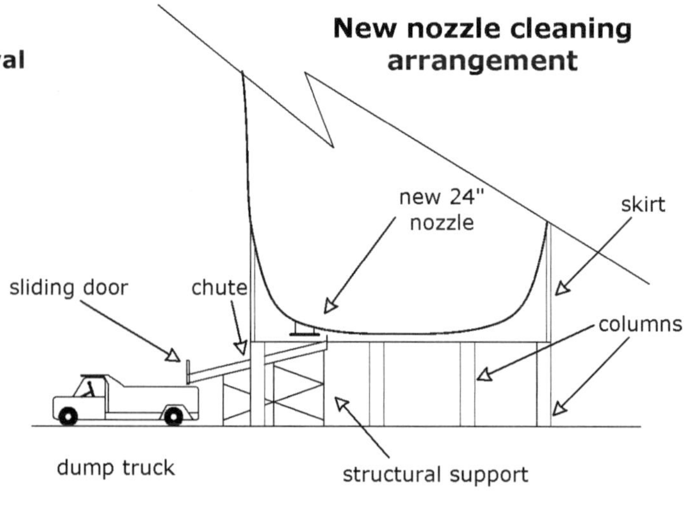

We had to design the nozzle, chute, and its structure, but these were relatively simple tasks. The chute had a sliding door at the truck end to allow control of the debris flow into the truck. The arrangement is illustrated in Figure 38.3.

38.7 Results

The vessel could be cleaned continuously during the shutdown. Instead of an extra 2 days for cleaning, we could shorten it by 2–3 days. Since the activity was on the critical path, the whole shutdown duration could be reduced by 2–3 days. The new cleaning procedure was worth 5–6 days of shutdown duration, on this occasion and 2–3 days for all future shutdowns as well. The effort required for the work would be reduced by about 80–100 man-days.

38.8 Lessons

1. Innovative solutions can be found when the objectives are clearly under stood by all the members of the team. Such innovations are not the exclusive preserve of design engineers or managers.

2. When we are receptive to ideas, especially those from the field supervisors or technicians, we enlarge our pool of brainpower. Some of these ideas may be quite brilliant, and add good value.

We must exercise change control procedures rigorously when modifying any asset or when changing an established procedure.

38.9 Principles

Innovation is like a tender plant; it needs nurturing and support. Novel ideas may sometimes be half-baked and need further development. The person suggesting the idea may not always be capable of developing it, and other resources may be needed. Innovation thrives in an environment where people are receptive and support one another. The leader's job is to point everyone in the right direction, so they not only understand the goals, but also recognize and support the contributions of team members when they come up with suggestions.

Chapter 39

Motor Maintenance Regimes

Listen, look, but keep them running?

The chain of habit coils itself around the heart like a serpent, to gnaw and stifle it.

William Hazlitt, Writer and Critic.

Author: Jim Wardhaugh

Location: 2.3.3 Corporate Technical Headquarters

39.1 Background

I left my Far Eastern idyll for cold dark Europe. Those who move the pawns in the corporate headquarters noticed my success in my assignment in the Far East. So they head-hunted me and I joined this elite group. The job was to help locations that were less well performing to make significant improvements. I met Vee Narayan, who was already in this group; later, Mahen Das also joined us.

As an electrical engineer, I had been extensively involved in previous lives for the implementation of computerized maintenance management systems (CMMS). So I became responsible for electrical maintenance and CMMS implementation.

An Electrical Engineers' Conference gave me the opportunity to prompt the collection of some data. In the Far East, I found that motor repairs were a significant ongoing manpower consumer and, hence, cost. I decided to put some effort into collecting data on this subject so that we could make informed decisions.

39.2 Motor Failure Data

A few of our bigger locations and the Institute of Electrical and Electronic Engineers (IEEE) had been collecting data for many years. We captured this data and collated them (see Tables 39.1 and 39.2). We found large differences between our numbers and those from IEEE. We can account for these in part by the difference in the ways of collecting data and in definitions. We believe that IEEE figures give a pessimistic view of overall failure rates that one might expect from a professionally-managed site.

Location	Offshore			Onshore			
Data source	IEEE	Author		IEEE		Author	
Voltage	HV	HV	LV	HV	LV	HV	LV
Failures per motor-year	0.1	0.1	0.07	0.07	0.08	0.07	0.05

Table 39.1 Motor Failure Rate in Failures per Motor per Year

	Location	Offshore		Onshore	
	Data Source	IEEE	Author	IEEE	Author
Component	Bearings	40	30	50	50
	Stator	12	23	23	23
	Rotor	12	7	8	4
	Others	36	40	19	23

Table 39.2 Percentage of Failures per Failed Component

The information shows that motors are generally very reliable. There are few electrical failures and most failures are bearing related. The most significant underlying causes of failure are:

- Defective components
- Poor installation/maintenance
- Poor lubrication
- Water ingress

39.3 Electrical Failures

Scrutiny of the data showed three prime causes of electrical failures:

- Catastrophic failure due to bearing collapse and rotor rubbing on the stator
- Water ingress due to cooler leaks or cleaning with high-pressure water hoses

- Breakage of connections (due often to inadequate bracing)

There did seem to be some deterioration of motors with age. This was most apparent in large motors with a high starting frequency. We found that the larger the machine, the higher it was stressed. Manufacturers had algorithms which could predict the end of useful life with reasonable accuracy. For this they needed service and operational data, such as frequency of starts. This might be worth doing for older machines in critical services.

39.4 Bearing Failures

Scrutiny of the data showed four main causes of (premature) bearing failure:

- Wrong (or inadequate) bearing installed
- Poor installation practice causing initial damage to the bearing
- Poor lubrication regime
- Poor alignment of driver and driven

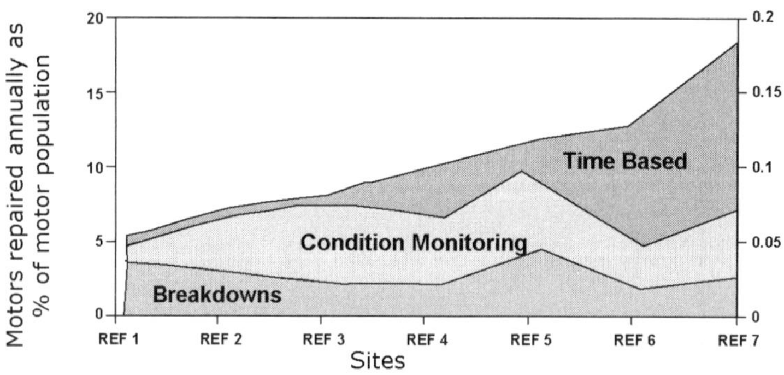

Figure 39.1 Seven Locations Compared

39.5 Performance of Seven Motors at Locations

We looked at the motor maintenance activities in seven of our companies in six different countries. Each was a company plant built to corporate standards, with most rotating equipment having an installed spare. However, there were a variety of maintenance strategies in place.

Figure 39.1 summarizes our findings. It gives the percentage of each site's inventory of motors removed to the workshop for significant repair each year. These percentages have been broken down by reason for removal:

- Breakdown (i.e., the motor had been run to failure)
- Condition monitoring had indicated imminent failure
- Time-based overhaul regime in place for some or all of the motors

Considering this, we found that:

- The large proportion of time-based overhaul activities of Location 7 did not seem to reduce breakdowns significantly.
- Location 6 did seem to be somewhat more effective, but arguably was still not cost effective.
- Location 5 had many breakdowns, even though a significant percentage of motors were repaired because condition monitoring was predicting imminent failure.
  - Their condition monitoring did not seem very effective in predicting and/or pre-empting failures.
  - The site had an extreme blame culture.
- Location 1 had minimized repair efforts by using run-to-failure as a default strategy.
  - The small amount of time-based maintenance was for a very few unspared furnace fans which were overhauled when the plants were shut down every four or five years.
  - This location had a fairly skeptical view of the merits of condition monitoring. They would keep motors running until imminent failure was very apparent.
  - What they also had found was that running motors less than 30 hp to failure did not result in significant additional consequential damage and cost compared to pre-emptive action.
- The proportion of breakdowns is fairly constant whether you do condition monitoring and/or overhauls or just let things run to failure.

39.6 Summarized Findings

From our review, we learned that electrical motors are very reliable. In more detail, we found that:

- Windings do not exhibit significant wear-out unless they are:
  - Too frequently started or
  - Large and highly stressed
  - Winding connections are prone to breaking if not well braced against movement.
- Bearings do wear out, but long life is a function of a few simple things:
  - Correct bearing selection (not necessarily the same as found in the machine)
  - Correct installation, using bearing heaters etc. to minimize damage
  - A good lubrication regime (correct type and quantity of lubricant)
  - Correct alignment
- Smaller machines (less than 30 hp) make up the bulk of the population and can be run to catastrophic failure without significantly increasing consequential damage to shaft or windings.
- In many locations, there is a high level of installed sparing so the conse-

quences of failure are low.

- Motor condition monitoring has to be quite cheap; else it is not an economical strategy. We concluded that:
  - Vibration monitoring could not be justified for most motors; it became viable (just) if you were already going to check the driven equipment.
  - Ultrasound was potentially useful if you had many close packed fractional horsepower motors.
  - Winding monitoring could not be justified for most motors; a special justification was needed for critical applications.
  - Winding monitoring is improving, so this would be kept under review.

39.7 A Maintenance Strategy for Refinery Motors

We carried out an evaluation of possible strategies and monitoring techniques. There were many magic bullets being advocated by credible universities, consultants, and large companies.

We concluded that for very large critical machines:

- A proprietary monitoring installation that continually monitors vibration, axial displacement, etc., should be the norm.
- Information should be centrally monitored.
- Alarm and trip parameters should be set after agreement with the manufacturers.
- We should do regular (annual) internal inspections, using a boroscope.

For the bulk of industrial motors, we concluded that regimes described briefly in Tables 39.3 and 39.4 were justifiable and should be the norm in most environments. Obviously in environments which are extremely arduous, ad-

Type of Inspection	Description	Frequency
External visual (do not shutdown)	▪ No visible unauthorized modifications ▪ Bolts, cable entries, gaskets, earths, cables, etc., OK. ▪ Not too dirty ▪ No unusual vibration, sound or heat ▪ Circuit identification OK	4 years
Internal (need to shutdown)	▪ Tightness of electrical connections ▪ Audit operation of standard industrial protection devices ▪ Test operation of all protection if special requirements	6 or 8 years or at planned plant shutdown or turnaround. Do vibration, etc., checks before plant shutdown.

Table 39.3 Periodic Inspection and Testing of Motors

Category of motor	Techniques suggested
LV motor - non critical	No routine monitoring unless in association with driver - monitor vibration before plant shutdowns.
LV motor - critical	Vibration monitoring
HV motor < 1000kW - non critical	Vibration/bearing temp. monitoring
HV motor < 1000kW - critical	Vibration/bearing temp. monitoring, Current spectrum
HV motor > 1000kW - non critical	Vibration/bearing temp. monitoring, Current spectrum
HV motor > 1000kW - critical	Vibration/bearing temp. monitoring, Current spectrum

Table 39.4 Monitoring of Motors While in Operation

ditional steps may be needed. See also Table 20.1 which gives a more general view.

If ultrasound and thermographic tools are available, a minimum effort periodic inspection could be beneficial. They are particularly useful if you have a large number of motors very small, and closely packed.

39.8 Lessons

- Buy reasonable quality motors which are inherently reliable enough for your application.
- Ensure good lubrication.
- Select correct bearings and install correctly.
- Run small spared motors to failure. Run to failure or at least imminent failure is a very respectable strategy for equipment where the consequential loss is low.
- Predicting and pre-empting failure, however cheap, is only cost effective if you can pre-empt catastrophic failure or major production loss. In general, pre-emptive actions must cost less (and probably significantly less) than the consequential loss due to failure.
- Condition monitoring can be costly and ineffective so you need to audit the effectiveness and cost effectiveness of the system.
- When motors fail, investigate to find root cause and try to eradicate repeat failure causes. (Think correct lubricant, correct lubrication regime, correct bearing, starting frequency.)

39.9 Principles

Preventive and predictive maintenance are generally sound strategies, but not universally applicable. Run-to-failure strategies are perfectly acceptable in many common situations. Maintenance strategies must be based on a rigorous understanding of the risks of failures, not on current fashions.

Chapter **40**

Boiler Feed-Water Pump Seals

I learned very early the difference between knowing the name of something and knowing something.
Richard Feynman, Physicist, Nobel laureate.

Author: V. Narayan

Location: 2.1.4 Large Petroleum Refinery

40.1 Background

The main construction contractor had not resolved and closed out some defects, as they needed research as well as coordination with vendors. One of the pending items (in the punch list of pending items) related to frequent failures of the mechanical seals of ten boiler feed-water (BFW) pumps used in the waste heat recovery systems in different units. We had two of these pumps P1409A and P1409B in our area.

Maintainers who have worked with boiler feed pumps will be aware that seals in this service worked in a harsh environment. Hence, they tended to fail more frequently than seals in other services. When the pump was running, there was a tiny flow of feed water through the neck bush to the stuffing box of the pump. This flow made up for the controlled leakage through the seal. Normally, the stuffing box contents will cool down due to heat loss by radiation. However, the flow through the neck bush provided heat continuously, so the stuffing box fluid remained hot.

40.2 Issues Relating to BFW Seals

In the case of the P1409A/B seals, some of the heat was dissipated by pumping the hot water in the stuffing box through external coolers. For this purpose, each seal had a built-in pumping ring. These rings act as mini-pumps themselves, and served to circulate a small quantity of water through an external cooler. The cooled water was re-injected near the seal faces, presenting the seals with a better operating environment (see seal cooler arrangements in Figure 40.1)

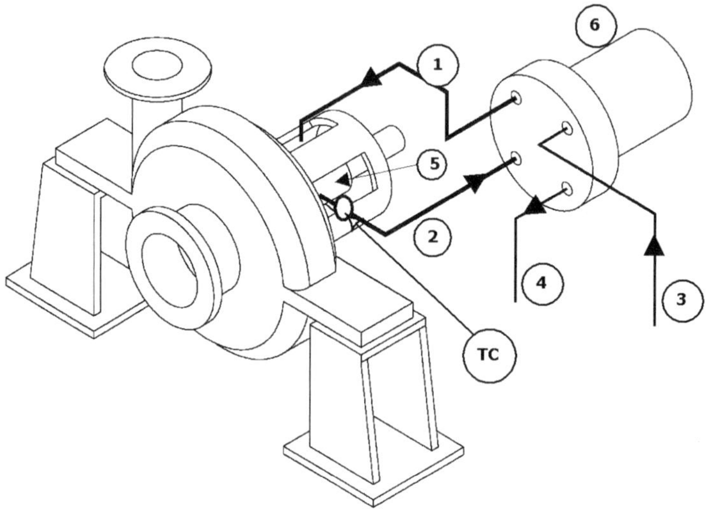

1 - Recirculation water - injection to seal
2 - Recirculation water - OUT
3 - Cooling water - IN
4 - Cooling water - OUT
5 - Mechanical seal
6 - Seal cooler
TC - Thermocouple

Figure 40.1 Seal Cooler Arrangement

The stuffing box pressure was the same as at the pump inlet, namely 5 bar gauge. At this pressure, the boiling point of water is about 150°C. When the pump was stopped, the hydrodynamic opening force was no longer available. So the fluid film between the seal faces was squeezed out. On starting the pump, it takes a short time for the hydrodynamic fluid film to be established. During this period, the seal faces are in effect running dry. This raises the temperature of the seal faces quite rapidly. As soon as the hydro-dynamic film starts forming, the hot faces convert the water film into steam. There was then a rapid rise in the pressure between the seal faces. As a result, the seal faces separate, allowing a large water flow that cools them rapidly. This quenching action can initiate seal face damage. By this time, a hydrodynamic film was established and this kept the seal faces relatively cool. There was no further heat to generate more steam, so the faces do not separate. At this stage, the hydrodynamic opening force was matched by the static seal closing forces, restoring the seal to steady state equilibrium conditions. Thereafter, the seal ran normally.

When starting the pump, if the stuffing box water temperatures were high, more steam would form, resulting in the faces springing further apart. The seal faces would tend to snap open and shut alternately, producing a chattering effect. This can result in impact damage on the seal faces, accelerating their degradation. It was therefore advantageous to keep the stuffing box fluid temperature as low as possible during pump starts. If this theory was correct, two actions were required to minimize seal degradation.

1. Minimize the number of seal starts, so that there are fewer dry run periods.
2. Keep the stuffing box fluid temperature as low as possible at start.

40.3 Measuring the Seal Face Temperatures

In order to establish our hypothesis, we wanted to measure the seal face temperatures during the starting and running conditions. Our instrument engineer suggested that the existing Distributed Control System (DCS) could be used to good effect. In the original project design, spare cable pairs had been laid, to allow for possible cable failures. He offered to connect our measuring devices to the DCS for the duration of the trials, using these spare pairs of cables. We arranged to measure the temperature of the hot water being pumped from the stuffing box to the cooler. For this purpose, we spot welded suitable thermocouple junctions to the hot water outlet pipes from the stuffing boxes to the coolers.

Using the DCS, we could get eight-hour, eight-minute, and two-minute printouts of the temperature of the water leaving the stuffing box. We were aware that there would be some errors in these measurements. Ideally, the water temperature at the seal faces was what mattered. However, we could only get temperature readings of a diluted mixture leaving the stuffing box. The measurement points were 2–3″ away from the seal faces. The water temperature was inferred from that of the pipe wall temperature, so there was potential for further error. The pipe temperature would be influenced by prevailing winds, rain, and the thermal inertia of the metal. Rain was generally not common in the Middle East, but prevailing wind speeds were variable and often quite high, so we had another source of error. We realized that these errors could be significant, but decided to carry on with the trials.

40.4 Testing the Hypothesis

According to our hypothesis, seal damage occurred principally at the time of starting the pumps. If true, we could accelerate seal failures by frequent changeovers. We discussed this procedure with Operations and they agreed to a daily changeover schedule for the trials.

Just prior to starting the standby pump, we planned to cool its stuffing box contents by using the cooled (seal) water from the running pump. For this purpose, we would connect the line from the cooler outlet of the running pump to its matching point on the standby pump. The flow through this line would be controlled carefully, as excessive flow would mean that the running pump would be starved and see an excessive rise in temperature. We designed and fabricated a new 1/2″ pipeline to connect the outlets of the two coolers. It had isolation valves at each end, and a 1-mm orifice plate to restrict the flow (see Figure 40.2)

40.5 Results

So far, we only had a theory. We would know if any of it was true when we started the trials. In the chart (Figure 40.3), we can see the result of the first trial carried out on March 8. At this time, the P1409A seal was brand-new,

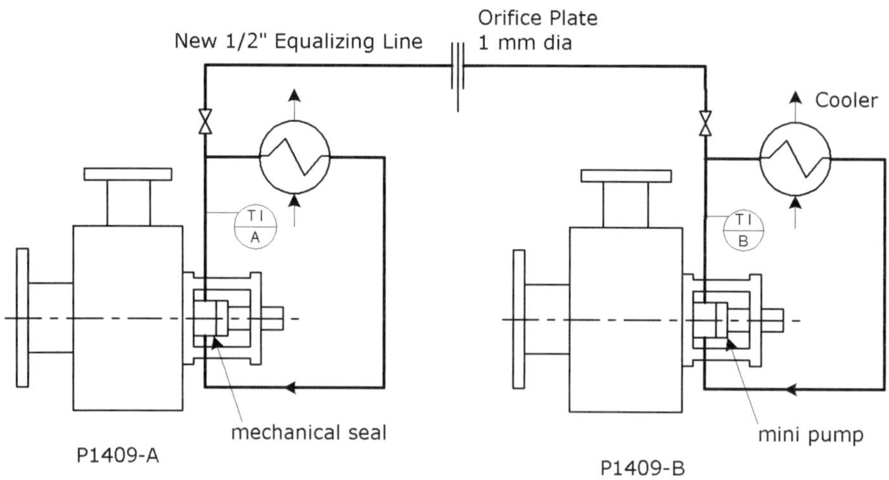

Figure 40.2 Temporary Seal Piping Arrangement

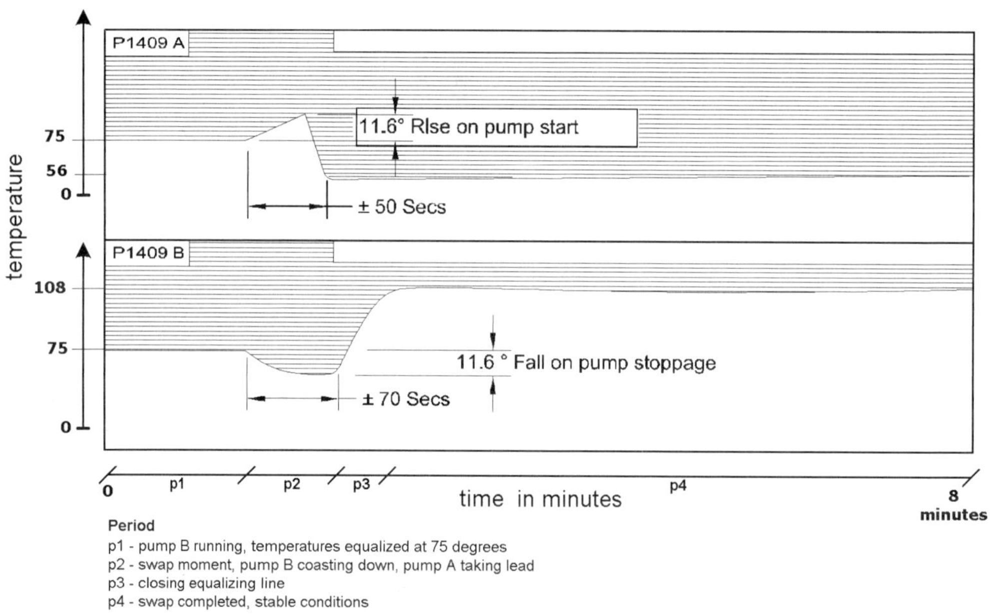

Figure 40.3 Trial 1 Results

while that in P1409B had been in-service for some time. Reading from left to right, the chart shows that P1409B was running and P1409A was about to start. The stuffing box water temperature of P1409B, which was normally at about 56°C, rose to 74°C. This was because the interconnecting 1/2" pipeline valves had been opened earlier, and the B pump seal's pumping ring was now supplying the A pump as well. In the stuffing box of the A pump, the water temperature, dropped from 108°C to 75°C.

This eight-minute printout shows the following:

1. The A pump was started about 100 seconds from the start of the print out. At this stage, the temperature of the water leaving the B pump's cooler was 75°C. Hence, the seals of both pumps were at 75°C.
2. During the next 45–50 seconds, the A pump water temperature rose to 86.6°C. This 11.6°C rise was due to the heat generated at the seal faces.
3. In the next three or four seconds, the A pump's seal pumping ring established a flow through its own cooler. The temperature then dropped to about 56°C. Thereafter, the temperature remained steady at this level.
4. Once the A pump started and the B pump stopped, the interconnecting pipeline valves were closed.
5. The B pump fluid temperature was about 74°C at the beginning of the chart. During the coast-down from full speed to zero, the heat generated by the seal faces due to viscous friction also fell.
6. The pumping ring discharge temperature dropped by 11.6°C, but this took about 75 seconds, as the pump took longer to come to a stop.
7. After the pump stopped completely, the temperature rose to reach the pump stuffing box temperature of 108°C, because at this time the interconnecting pipeline valves were also closed and there was no supply of cool water.

This chart shows that our hypothesis was not far off the mark. Note that the viscous seal friction produces a temperature rise of 11.6°C. The chart shows that the pump seal ran for about 45 seconds before its seal cooler became effective. Once this happened, the temperature dropped to 56°C, and remained there as long as the pump was running. Similarly when the pump was stopped, there was an initial fall in temperature by 11.6°C, confirming our evaluation of seal friction heat generation. Once the pump came to a halt,

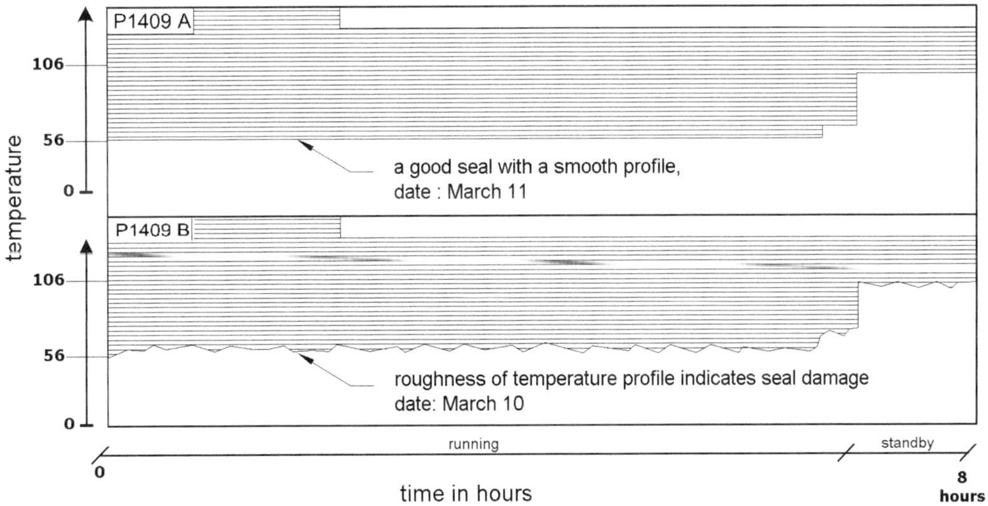

Figure 40.4 Seal Performance on March 10–11

the temperature rose gradually till it reached the pump suction conditions of 108°C.

The next chart shows the performance of the P1409B on March 10 and that of P1409A on March 11 after the changeover. As you can see, the B pump temperature profile was quite rough, while that of the A pump was smooth—note that these are 8-hour charts. The roughness in the B chart was because the seal faces were already damaged, causing erratic hot spots. As a result, steam is formed, leading to occasional seal separation. There was a clear rising trend in the temperature and the seal was close to failure. In any event, the B pump seal failed on March 12. The trial shows that the temperature profile was a good indicator of seal condition.

The daily change-over definitely accelerated the seal wear quite dramatically. This supports the hypothesis of seal damage due to dry running during pump starts.

40.6 Lessons

1. Seal damage occurs mainly during pump starts; if you want seals to last long, minimize pump starts.
2. Keeping the stuffing boxes and hence the seal faces cool during starts helps prolong seal life, especially when the liquid is near its vapour point.
3. With modern Process Control Systems, it is technically feasible to monitor seal face condition. However, this may only be economically feasible with the pumping-ring design.
4. When the seal is in good condition, the fluid temperature and, by inference, the seal face temperature is constant, with a smooth profile.
5. Once seal faces start getting damaged, this profile became jagged and rough and the temperature starts rising.
6. Using this type of data, it is possible to predict the time of failure of such seals. This knowledge can be used to plan the timing of seal re placement accurately, and maximize their service life, reducing down time and costs.

With improving technology, it is possible for temperature sensors to be embedded in stationary seal faces, which can then be monitored. This will add to the growing list of condition monitoring techniques that will become available. Seal failures account for over 20% of pump failures. Predicting their time of failure can lead to significant reductions in downtime and costs. The method can be universally applied and is not restricted to this particular seal design.

40.7 Principles

Having a theory or hypothesis is not much use unless we can verify its validity by field trials. When we conduct such trials, potential errors have to be identified, and results corrected if they are significant or introduce bias.

Understanding exactly what happens during a failure can guide us in designing better components and equipment.

Chapter 41

Cooling Water Pump Failures

Kites rise highest against the wind, not with it.
Winston Churchill, Politician, Author, Statesman.

Author: V. Narayan

Location: 2.1.4 A Large Petroleum Refinery

41.1 Background

When carrying out refining processes, large quantities of heat have to be removed. High-grade heat can be reused for product-to-product heat transfers, but low grade heat is invariably discarded. In this refinery, we used air cooled heat exchangers (fin-fan coolers) to discard low grade heat. Final cooling to ambient temperature was sometimes required after the air cooling. We used sea-water coolers for this purpose. A Public Utility Company supplied the cold sea water. Once used in the coolers, the warm cooling water was returned to the sea. The Utility Company provided elevated concrete supply and return channels, and we had to pump the warm water up about 30 feet. We had three large vertical cooling water pumps to pump out the warm water. Two of these were needed to cope with the volume involved, while the third served as a spare unit.

A concrete channel was used to transport the water from the process plants to the pump-house. This was about 12 feet wide and 10 feet deep, with the top (covered with removable slabs), about 6″ above ground level. From the channel, water flowed into an open concrete reservoir through an inverted weir. There was a 'waterfall' effect, as the drop was about 12 feet. This caused a lot of turbulence, and there was plenty of 'white water' in the reservoir.

The pumps' suction pit was oriented perpendicular to the inlet channel, so the water changed direction by 90° in the reservoir. At the entrance to this pit, there were trash racks to catch large debris such as wood pieces. Figure 41.1 shows the layout of the cooling water pump house.

These pumps suffered from a series of failures, resulting in their inability to meet their specified flow requirements. Eventually all three pumps had to operate to meet the capacity requirements. If a pump came out for repairs during this period, we had to reduce the refinery throughput. This situation was clearly unacceptable.

41.2 Internal Inspection Results

A general arrangement of the pump can be seen in Figure 41.2. Internal inspection showed that the pumps suffered severe cavitation damage. There was extensive damage to the aluminum-bronze suction bells (inlet cones).

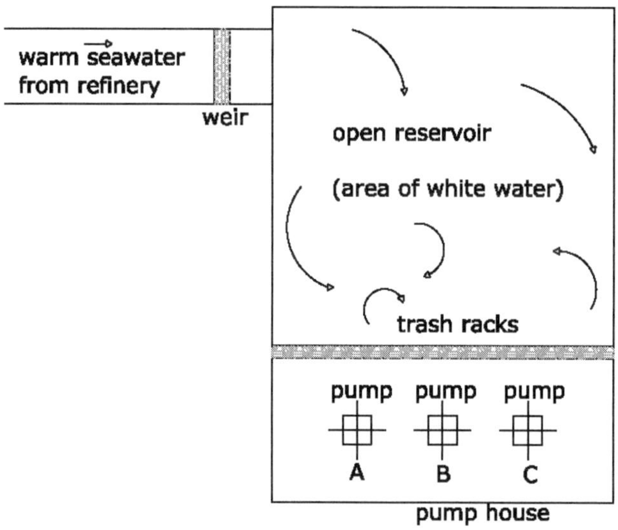

Figure 41.1 Cooling Water Pump House Layout

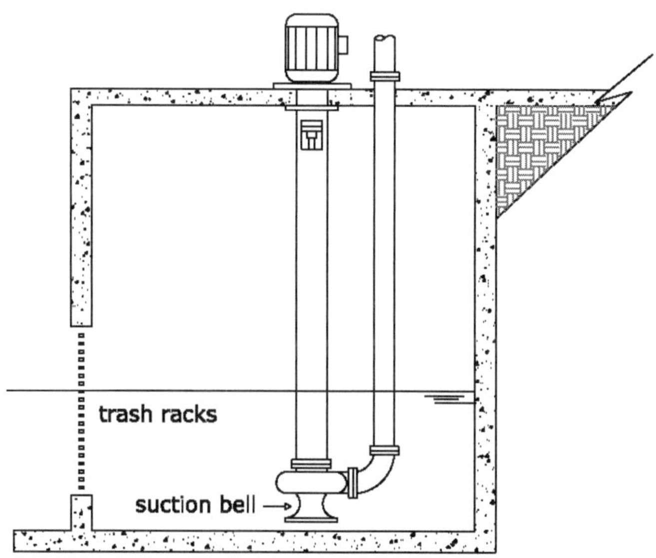

Figure 41.2 Pump House Cross Section

The impellors and wear rings, made of monel metal, were, however, largely unaffected. Damage to the suction bells meant that we had to purchase new bells for all three pumps. The vendor needed 6 months to supply these, while we estimated that the bells would last only 3–4 months. Cavitation by itself had a process impact, resulting in loss of capacity.

41.3 Analysis

We sought assistance from the pump vendor and design contractor to analyze the problem, so that we could identify the physical causes of damage. There was no doubt that cavitation was the cause of damage to the suction bells, as all three pumps produced the characteristic 'pinging' noises associated with this process. The suction bells had many hemispherical cavities with metal loss up to 1/2″ in diameter. This confirmed the cavitation theory.

The white water indicated significant aeration, which could account for loss of capacity. The vendor had already researched this subject. Their test data showed that 1% aeration could reduce pump flow by about 8%. We wanted to establish the level of aeration to see if we could explain the measured loss of flow. Our instrument engineer came up with an innovative way to measure the degree of aeration (see Appendix 41-A). Using his technique, we measured the level of aeration in the reservoir at 1–2% over the depth of the water. With this information, we could attribute loss of capacity as being entirely due to the white water. Bursting of the air bubbles in the suction bell was the most likely cause of the pinging noises and metal loss in the bells. Loss of capacity and suction bell damage could both be the result of the high level of aeration, seen as white water.

The white water itself originated in the waterfall. The water was clear till it reached the weir, and then became frothy (see Figure 41.3). The next ques-

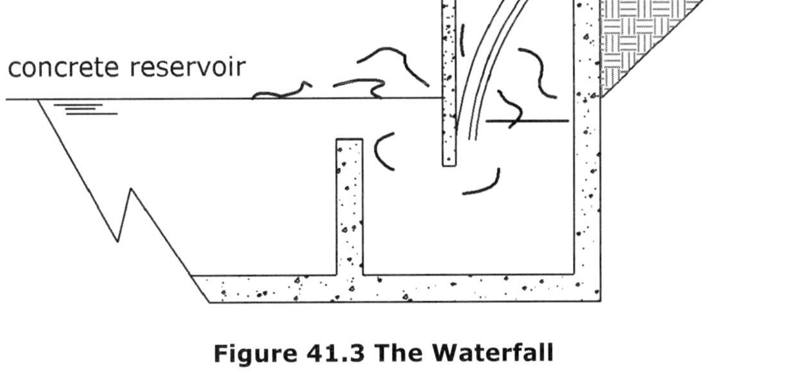

Figure 41.3 The Waterfall

tion was how to minimize or eliminate this aeration without disturbing the weir.

41.4 Temporary Solution

We wanted a temporary solution within two months, which would allow us some breathing space to find a more lasting solution. A refinery shutdown was scheduled in 18 months when a permanent solution could be implemented.

The most urgent task was to repair the suction bells, as there was no possibility of getting spare parts in time. Many years earlier, in a different refinery, I had experimented with the use of 'plastic bronze' for doing 'cold' repairs to sea water pumps. Plastic bronze is a two-part epoxy resin, with metallic filler. As in this case, the pump casings had suffered severe cavitation damage in that refinery as well.

In the previous location, prior to the use of plastic bronze, we used to fill up the large (ca.3/4" dia.) hemispherical cavities in the casing by welding, using bronze electrodes. We had to take precautions to prevent casing distortions that could result from large localized heat inputs. Our attempts to repair the damage with plastic bronze were very successful there, so we decided to try that option in this case as well.

41.5 Solution: Trial Repair of Pump Bell

One pump was in the workshop with a severely damaged suction bell, awaiting repairs. We grit-blasted the bell, so that the surface was clean and had a rough texture to ensure the epoxy bonded with the metal. We added chopped fiber-glass to the epoxy mixture and applied it in layers of about 1/8", allowing plenty of curing time between layers. We laid a clear plastic sheet over the final coat and smoothed the surface. Once the epoxy dried, we could peel off the plastic sheet, leaving a smooth, glassy surface. Some minor machining was required to clean up the flange face and bolt holes. The whole job took four days to complete. Meanwhile we repaired the rest of the mechanical parts of the pump. We reinstalled the pump and found it mechanically satisfactory, though the capacity loss problem remained. Over the next three weeks, we did the same type of repair to the remaining two pumps.

41.6 Solution: Aeration of Water

The white water posed a much more difficult challenge. Our hypothesis was that if we restricted the flow downstream of the waterfall, the level would build up in the box section, reducing air entrainment during the free fall and, hence, the frothing.

The first idea was to hang a sheet of 1/2" thick rubber at the box outlet. It would have a steel flat along the horizontal edges to give it some stiffness, but would allow it to flex. A simple Strength-Weakness-Opportunities-Threat or SWOT analysis showed there were some threats to be managed. First, the sheet could break into pieces and enter through the trash rack into the pump suctions. Second, the horizontal force of the water might be excessive. This might destabilize the crane that would be used for hanging the sheet. The work area also posed some hazards to the people who were to execute it. We presented the proposal along with the SWOT (and our proposed actions to mitigate them) to the refinery management team. After getting their approval we did the first trial.

This was a complete failure. The force of the water was so large that the sheet was lifted till it was at about 45°. The sheet could not restrict the water flow at all, flaying about so wildly that we had to pull it out quickly.

We went back to the drawing board. The new design was a horizontal steel plate barrier, with a 3″ gap along the four edges and a few vent holes in the middle. We strengthened the plate with a 6″ pipe frame and decided to hang it in place using chain blocks (see Figures 41.4 to 41.6 showing the installation sequence). We checked the design calculations to ensure that the plate would withstand the expected loads and bending moments with a good safety margin. We checked the concrete channel walls to see if they would take the additional loads from the chain blocks holding the plate. We had to cut four holes 6″ diameter, through the top slab, for the wire slings. An external consultant verified the wall and slab designs, to confirm that these changes were acceptable.

The SWOT on this design also exposed several threats, including the possibility of one sling alone breaking due to uneven impact loading. If this happened, the plate could get stuck at an angle in the box, perhaps blocking flow excessively. To resolve this potential problem, we decided to attach a second set of longer slings (with extra slack), to the steel frame.

The second trial was done in the presence of some members of the management team. Getting the plate in place was quite tricky, but we had a very competent and imaginative mechanical supervisor in charge. When the plate was in place, it did restrict the flow, but not adequately. The white water be-

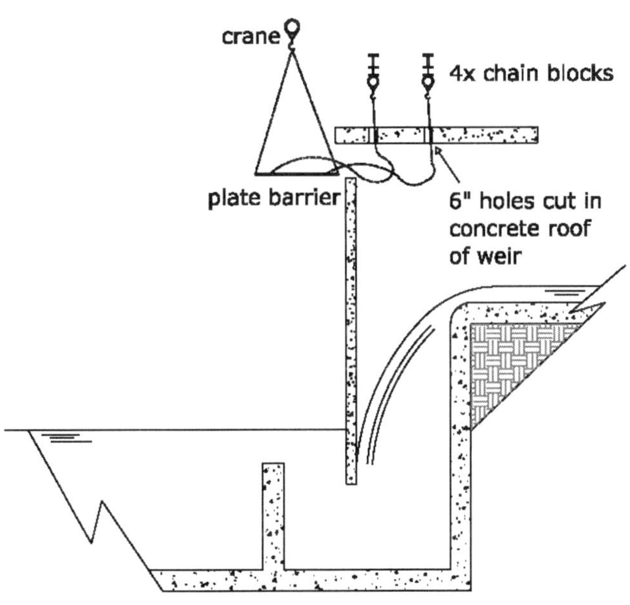

Figure 41.4 Preparing the Plate Barrier for Installation

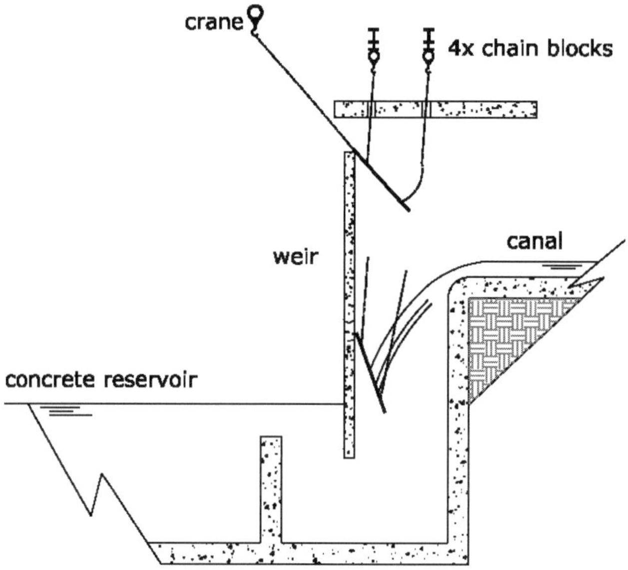

Figure 41.5 Progress of Installation of Plate Barrier

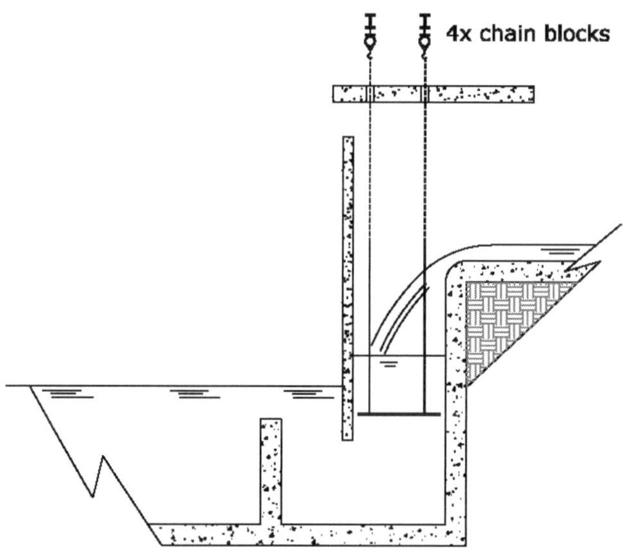

Figure 41.6 Plate Barrier in Place

came less frothy, but it was obvious that the flow passage was still too large. In a sense, this trial was also a failure, but we were making progress. We pulled out the plate and modified the edges, so that the gap from the walls of the box was reduced to about 2″.

41.7 Result of Trials

This time, the design worked perfectly, with the water level building up above the plate and eliminating the waterfall effect. The water in the reservoir became calm and clear. The pumps stopped their pinging noises and their capacity was back to design levels. This meant we could go back to two-pump operations.

41.8 Change Control

Good communication and coordination with relevant operations and process design staff helped fast track the change-control activities for the temporary solution. The process technologist and utilities manager supported the work at all stages, and took an active role in vetting the mechanical and structural design. We kept the management team informed at every stage, as this was a problem that had the potential to shut the whole refinery down for an extended period.

41.9 Long-Term Solution

This required major engineering effort. Significant design changes were envisaged, based on the results of these trials. The plan was to redesign the overflow from the weir so that aeration was eliminated. By the time all this happened, I had already moved to a new location, so I could not keep track of subsequent actions.

41.10 Lessons

1. It is necessary to identify root causes to solve serious reliability problems.
2. The participation of the pump vendor and design contractor enabled quick analyses. Vendors and design contractors are a great source of knowledge, and should be used when available.
3. Temporary solutions must also be properly engineered. We have to apply change control procedures, even if changes appear insignificant. We can fast-track these with good communication and cooperation.
4. Maintenance engineers can make their jobs more interesting by methodically addressing serious reliability problems. Solutions may sometimes be innovative, but analysis must always be thorough and based on facts.
5. Given the freedom to explore, people often come up with innovative solutions. In this instance, we used an unusual way to measure aeration, another to repair the suction bells and yet another to stop the aeration. Even the way in which the plate was inserted into the weir box required creative ideas, in this case from the mechanical supervisor and crane operator.

41.11 Principles

In solving difficult reliability problems, tap knowledge and skills wherever they reside. Innovation and creativity are available at every level in the organization. These talents can flower or be smothered, depending on how we deal with people.

Analyze risks and apply change control procedures before implementing temporary solutions. These need the same quality of design and engineering as permanent solutions.

Difficult challenges make a maintainer's life interesting.

Appendix 41-A

Measuring Aeration in the Cooling Water

Instrument engineers will be familiar with the use of pressure readings to measure the height of liquid columns. One application of this principle is the use of bubbler tubes. Low pressure air was supplied at the top end of a small bore tube with its lower end immersed in the liquid whose density we wished to measure. The depth of the lower end below the liquid surface was known in advance. At the air supply end, there was a flow regulator, flow indicator, and high-resolution pressure measuring device, usually a water manometer or differential-pressure (DP) cell. Such an arrangement can be seen in Figure 41-A.1.

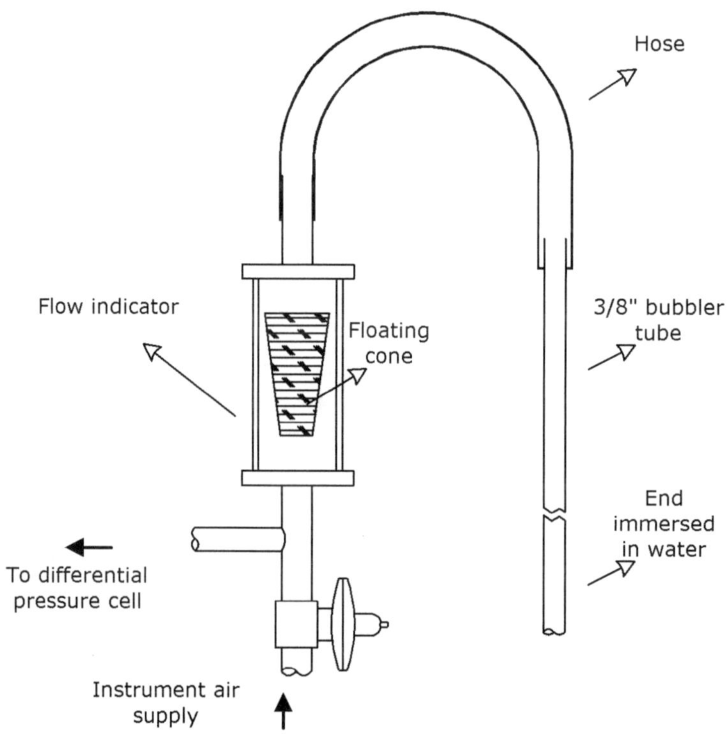

Figure 41-A.1 Bubbler Tube Arrangement

Using the air flow indicator, the air flow was reduced to the point when there was nearly no air flow, with a bubble of air escaping from the bottom of the tube every 10 or 20 seconds. The DP cell reading tells us the static pressure at the bottom of the tube. The depth of the tube below the liquid surface was already known from earlier calibration. From this we could compute the density of the liquid at that point, using the equation

$$P = \sigma \times H, \text{ where}$$

P is the pressure measured by the DP cell,
σ is the liquid density at the bottom of the tube, and
H is the known depth of the immersed part of the tube

When sea water is aerated, it becomes less dense. If we know the depth of the water below the surface, we can use this method to measure the level of aeration by measuring the apparent density.

The measuring device fabricated by the instrument engineer is illustrated in Figure 41-A.2. The external 2″ pipe protects the measuring tubes from the

Figure 41-A.2 Bubbler Tube Measuring Arrangement

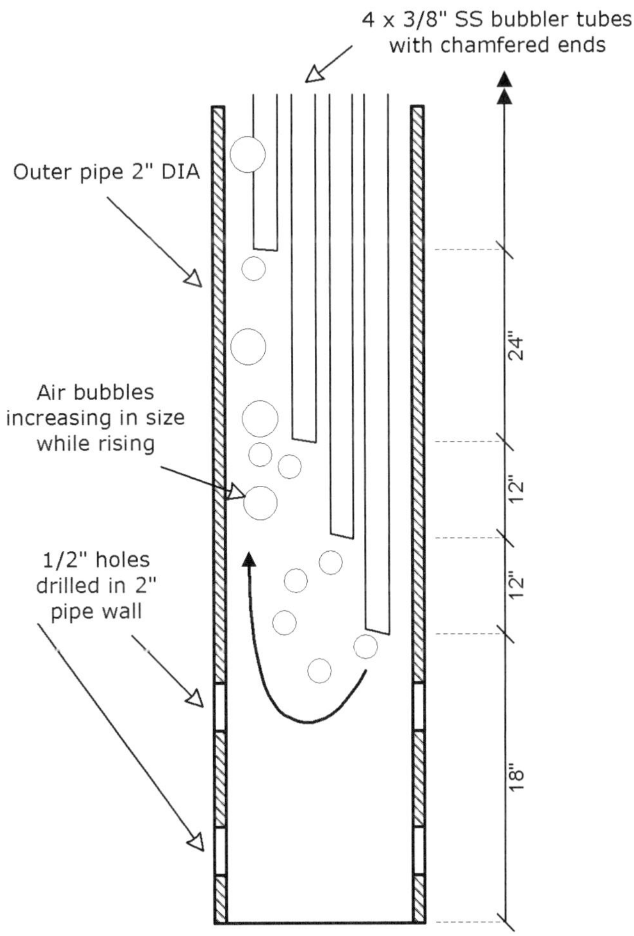

turbulence of the frothy water in the reservoir. This pipe rested on the floor of the reservoir. Several 1/2″ holes were drilled in the bottom 18″ of the pipe, so that in effect it was 18″ clear off the floor. The four 3/8″ bubbler tubes were arranged in different heights, as shown in the figure. Their ends are slightly chamfered so that the release of air bubbles is controlled, and bubbles are released one at a time. This bubbler-tube stack arrangement allowed us to measure the density and estimate the level of aeration at different depths.

Water temperature also affects its density, so the device was first placed in the channel just before the weir, where it was calm and clear. This allowed us to calibrate the device at the actual water temperature. There were some minor errors, caused, e.g., by the cooling effect of the waterfall, but these were not significant.

Chapter **42**

Heater Outlet Flue Gas Dampers

The beginning of knowledge is the discovery of something we do not understand.

Frank Herbert, Science Fiction writer.

Author: V. Narayan

Locatio: 2.1.3 Petroleum Refinery

42.1 Background

The Crude Distillation Unit, capacity 40,000 barrels per day, had two fired heaters with refractory-lined steel chimneys. There were dampers to control the draft at the flue gas outlets from the heaters. These were operated as a set and positioned manually, using chain wheels. The hollowed-out cast-alloy dampers (high chrome nickel steel) were 51/2′ long, 24″ wide, and 5″ thick, weighing about 350 lbs. each. A longitudinal section showing the dampers in an open position, and cross sections in closed and open positions, can be seen in Figure 42.1. The bearing assembly details are in Figure 42.2.

These dampers had a long history of failures. Usually, they would get stuck in one or the other position within a few days after replacement. Once this happened, they could not be rotated, making the heater draft control very difficult. The problem was known for more than ten years and there had been several unsuccessful attempts to resolve it.

42.2 Inspection Results

The ball bearings, on which the damper shafts were mounted, showed clear signs of overheating, some with burn marks. All the grease had melted and burnt away.

Failed dampers sagged 2–3 inches near the center due to their self-weight and as a result of exposure to the flue gases at 650°F. There was a 3″-thick refractory lining on the walls of the flue gas duct. Once the dampers sagged, their ends rubbed against the lining, preventing free movement. Rub marks were visible on the lining, confirming this observation.

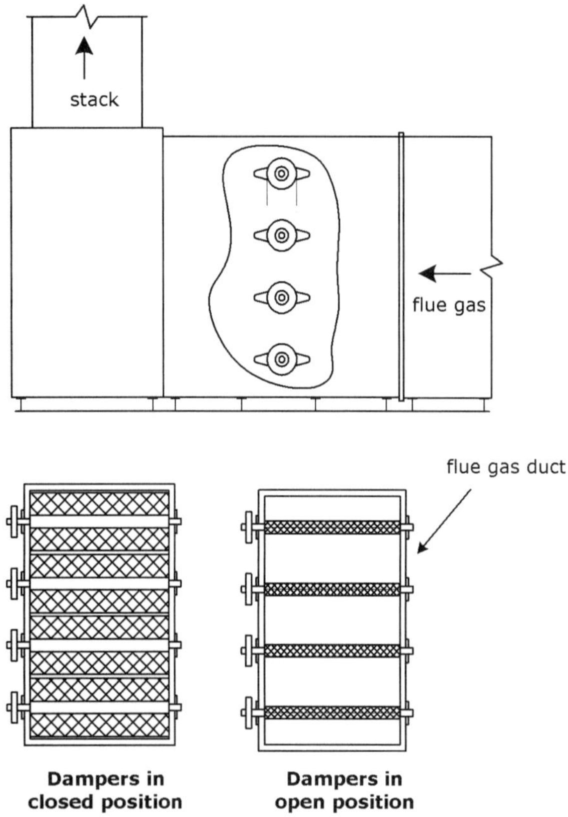

Figure 42.1
Old Design of Dampers

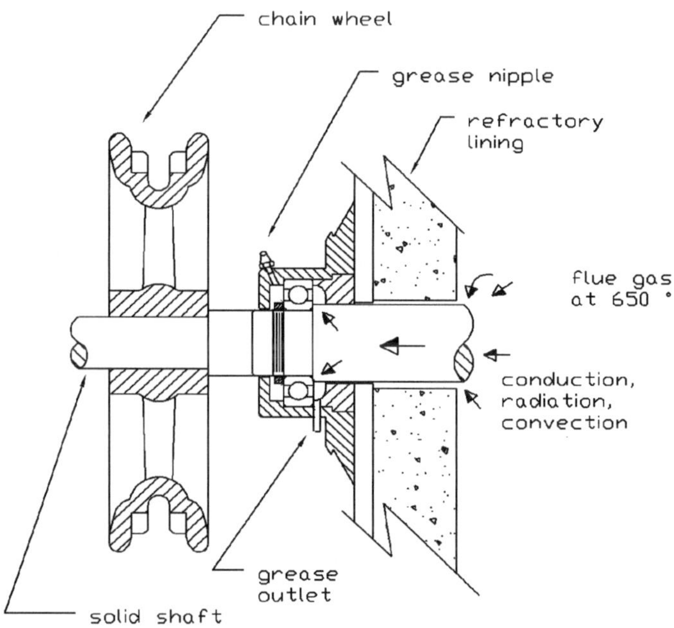

Figure 42.2
Old Design: Bearing and Pulley Arrangement

42.3 Analysis

The bearings, which were mounted outside the duct, became very hot during operation. The damper shafts were at the flue gas temperature. The solid shaft ends conducted heat continuously to the bearings. We reviewed the design and came to the conclusion that the bearings had to be kept cool and that damper sagging should somehow be avoided.

42.4 Solutions

We asked the process technologist and operations staff how leak-tight the dampers should be when closed. They said that a small gap of, say, 3″ along all four edges of each of the dampers would be acceptable.

In order to reduce the weight of the dampers, we selected a box frame section. We planned to use 1/8″ thick 18/8 stainless steel (AISI 304) sheet metal to fabricate an open box design. Two sheets would form the top and bottom of the box. So as to provide some stiffness, we decided to provide a few cor-

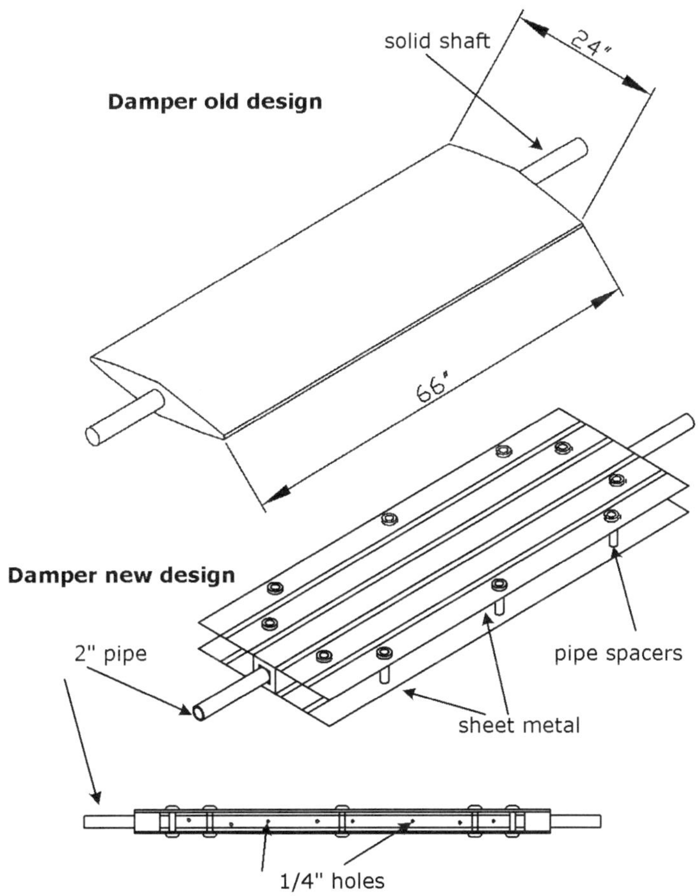

Figure 42.3 Perspective View of Old and New Design of Dampers

Figure 42.4 shows longitudinal and cross sections of the new damper design.

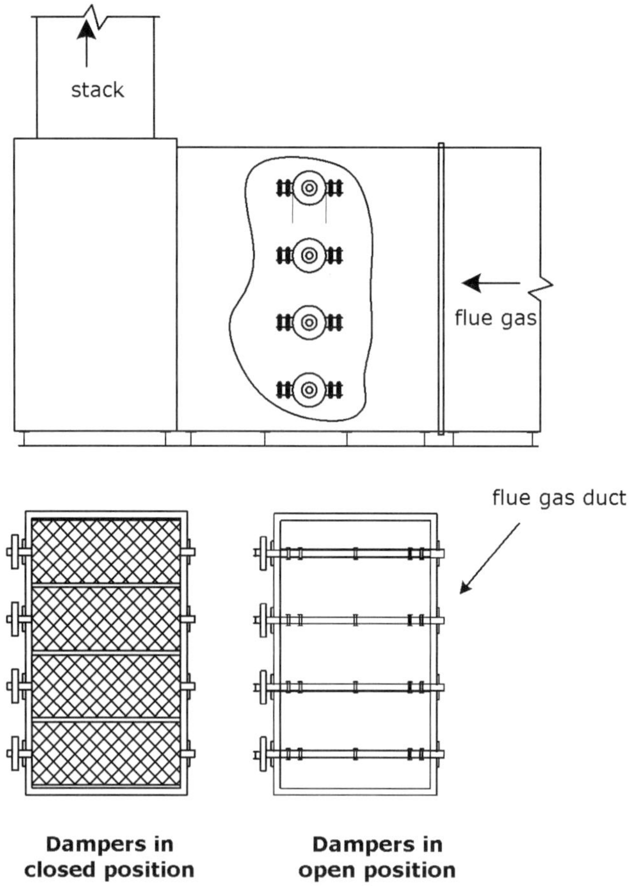

Figure 42.4 New Design of Dampers

rugations on the sheets, along the length. The box section had open sides and ends. The two sheets would be bolted together 4″ apart, using machined pipe spacers. We planned to use a 2″ stainless steel pipe as the axle on which the new damper would be mounted. The pressure inside the duct was considerably lower than outside the duct, due to the powerful suction effect of the chimney. A few 1/4″ holes drilled through the wall of this pipe-axle would allow outside air to leak into the duct from the open pipe ends. The air flow would help cool the bearings. Roller bearings would replace the existing ball bearings, as they were more suited to the small and infrequent movements of the dampers.

The new dampers were expected to weigh less than 120 lbs. As their section was quite light and stiff, it was unlikely that they would sag. Figure 42.3 shows perspective views of the old and new designs.

In this design, the bearing is mounted on a small step machined on the pipe, while the pulley is mounted on a further step on the pipe, as can be seen in Figure 42.5.

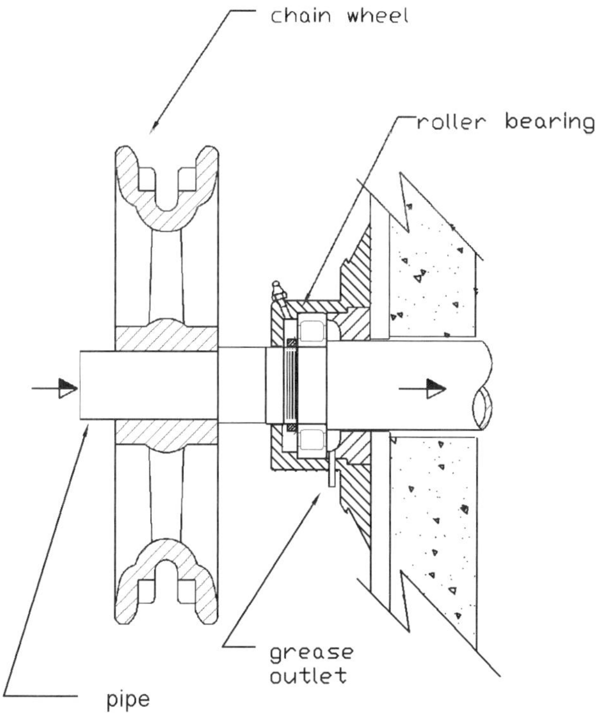

Figure 42.5 New Design: Bearing and Pulley Arrangement

42.5 Change Control

By involving the process technologist and operations in the design of the new dampers, potential problems were identified in time. Rigging activities inside the duct would be quite demanding, so we took advice from the rigging foreman in sequencing the work. We applied the normal change control procedures to obtain approvals.

42.6 Results

The new design of dampers performed satisfactorily, allowing operators to control the draft efficiently. Contrary to the experience with the old design, the dampers worked right through to the next planned shutdown. Repair of dampers was no longer an item on the shutdown work list. The cost of fabricating and installing two sets of new dampers was less than 30% of the cost of one set of original dampers. During the subsequent shutdown, inspection showed that the dampers were in good condition without any sagging. The bearings were also undamaged and did not need replacement.

42.7 Lessons

1. The cause of damper failure had to be established before attempting a solution. Once this was done, innovative solutions were required

2. Problems that have remained unsolved for many years offer interesting challenges.

3. The change control process was speeded up by working closely with operations and the process technologist.

42.8 Principles

No amount of maintenance can cure an inherently poor design. Some reliability problems will require redesign.

Solving problems that affect operability brings credibility to maintainers.

Chapter 43

Laboratory Oven Failures

I decided that it was not wisdom that enabled [poets] to write their poetry, but a kind of instinct or inspiration, such as you find in seers and prophets who deliver all their sublime messages without knowing in the least what they mean.

Socrates (469 BC – 399 BC).

Author: V. Narayan

Location: 2.1.1 Pharmaceutical Plant

43.1 Background

In this location, the Research and Development (R&D) department also managed the Quality Assurance (QA) program. In line with QA requirements, they tested product samples by developing cultures (in Petri dishes). The process of developing cultures requires a constant temperature of about 30°C for several days. There were eight ovens in the laboratory for providing this controlled environment.

The ovens had a large cavity, about 3′ wide, 3′ deep and 30″ high. A set of electrical heating elements along the side walls provided heat. A bimetallic thermostat turned the heating elements off when the temperature exceeded set limits. A small fan 8″ in diameter was mounted at the center of the ceiling of the oven. It rotated very slowly, gently wafting the warm air from the sides down the center of the oven. The intention was to produce a toroidal flow path to help distribute the temperature uniformly in the oven (see Figures 43.1 and 43.2).

Petri dishes, about 4″ in diameter, containing the culture samples were placed on wire trays. The latter allowed free flow of air. These trays could slide out once the oven door was opened. There were seven trays in each oven, spaced about 3″ apart. The top tray was 9″ below the oven's ceiling while the fan blades were 2″ below the ceiling. In placing the dishes, the laboratory technicians took care to ensure that there was a uniform airflow around them.

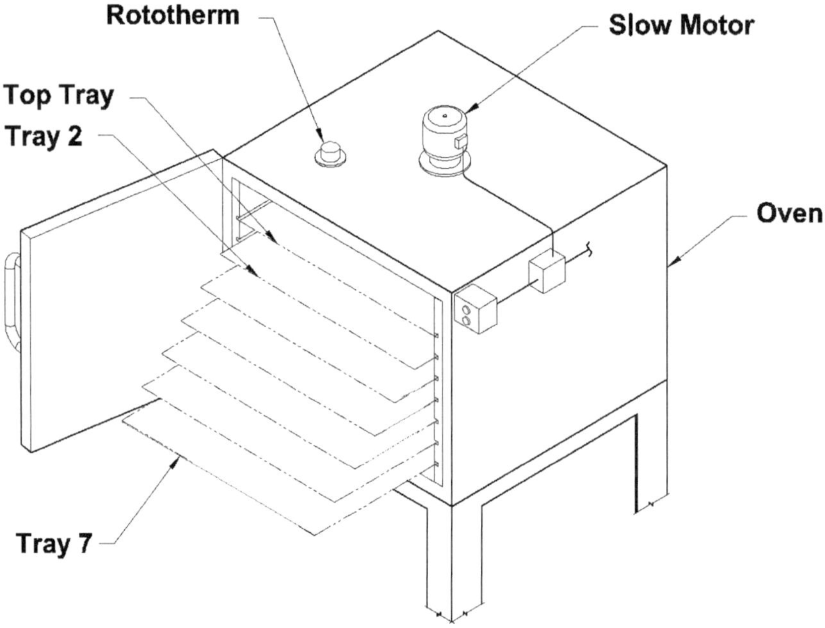

Figure 43.1 Perspective View of Oven

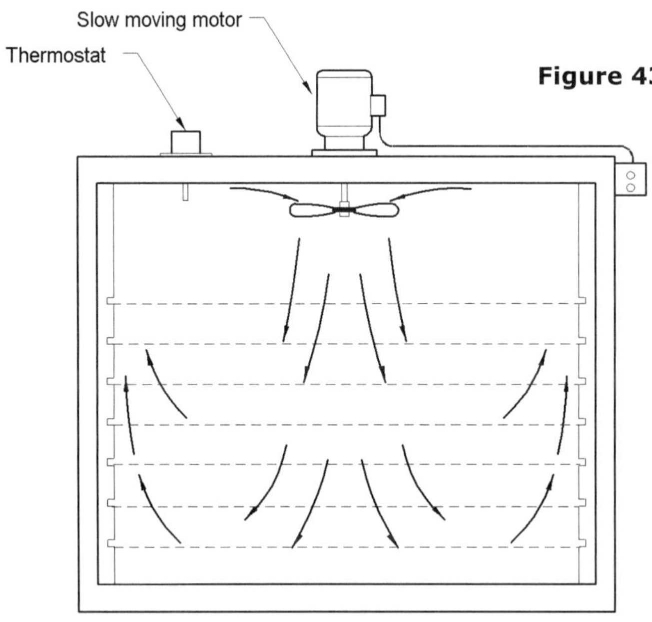

Figure 43.2 Original Air Flow Paths

43.2 Problem Description

Culture samples are very sensitive to temperature variations. Processing requires several days and cultures could be damaged if the temperature is not kept uniform during the whole period. If an oven fails and the samples are overheated, the entire batch has to be abandoned and a new batch started.

Over the past three or four years, the laboratory had lost a number of batches due to such failures. These were mainly due to bimetallic thermostat failures, but also due to uneven heating inside the oven or incorrectly distributed Petri dishes. Various unsuccessful attempts had been made in the past to resolve these problems. Different designs of thermostats were tried, but the new ones were no better. Changing the fan speed or pitch of the blades did not seem to make any difference. Training the laboratory technicians to distribute the Petri dishes did help, but only as long as the ovens themselves performed well. The thermostats continued to fail randomly, and the temperature distributions inside the ovens were still quite variable. At the time of this episode, the laboratory had a large backlog of cultures to process. I knew very little about laboratory ovens, so this was going to be an interesting challenge.

43.3 Analysis

Lack of knowledge can sometimes be an advantage. I spent some time reading the manual, talking to the lab technicians and with my staff. The records showed that there had been 21 oven failures in the past year. Since there were eight ovens in the laboratory, this meant that on average there were over 2.5 failures per oven per year. The maintenance supervisor and electricians had identified two distinct problems. One related to the distribution of airflow inside the ovens; the other was the performance of the bimetallic thermostats. Of the 21 failures recorded, 7 related to airflow distribution and 14 to failure of thermostats.

The thermostat contact remained in the closed position as long as the oven temperature was below the set point. This allowed the solenoid to remain energized, allowing the electric heating elements to continue providing heat. When the oven temperature exceeded the set point, the thermostat contacts

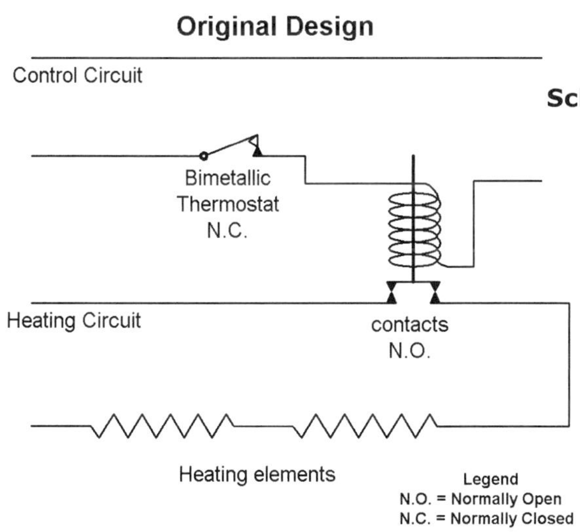

Figure 43.3
Schematic Control Circuit Diagram

opened. As a result, the solenoid switch was de-energized, in turn removing the power supply to the heating elements (see Figure 43.3). The solenoid switches had never failed, so we decided to focus on the two known problems.

As far as the airflow distribution was concerned, it was clear that the design intent was not being achieved. We reviewed the design and discussed the possible options. There was evidence that the airflow pattern was not well controlled.

The thermostat was located about midway on the ceiling on one side of the oven. The temperature it sensed was not very representative of the temperature in the oven. This location was clearly not optimal.

43.4 Solutions

We had to guide the upward flow of warm air along the walls and not allow it to mix easily with the air flowing downwards from the fan. We could achieve this by introducing a horizontal baffle or false ceiling at the plane of the fan blades. For this purpose, we designed and fabricated a sheet metal false ceiling. This was 4″ smaller than the oven's ceiling, leaving a 2″ gap along the four edges to allow upward air flow. We aligned this false ceiling with the plane of the fan blades, using twelve pieces of two-inch long pipe spacers and machine screws. The idea was that the warm air would be forced to travel vertically upwards along the walls and then enter the space between the oven's ceiling and the false ceiling. The fan would then blow the warm air down the center of the oven (see Figure 43.4) to produce the desired air circulation path.

An electrician suggested installing a second bimetallic thermostat. This was an intuitive suggestion, not one based on any analysis or knowledge of reliability engineering. After some discussion, we decided to install two thermo-

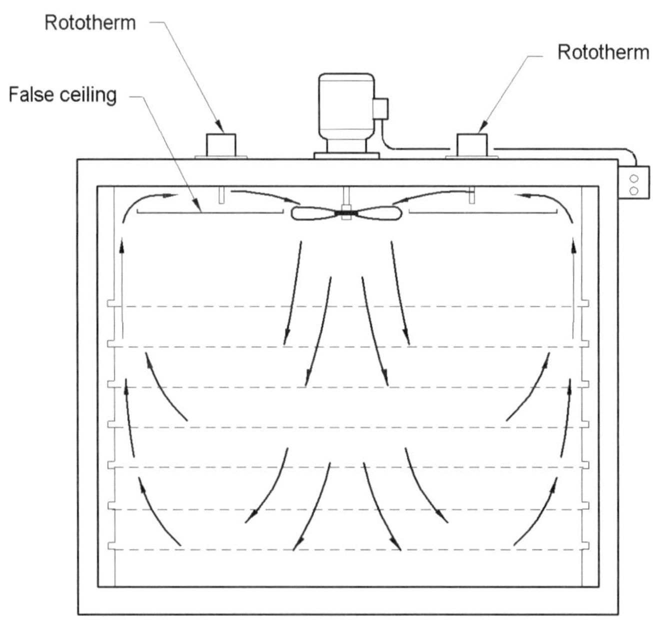

Figure 43.4 Modified Air Flow Paths

stats on the underside of the ceiling, where there was a controlled airflow pattern between the ceiling and the new false ceiling. The electrical circuit was wired so that both sets of thermostat contacts were in series. This meant that if the contacts of any one of the two thermostats opened, the power supply to the solenoid and, hence, heating elements would be cut off. Even if only one thermostat failed (i.e., its contacts did not open when the temperature was above the set point), it did not matter, as the second thermostat would be able to break the circuit. A schematic circuit is shown in Figure 43.5, and a discussion of the theory is given in Appendix 43-A.

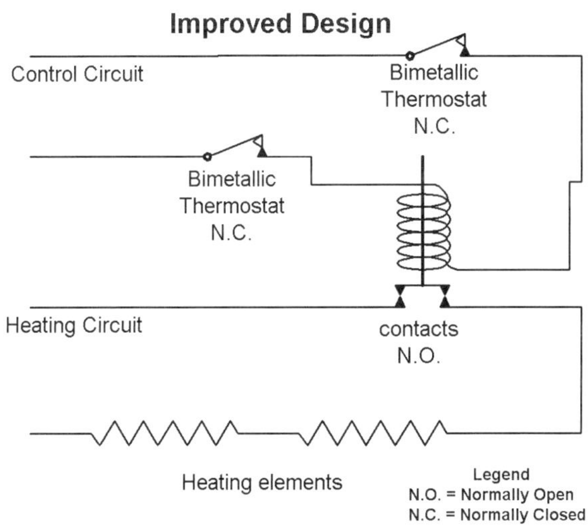

Figure 43.5 Modified Schematic Circuit Diagram

43.5 Trials

We implemented these modifications on a trial basis in one oven. We processed batches in all eight ovens, seven of which were of the original design. In a period of two months, the trial oven did not fail at all, while there were three failures (all due to thermostats) in the remaining seven ovens. This was a vast improvement over the earlier performance, so we decided to modify the remaining seven ovens.

43.6 Results

The results were satisfactory, and the laboratory was very pleased. We continued to have some failures of the ovens, but these had dropped to about two incidents per quarter. There were no failures attributed to airflow distribution.

While this work was going on, there were a number of production-related issues which needed resources and effort. Initially, it was difficult to justify the time and effort required to work on the laboratory ovens. However, it was clear that the knock-on effect of delaying the work on the ovens would eventually bring all production to a halt. From an implementation point of view, it was not complex and the time required to do the modifications was in days rather than weeks or months. The costs involved were quite low, so from all aspects, it looked a good candidate. In the event, this judgment proved right and with this success the maintenance department's profile rose significantly.

43.7 Lessons

1. Production department pressures can sometimes overwhelm maintenance people to the point where other departments may be ignored. In this instance, such actions would ultimately have resulted in production losses. An assessment of the overall impact can drive an apparently unimportant problem up the criticality ranks. Such overviews can be crucial.
2. Ideas are not the prerogatives of qualified engineers or managers. Often, the best ones come from the shop floor, so we have to listen to the workers!
3. Analysis, using factual data, a receptive mind, and common sense can produce results that may compare with those of experts with knowledge of theory (see Appendix 43-A).
4. Credibility of the maintenance department is valuable currency—hard to earn and easy to lose. Solving tricky problems that have vexed people for some time is one way to build credibility.

43.8 Principles

One of the pleasures of working as maintainers is the opportunity it gives us to apply innovative ideas. It offers fertile soil to sow creativity and reap the fruits of significant business benefits.

If maintainers understand basic reliability engineering concepts, they can apply it to their advantage in their work.

Appendix 43-A

43-A.1 An Explanation of the Underlying Theory

Reliability Block Diagrams (RBD) can be used to understand why two thermostats in series worked better than one. The following discussion is to assist those who are not familiar with the principles involved.

Reliability is the probability that at any given point in time, an item will survive a further stated period of time. It is also called survival probability, and expressed in percentage terms. An item can be in one of two states, i.e., either it is working or it has failed. The probability of its being in a working state, or its reliability at time 't' designated R(t), and read as R of t, plus its probability of being in a failed state or F(t) is 100%, or

$$R(t) + F(t) = 1$$

A Reliability Block Diagram (RBD) is a pictorial representation of a system, which mimics it mathematically. The items are represented as blocks and are linked to each other following the logic of the flow scheme.

Series RBD systems represent situations where every component item has to work for the whole system to work. For example, your car needs all of the following systems to work, for it to function.

- power system (engine)
- suspension system (chassis, springs, shock absorbers)
- steering system (steering wheel, steering gears and linkages)
- safety systems (brakes, lights, emission control)
- transmission system (clutch, gearbox, drive shaft, differential, wheels, and tires)

Any one system by itself is not enough; it needs all systems at the same time. Such a system is called a series system.

A system with three sub-systems is represented in Figure 43-A.1:

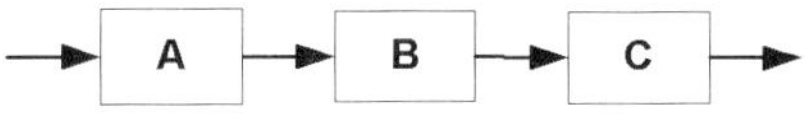

Figure 43-A.1 Series RBD with three sub-systems

For the above system to work, all three sub-systems must work. The logic operator is AND. It can be shown that

$R_{SYSTEM} = R_A \times R_B \times R_C$ where,

R_{SYSTEM} is the reliability of the whole system

R_A ,R_B, R_C designate the reliability of each of the sub-systems, A, B and C

Since the reliability of each sub-system can never exceed 1, the system reliability falls rapidly as the number of sub-systems rises.

Parallel RBD systems consist of a number of sub-systems where any one of them is enough for the whole system to work (see Figure 43-A.2 below). This is the case with our two-thermostat design, where if any one of them works, we will have a successful isolation of power to the solenoid. The logic operator in the model is OR, since it is enough if A or B or C operate for system success.

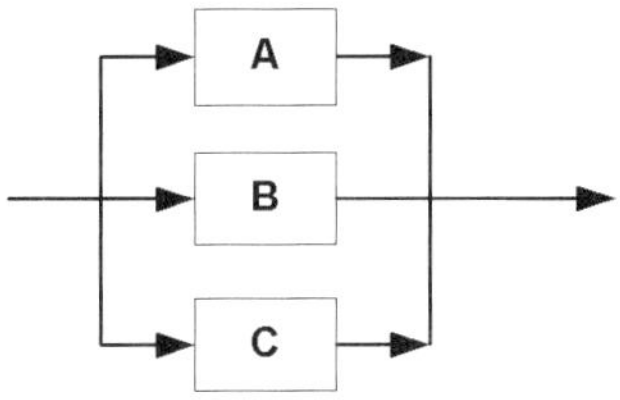

Figure 43-A.2 Parallel RBD with three sub-systems

It can be shown that

$$F_{SYSTEM} = F_A \times F_B \times F_C \text{ or,}$$

$$(1 - R_{SYSTEM}) = (1 - R_A) \times (1 - R_B) \times (1 - R_c) \text{ where,}$$

F_J is the failure function of element J and $R_J + F_J = 1$

In the case of parallel systems, the more the elements, the higher the system reliability. You can see why redundancy helps improve system reliability.

The same physical layout may result in a series or parallel RBD, depending on the failure mode we are considering, or the design capacity of each item.

As another example, consider two safety relief valves on a steam boiler drum. These valves will lift when the pressure exceeds the set point. Are the valves in series or parallel? If each valve is capable of relieving the pressure on its own, i.e., it has 100% capacity, you note that they are in an OR configuration, so they are in parallel. If each valve can relieve only 50% of the flow, both valves need to operate at the same time, so it is an AND configuration. In this case, the valves are in series, in an RBD. In both cases, if the failure mode we are considering is "leakage to atmosphere," the leakage of any one valve is enough to result in a leak to atmosphere. For this failure mode, the logic is OR, whether the valves are of 100% or 50% capacity.

Various combinations of series and parallel configurations are possible. Another configuration is the Bridge circuit, with 5 elements. Readers interested in these more complex arrangements should refer to the texts in the reference list.

43-A.2 Thermostat Failures

We will use an RBD to explain the performance of the two designs. The RBD of the old arrangement is shown in Figure 43-A.3.

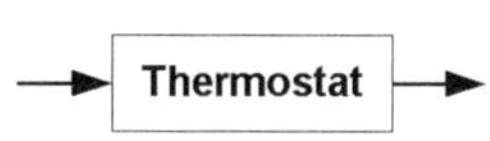

Required outcome: solenoid cuts off heater power if thermostat opens

Figure 43-A.3. RBD of the original arrangement

With two thermostats physically in series, if the contacts of either thermostat A or B opens on demand, the solenoid will be de-energized and the power to the heating elements cut off. This is the desired outcome or 'success.' The thermostat has failed if its contacts remain closed when the temperature rises above the set point. If both fail to open, we have a system failure. If either thermostat A or B works, we will successfully cut off the power to the heating elements. In this case, the logic operator is OR, so in the RBD the two thermostat elements are shown in parallel. This is shown in Figure 43-A.4.

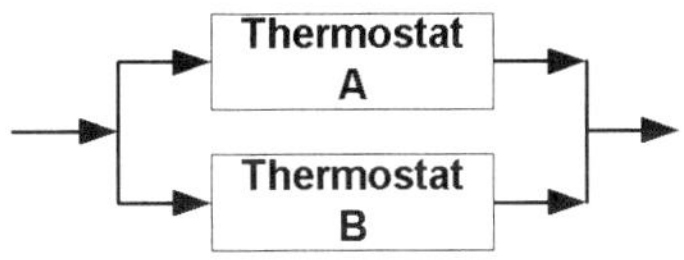

Thermostat A or B opens solenoid, loss of power is the desired outcome or success

Figure 43-A.4 RBD of the new arrangement

The mean time between failures (MTBF) and failure rate of the thermostats is computed thus:

Number of oven failures recorded in one year or 8760 hours = 21

Of this,
Number of failures of thermostats (fail to open on demand) = 14
Number of ovens in service during this period = 8
Thermostat failure rate (fail to open on demand)
= 14/(8x8760) = 0.0002/hour
MTBF for this failure = (8x8760)/14 = 5006 hours

In this case, we are assuming a failure distribution called the exponential distribution, described by the following equation.

$$R(t) = e^{-\lambda t}, \text{ where}$$

$R(t)$ is the reliability of the item at time 't',
'e' is the natural logarithm base,
λ is the failure rate, in this case computed as 0.0002 failures per hour

Thus when the operating age is 100 hours, if we use this value of 't' in the above equation, we can compute the reliability as 0.9802. Similarly when the operating hours are 500, the reliability is 0.9048.

In the modified design, when we have two thermostats, we have a parallel configuration. For this system,

$$(1 - R_{system}) = (1 - R_A) \times (1 - R_B)$$

Table 43-A.1 compares the reliability of the original system with one thermostat and that of the modified design with two thermostats. R_A is the reliability over time of the single thermostat design, while R_{system} is that for the new two-thermostat design.

For a simple system with one thermostat, when the elapsed time is equal to the Mean Time Between Failures (MTBF), we expect a reliability of approximately 37%. In our case, the MTBF is 5006 hours and R_A at this age is about 0.37 (See Table 43-A.1).

	Reliability with 1 thermostat	Reliability with 2 thermostats
Hours	R(A)	R(Sys)
100	0.9802	0.9996
200	0.9608	0.9985
500	0.9048	0.9909
1000	0.8187	0.9671
1500	0.7408	0.9328
2000	0.6703	0.8913
2500	0.6065	0.8452
3000	0.5488	0.7964
3500	0.4966	0.7466
4000	0.4493	0.6968
5000	0.3679	0.6004
6000	0.3012	0.5117
7000	0.2466	0.4324
8000	0.2019	0.3630
9000	0.1653	0.3033
10000	0.1353	0.2524

Failure rate of single thermostat is 0.0002/hr

Table 43-A.1 Comparison of Reliability Old/New Designs

The corresponding age for the two-thermostat system is about 8000 hours, when the reliability is about 37%. You can see that the modified system has a considerably higher level of reliability at any age. This means that the chance of the system failing and allowing the samples to overheat with the new design is much lower than if there were only one thermostat in place.

Further details about RBDs can be obtained from texts on Reliability Engineering (see the first two items in the list of references). A simplified explanation is given in Chapter 3 of the last reference item for this chapter. Knowing more about these will help you predict performance of design changes before the event. As you will agree, this approach is a lot better than the hit-or-miss method we used in our case.

Additional Reading

1. Hoyland, A. and M.Rausand. 1994. System Reliability Theory. 18-72. New York: John Wiley and Sons, Inc. ISBN: 0471593974
2. Davidson, J. 1994. The Reliability of Mechanical Systems. 22-33. London: Mechanical Engineering Publications, Ltd. ISBN 0852988818.
3. Narayan, V. 2004, Effective Maintenance Management: Risk and Reliability; Strategies for Optimizing Performance. Chapter 3. New York: Industrial Press Inc. ISBN 0-8311-3178-0.

Chapter **44**

Pump Reliability

0.1% of water in oil can reduce bearing life by 70%: Mark Barnes, "Water - The Forgotten Contaminant", Practicing Oil Analysis Magazine. July 2001.

If the motor is offset misaligned by 10 percent of the coupling manufacturer's allowable offset, then one can expect a 10 percent reduction in inboard bearing life.
Wesley Hines et al, Maintenance Technology, URL: http://www.mt-online.com/articles/04-99ma.cfm

Author: Jim Wardhaugh

Location: 2.2.2Large Complex Oil Refinery in the Far East

44.1 Background

A recent benchmarking study and follow-up review revealed that our reliability was poor. The overall Mean Time Between Failure (MTBF) of rotating equipment in our refinery was about ten or eleven months. Locations with high reliability were getting an MTBF of about four or five years, so we had a long way to go. The Engineering Manager tried to address this issue, but previous attempts ended in failure. This time he was determined to succeed, so he gave me the job of identifying the issues and putting together a strategy to bring top performance.

I chatted with a number of people in the location as I don't find formal interviews very effective. These chats threw up a number of interesting comments:

- We had about 3000 pieces of rotating equipment. Most were centrifugal pumps so the problem could immediately be refined to "Poor reliability of centrifugal pumps."
- The rotating equipment advisory group was competent, felt a responsibility for pump reliability, but had little influence on events.

- They felt that bunches of incompetents operated and maintained pumps.
- A set of silos was in existence, typified by:
- Production and maintenance had unclear roles in operating, maintaining and lubricating pumps
- Little knowledge of best practices or consequences of bad practices
- No coherent drive for improved reliability
- Repair service rather than failure elimination
- Facts (as opposed to opinions) were difficult to retrieve on reasons for failure, and actions taken.

Based on these findings we agreed on a multi-pronged approach to pump reliability improvement and a framework for action. What we wanted to achieve is well captured in Ron Moore's words, "Fixed forever as opposed to fixing forever."

44.2 Pump Reliability Improvement Framework

The framework aimed to put the vital basics in place and create clear roles and responsibilities. The elements of this framework were:

- Each plant manager would be jointly responsible with the area maintenance engineer for pump performance in his plant area
- They would identify the top 10 poor performing pumps and paint them yellow We called these yellow pumps bad actors.
- They would identify a remedial action plan for each yellow pump.
- They would revisit start and stop procedures, using the rotating group as advisors, train operators, and carry out regular audit start-ups by production supervisors, and trainers.
- Production operators would conduct formal tours of their plants to check running pumps for correct suction and discharge pressures, unusual noises such as cavitation, unusual vibration, etc.
- Rotating equipment advisors would review the type of lubricants used and rationalize where possible on a few best performing lubricants.
- Maintenance would set up and manage best practice lubricating oil and grease handling facilities avoiding cross contamination or water ingress:
  - Containers in dry area under cover, stored on side
  - Filled under clean, dry conditions, etc
  - Pumps with direct connection to barrels, etc.
  - Maintenance would form a two-man machinery-care team which would visit each pump monthly, along set routes in the site. The team would:

- Measure vibration, temperature, motor current, etc.
- Identify any obvious problems such as excessive vibration, cavitation, etc.
- Clean up around the pump if needed.
- Put back missing guards or bolts.
- Check lubrication system for contamination, top up and/or change lubricant.
- For large critical machines, take regular lubricant samples in scrupulously-clean sample bottles for laboratory analysis.

- We would modify the Computerized Maintenance Management System (CMMS) to allow easy entry of failure history data by technicians, including:
  - Reasons work was requested on a pump (breakdown, condition monitoring predicting failure, etc.)
  - As-found condition
  - Causes of the problem
  - Work done
  - Test results
  - These entries were vetted by the rotating equipment group for completeness and transparency.
- During workshop overhauls, the workshop supervisors (with advice from rotating group) would scrutinize the following aspects:
  - Suitability of pump design for present-duty conditions
  - Underlying causes of failure
  - Examine if they could beneficially replace existing components with more modern versions
  - Examine if the pump could be economically made more resilient to failure
  - Use findings from another European location on premature failures of mechanical seals as a check list; see Appendix 44-A
- All new installations and re-installations of pumps would put a focus on:
  - Solid foundations with minimal shimming (using only stainless steel shims)
  - Accurate laser alignment
  - Pipe alignment to minimize stresses
  - Capture of acceptable base-line vibration and noise signatures.
- We would implement a modern vibration monitoring system. See Figure 44- 1. It would have the following features:
  - We would set up routes around pumps on site in a PC, down-load data to a portable hand-held data collector-based system, and use this to capture vibration spectra of pumps.

- Paint spots on each pump would identify vibration capture points for consistency and repeatability
- We would install studs on a number of pumps, in direct contact with the bearing, to evaluate benefits.
- An expert system in the PC would identify potentially failing pumps.
- Vibration trend data would be uploaded from the PC to refinery CMMS to give trend information to interested parties.

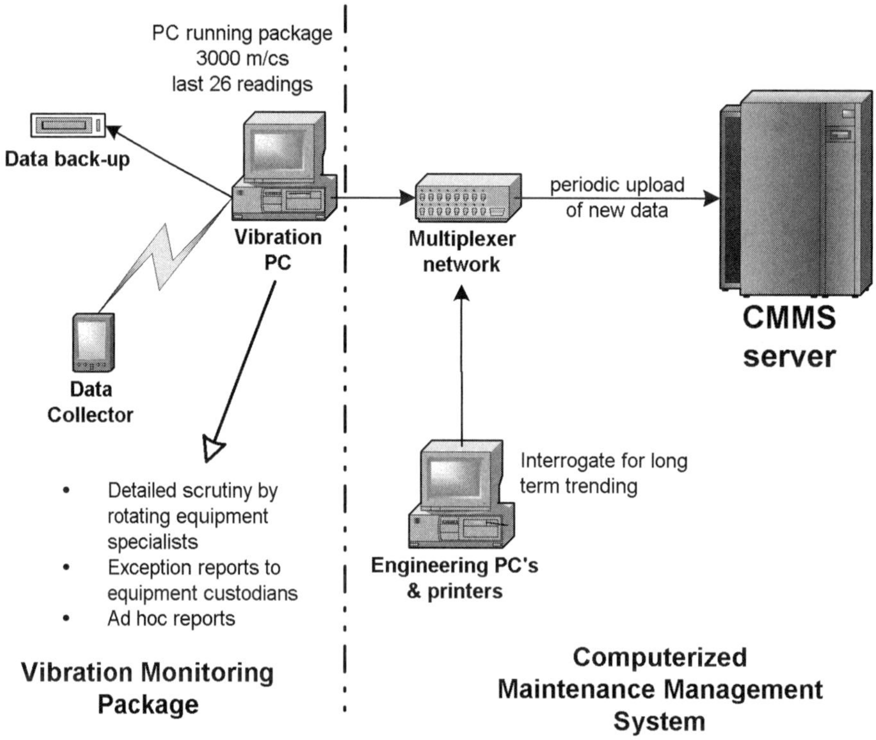

Figure 44.1 Vibration Monitoring Package Integrated into CMMS

44.3 Results

We had a number of initiatives running on the site in our drive for better performance. A concern that we might have too many, and efforts would be dissipated is always there. We went to a lot of trouble to explain what this pump initiative was all about, and in aligning efforts into a coherent whole.

We focused people's attention at all levels on the bright new tomorrow where there would be fewer pump failures. This took attention away from the poor past performance. We minimized blame and finger pointing.

We got the basics in place within a few months and felt confident that we

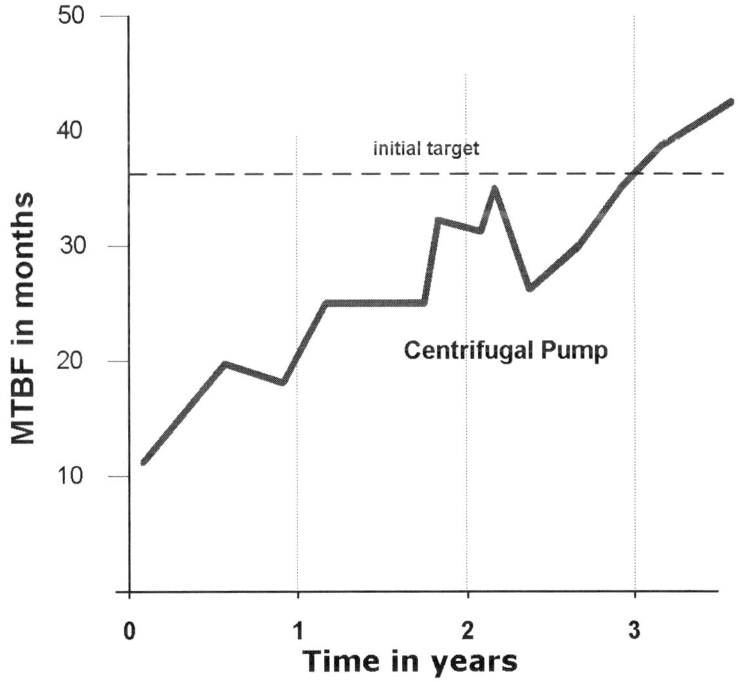

Figure 44.2 Centrifugal Pump MTBF Improvements

were building on a sure foundation.

Reliability results never come overnight, but an improvement trend became obvious quite quickly. The reduction in repeat failures was particularly welcome. With time the results worked through into the MTBF figures, as shown in Figure 44.2. After three years, we exceeded the initial target MTBF of 3 years.

44.4 Lessons

- Fragmented activities, however well-meaning, produce little performance improvement
- Alignment and coherence is vital
- Concentration on the basics is a must

44.5 Principles

- Move away from a repair focus to a reliability culture.
- Institutionalize failure elimination at all levels in the organization.

Appendix 44-A

Mechanical Seal Unreliability

A European location had a problem with shorter-than-expected life from their mechanical seals. In particular, 20% of seals failed within 10 weeks of putting into operation, and average life was less than 2 years.

The location's rotating equipment group and the seal vendor carried out a joint review of 330 seals as they failed over a period of months.

They found three problem areas: Selection, Operation, and Fitting. Figure 44-A.1 summarizes the principal causes of failure.

44-A.1 Selection Problems

40% of seals reviewed would benefit from re-selection, i.e., a better type was available. In particular, areas that could be beneficially addressed included:

- Abrasive wear of seal faces by using harder materials
- "Hang up" of seal due to solids building up on the shaft
- Materials incompatibility or chemical attack
- Vulnerability to dry running by better design or harder faces

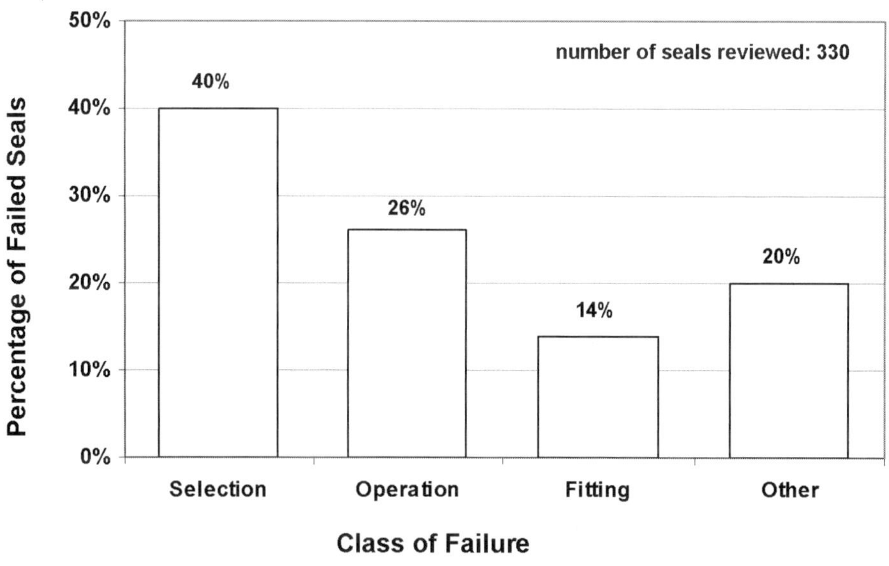

Data gathered from the analysis of CMMS failure data

Figure 44-A.1 Classes of Mechanical Seal Failure

44-A.2 Operation Problems

26% of the seal failures reviewed were classed as caused by poor operation. Particular issues were:

- Poor start-up procedures
- Dry running
- Inadequate venting
- Cavitation
- Changes in duty for which pump was inadequate

44-A.3 Fitting Problems

14% of seal failures were classed as caused by fitting problems. Particular issues were:

- Mechanical damage while fitting
- Incorrect assembly
- Misalignment
- Dirty surroundings
- Production demands causing rushed work or the use of fitters of inadequate competence

Chapter 45

Book Summary

Author: V. Narayan

My co-authors and I were fortunate to be able to see the seeds of maintenance thought blossom into an understanding that it was an organized and structured process to manage degradation. The latter part of the 20th century saw modern philosophers coming into full bloom, as incarnations of Plato, Socrates, Aristotle, and Confucius. Many of their theories and philosophies were applicable to maintenance management. We were able to see and experience these changes, learning from them as we went along on our respective journeys.

When we started our careers, the purpose of maintenance was generally believed to be able "to fix it when it breaks down, as fast and as cheaply as possible". Managers of a business saw it as a necessary evil. In the latter half of the last century, there was a spurt in the growth of risk based approaches to maintenance. With the establishment of Reliability Centered Maintenance (RCM) in the airline industry, a logical way of deriving the right tasks to do, and the right frequency to do it, revealed that maintenance was in fact a science, with laws similar to those in physics, chemistry, biology, and mathematics. Following on the heels of RCM, similar processes like Risk Based Inspection and Instrumented Protective Functions followed, making the management of reliability a scientific process. The purpose of maintenance now was beginning to be seen as "the activity to ensure that the reliability and integrity of assets is achieved at an acceptable cost."

The three of us have tried to show you what we have experienced and learned, both in respect of the philosophies and the processes. We have explained these using anecdotes about events in our working lives. These are true as far as we can trust our memories. Our explanation of how we handled the situation may differ from the way you would in your own situation with the constraints you face. At least you now have a point of reference.

There is a great deal of emphasis in industry about the need for problem solving abilities. While granting that this is true, we suggest that discovering opportunities to exploit is even more important, mostly because it also makes our organizations more profitable and our life more enjoyable. Having found the opportunity, we still need problem solving abilities, but the second approach puts us in a proactive frame of mind.

A manager does things right, a leader does the right things, so said Warren Bennis, one of our great current day philosophers. Steven Covey, another great thinker, gives a number of examples to drive home this point. Like Martin Luther King with his "I have a dream," leaders have a vision of the future. In Chapter 3, we demonstrate how we can create such a vision for our own factory or facility. Visions result in analysis of how to get there, leading to missions and objectives. You can see specific examples of these in Chapters 3, 4, and 5.

The customer is king, but who is this person? Identifying them and their expectations helps us set clear objectives, as shown in Chapter 4. In Chapter 5, we see how a transformation takes place in an organization steeped in somnolence. Mahen and his GM turned the place around, stating and sharing visions. Communication is the key to success, and we see examples of how 'they did it.' In Chapters 5 and 6, we see the importance of looking at the outside world and comparing ourselves with the best. Benchmarking can reveal our shortcomings, sometimes quite dramatically, and often we will go into denial. Jim describes such reactions at his refinery, and how they fought their way out of it, using the ideas of John Kotter, a Guru on change management.

Knowledge is a precious thing, to be guarded as a treasure. In their eagerness to reduce numbers, companies sometimes throw out the baby with the bath water, farming out work to contractors willy-nilly. This may lead to loss of core competencies, one of which is to be able to make proper cost estimates. In Chapter 7, Mahen explains how to recover from such situations.

For those readers who are less familiar with benchmarking, Jim leads us through the mechanics of the process in Chapter 8. He also explains what they did with the knowledge, to make real improvements in maintenance and reliability performance. There are recipes and practical advice for those who wish to embark on a similar improvement path.

One thing we often learn from benchmarking is that we are overstaffed. Traditions and featherbedding practices prevail in many factories, resulting in too many layers and too many people. In Chapter 9, Jim tells us how he broke that stranglehold in a strongly-unionized environment. Too many people can cause problems in many ways including high costs, inflated support infrastructure, management complexity with slow decision-making, and poor productivity, Most of the extra people do little useful work. Worse still, they stop the good workers from producing as well.

Having too many people also leads to empire building, a silo mentality, and turf wars. In Chapter 10, we see how one location developed a team approach to problem solving and began to demolish silos. Using the degradation management process to facilitate a change in attitudes and behavior, they used it as a first step in a reliability improvement program.

People are our greatest assets – but they are liabilities unless trained to meet the required competence levels. In Chapters 11 and 12, Jim and I deal with training and competence, and what we did in our particular situations. In Chapter 13, Jim explains how he used a different approach in a brand new refinery. He tells us how they applied the Scandinavian model in selecting and training people doubling up as operator-maintainers.

Motivating people to work effectively is never easy. Peter Drucker, the great management Guru, remarked that "So much of what we call management consists in making it difficult for people to work." Imagine how de-motivating it must be to have to continually surmount hurdles and get authorization to do work which, in your view, is largely unnecessary. All too often, workers are treated as 'hands', not whole people with brains as well. This is not just de-motivating; it also makes poor business sense. Getting an element of fun into jobs can sometimes help cure this disease and what better way to do this than to energize their creativity and trust individuals to influence events which affect them? In Chapter 14, we discuss how we managed to get a reliability culture in place using this approach.

We started with the problems of overstaffing. In many poor countries, social security is weak or non-existent and safety nets don't exist. The loss of a job can be really bad news for the individuals affected and their families. How do we deal with surpluses in such situations? In Chapters 15 and 16 we discuss two different scenarios. In one, we have the traditional problem of dealing with excess staff and how one company managed it. In this case, I was only an observer, so I had no role in its success. The next chapter deals with an individual who was encouraged to leave, even though his failure was entirely due to management errors. I did have an active role in rehabilitating this man and making him a useful team member, as discussed in Chapter 16.

So far we have discussed two important elements that lead to success in business: Leadership and People. The remaining chapters relate to elements adapted from the well known Continuous Improvement Cycle (Plan-Do-Check-Act), often attributed to Edward Deming, but in fact first stated by Walter Shewhart.

In Chapter 17, dealing with Integrated Planning, Mahen makes the case for team-work and joined-up thinking. This is built on a foundation of a well-mapped business process, clarity of roles and responsibilities, and a clearly communicated vision. Tools such as Critical Path Planning can bring large benefits in managing large projects such as shutdowns. In Chapter 18, Mahen explains why success comes only when they are used properly, i.e., updated, monitored, and controlled regularly.

Plant shutdowns reduce revenues and cost large sums of money. In the next chapter, Jim takes us through some overall strategies that can help the business, using results from benchmarking studies to support his case. He then lists some detailed recipes that worked well for his company. Top performers put a focus on plant integrity and production availability. This promotes extended intervals between shutdowns and shorter shutdowns.

In 1958, Professor Cyril Northcote Parkinson stated his well known law "Work expands so as to fill the time available for its completion." The events in Chapter 20 show how people can be very busy and achieve very little. Jim uses a common sense approach to challenge and eliminates unnecessary work in a conservative and risk-averse environment. In the next chapter, he deals with the question of how to use the spare time available to process plant operators to do maintenance work. There are two stories, one illustrating how not to do it and the other showing a way forward. The local culture and emotional make-up matter and one has to mold the improvement steps to suit.

In his book, A Force for Change (1990), John Kotter makes a key statement that "Leadership produces change. That is its primary function." In the events described in Chapter 22, my boss made a seemingly impossible demand, asking for a five-to-six fold increase in the prevailing tempo of machine shifting activity. All of us around the table were pretty shaken up, well out of our comfort zones. Some were even in denial, but I could see there was an opportunity. In this case we used Method Study to find a solution. Clearly, working smart is better than merely working hard, and the first step is to identify the problem clearly before seeking solutions. We also learned how to manage change, using teamwork and good communication.

Infrastructure maintenance is costly, difficult to execute, and not easy to justify in the short term. Faced with a major coating breakdown scenario in a humid and saline environment, as described in Chapter 23, doing nothing was not an option. We definitely needed a master plan and a clear policy to guide us. Seemingly intractable technical problems sometimes have elegant solutions, but commercial issues are often more difficult.

An often forgotten goal is that we have to keep the money machine running. If we have to shut it down, we must find the best time to do it, so that losses are reduced. When we do shut it down, we must try to do the work quickly so the machine is back at work as soon as possible. In Chapter 24, Mahen describes a dynamic refinery which took scheduling to new heights. They converted their long-term plan into a 104-week schedule for base-load work, complete with resource identification and leveling. This needed a clearly mapped business process, strict controls on priority setting, addition of new work to the weekly plan, and good team work. Some years earlier, I faced a problem of poor priority setting in a different refinery. We applied a different process there, as described in Chapter 25, and that too worked quite well.

If you are a maintenance practitioner, at least once in your lifetime you will have faced the problems caused by arbitrary budget cuts. Infrastructure items are invariably the worst hit, for reasons Mahen explains in Chapter 26. If a serious incident occurs as a result of years of neglect, you are faced with large bills, not just to recover from the incident, but also for the rest of the iceberg that lurks nearby. We need a logical way to re-prioritize the planned work so that funds become available to schedule the new work. The chapter shows a way that requires teamwork and a suitable process.

Using IT systems to de-layer the organization and delegate the task of scheduling routine work to the technicians helped improve workflow and productivity significantly, as Jim describes it in Chapter 27. The technicians were empowered to schedule their work; this meant they could request spares, scaffolding, cranes, etc., for the jobs they would do next day or week. They worked within a framework of rules on priorities and backlog, and their actions were transparent.

In a high hazard industry, it is essential to manage Technical Integrity well. We have to ensure that trip devices and other protective systems work on demand. The way to do this is to test these devices; but these tests can cause production losses or other consequential damage. A commonly applied solution is to use limited functional testing, in Chapter 28, we examine the dangers of relying solely on such tests.

Jim has shown earlier that poor performers have lengthy shutdowns because they do too much unnecessary work. In Chapter 29, Mahen looks at the whole business process. In the execution phase, emerging work can throw a well-made plan out of gear. Mahen explains the use of a hurdle as a precondition to accepting such work. Such controls ensure that shutdowns do not stretch in duration or costs. In another example in Chapter 30, Mahen explains how with good teamwork, we could find an answer to a problem that appeared at first sight to be beyond solution. Such jobs bring out the creative juices and make the life of a maintainer more interesting.

Plant operators are only utilized formally for about 75% of their time. The remaining 25% is unstructured. This time can thus be used for doing interruptible maintenance work. However, operations managers are reluctant to accept this evaluation. Mahen tells us in Chapter 31 how he conducted an evaluation by plant operators themselves. There is a list of typical maintenance activities that operators can perform in this spare time.

In the next chapter, Jim explains the difficulties in controlling overtime work. Using IT systems, the refinery made the whole overtime scenario transparent. This helped identify the real source of the problem to which they applied some interesting and imaginative solutions. Continuing along the same vein of making processes transparent using IT systems, Jim explains in Chapter 33 how they solved another seemingly intractable problem, that of contractor numbers. These systems put facts on the table. We are pretty good at finding an agreed way forward when we have facts, but very poor when we have only opinions.

This brings us to the final part of the book, dealing with the 'analyze' phase. In Chapter 34, we discuss how reliability engineering can help designers build leaner, cheaper, and better plants, and demonstrate that the design objectives are met. This link between designers and reliability engineers is, at best, tenuous. We can make significant capital and operating costs savings by strengthening this relationship.

We have lots of data, but no information to help make better decisions! We all collect mountains of data, but only some of us use them effectively. In Chapter 35, we can see how one company harnessed this information to advantage. There are some guidelines and definitions to help others along this path.

Independent observation of activities in a shutdown can help identify scope for improvements that 'insiders' cannot always see. In Chapters 36 and 37, we have two examples when such reviews identified significant improvement potential. Both locations adopted most of the recommendations and reaped the benefits. In an environment where people feel free to challenge existing practices, there will always be progress. The civil engineering supervisor suggested a change in the way we cleaned a vessel in the Fluid Catalytic Cracker Unit shutdown. He received the full support of the team and this led to a very significant reduction in duration and costs. These events are discussed in Chapter 38.

There is a school of thought that believes that all failures can and should be prevented. According to them, run-to-failure as a maintenance strategy is inherently flawed and is poor practice. In Chapter 39, Jim puts some hard evidence in front of us to show that for spared electric motors, especially the smaller ones, run-to-failure is often the most cost-effective strategy.

Supporters of condition monitoring will be pleased to read Chapter 40, as we discuss a way to predict seal failures. Alas, while the experiment shows this to be true, in practice, this can only happen with some additional features in seal design. More importantly, the experiment showed the effect frequent starts have on seals, and how we can make changes in operational philosophy to make reliability improvements.

A maintainer's life can be very interesting, as we can see in Chapters 41 and 42. Both situations presented seemingly intractable problems, and needed imaginative solutions. When the modifications were completed, the results were so dramatic that there was no doubt that we had won the day. The problem described in Chapter 43 was solved intuitively, without any knowledge of reliability engineering. If I had known then what I learned years later, that problem could have been solved much earlier.

Human reliability drives reliability performance of equipment. Jim explains in Chapter 44 the steps they took to improve pump performance. These included accountability, surveillance, focus (yellow pumps) and education, all aimed at people. Then they got the basics right; keep the equipment clean, dry, properly lubricated, aligned, and with tight bolts. Just taking these steps can get rid of 50–60% of failures, so they had clearly hit the two most important aspects. The results were as expected, a sharp improvement trend in pump reliability.

We hope we have demonstrated that we have not spent our 100+ years in vain, and that you can take away a few lessons to apply in your own situation. And I hope you had as much fun reading the book as we had in writing it.

Glossary

A list of terms used in this book and their meaning in the relevant context, is given below.

Annualized Turnaround Cost	Actual turnaround costs divided by the turnaround cycle in years.
Asset Replacement Value	Investment needed to replace a production plant (asset) in its same location.
Austenitic stainless steel	Austenitic stainless steels are alloys containing nickel and chromium. They have high oxidation and chemical resistance and are ductile.
Availability	1) The ability of an item to perform its function on demand and under given conditions. For non-repairable items, it equates to the reliability of the item. 2) The proportion of a given time interval that an item or system is able to fulfil its function under given conditions. 3) Availability = 100% minus (annualized turnaround downtime plus a two years average for routine maintenance downtime)
Average Run Length	Mean on-line time of a process unit between stops.
Axial Displacement	Displacement of the rotating element from its axial position per design.
Backlog	Requested work hours divided by resource hours available per week. Often approximated by using request numbers rather than man-hours.
Bath-tub Curve	A failure curve which mirrors that of human mortality, i.e., infant mortality, constant failure rate wear out.
Battery Limit	Perimeter of a plant facility.
Bean Counter	An (uncomplimentary) term for an accountant.
Benchmark	The systematic comparison of organizational processes and performances in order to create new standards and/improve processes.
Bitumen Blowing Unit (BBU)	A plant to manufacture bitumen from vacuum distillation residue.

Blow Down Valve A valve in a plant actuated to reduce the process inventory rapidly.

Brainstorming A process to generate ideas in an open-ended way. Usually used in problem-solving.

Breakdown Failure resulting in an immediate loss of function.

Business Model An algorithm which defines in simple, easy-to-understand terms, how the business will be run and what is important to bring business success.

Business Processes Activities carried out to run the business.

Catalyst An agent which speeds up chemical reactions without itself being affected.

Catalyst Regeneration Reactivating worn-out catalyst.

Change Management Managing the necessary actions to achieve a new way of working.

Combined Cycle Power Plant A power plant designed to maximize thermodynamic efficiency.

Competence The combination of knowledge, skills and attitudes necessary to carry out a job to the required standard of performance.

Complexity In a petroleum refinery, equivalent capacity divided by crude intake capacity. Each type of unit has a defined complexity number with crude distillation (the simplest) being taken as 1.

Compliance The ratio of completed preventive or corrective maintenance work orders to that planned, in a given time period. Work is considered complete as long as it is done on the scheduled date or within a stated tolerance band around this date.

Condition Based or On-Condition Maintenance Maintenance initiated as a result of knowing the condition of an item as a result of inspection or by routine or continuous monitoring of performance.

Condition Monitoring The continuous or periodic measurement and interpretation of data to indicate the condition of an item and thus determine the need for maintenance.

Conformance Proof that a product or service has met the specified requirement.

Control Loop A term to describe the connection of a sensor (say, measuring flow), a controlling device which aims to keep some parameter (in this case flow) at its "set-point," and the final element (usually a control valve or similar).

Corrective Maintenance 1) The maintenance carried out after failure has been initiated and intended to restore an item to a state in which it can perform its required function.
2) Any non-routine work other than breakdown work required to bring equipment back to a fit-for-purpose standard and arising from:
- defects found during the execution of maintenance routine.
- defects found as a result of inspection, condition monitoring, or ob servation.

Critical Path Planning A process to determine the sequence of activities in a project that takes the longest time (thus determining the project duration) with a view to minimizing this period.

Crude Distillation Unit A process plant used for distilling crude oil.

Cryogenics A branch of physics and engineering which deals with the production of very low temperatures (say below -150 degrees Centigrade) and examines the behavior of materials at those temperatures.

Decision Matrix A matrix usually with axes of frequency of occurrence and consequences on which events can be plotted to determine their importance.

Defect An adverse deviation from the specified condition of an item.

Diagnosis The art or act of deciding the nature of a fault from its symptoms.

Downtime The period of time during which an item is not in a condition to perform its intended function.

Dry Film Thickness Thickness of a paint film when dry.

Emergency Shut Down System A system or valve that is used to shut down a process plant in an emergency.

Emergent Work Work in a project which cannot reasonably be anticipated. Work which can be anticipated is called contingent work.

Entrainment A process by which liquid particles are carried away by flowing air or gas streams.

Equivalent Distillation Capacity (EDC) Equivalent Distillation Capacity(EDC) of a unit is the unit capacity multiplied by the unit complexity factor.

Equivalent Maintenance Personnel	The annual sum of own and contractor man-hours used doing routine maintenance, plus annualized turnaround man-hours, and all overtime manhours to get the average annual manhours consumed by maintenance. This number is then divided by the annual work hours of an individual maintainer. For standardization reasons this is taken as 2080 (52 weeks x 40 hours). This gives the number of maintenance personnel used ignoring sickness, vacation periods etc.
Failure	Termination of the ability of an item to perform any or all of its functions to the specified performance standards.
Failure Cause	The initiator of the process by which deterioration begins, resulting in functional failure.
Failure Effect	The consequence of a failure mode on the function or status of an item.
Failure Mode	A specific single event that leads to functional failure.
Failure Modes & Effects Analysis (FMEA)	A structured qualitative method involving the identification of the functions, functional failures, and failure modes of a system, and the local and wider effects of such failures.
Fatal Conceit	A flawed concept that a single mind or a single committee can some how do things better than the spontaneous, unstructured, complex, and creative forces of the market (based on a book written by the Nobel laureate economist F.A. Hayek).
Fault Tree Analysis	A structured and logical process for identifying the contribution of various causes of an unwanted event.
Fin-Fan Coolers	A heat exchanger (with tinned tubes and a large-bladed fan) used to cool process fluids.
Fluidized-bed Catalytic Cracker (FCC)	A process unit where long-chain hydrocarbon molecules are broken into high value short chain molecues by removing carbon molecules using a fluidized catalyst.
Function	The role or purpose for which an item exists. This is usually stated as a set of requirements with specified performance standards.
Functional Failure	The state when an item ceases to perform its function to the specified standards.
Gantt Chart	A widely-used planning chart; the Y-axis represents the activities while the X-axis represents time. The chart tells what is to be done, when, and by whom. It groups related tasks together, and identifies task that needs to be done before another can start.

Gas-Oil	A distillation product used for blending into diesel fuel oil.
Guru	A world-renowned thinker who has crystallized management ideas in an effective way.
Heat Exchanger	Equipment used to transfer heat between two fluid streams.
Hidden Failures	A class of failures which the operator will not knowabout under normal operating conditions. A second failure or other event is required before a hidden failure has any consequence.
High Vacuum Unit (HVU)	A process plant where distillation is carried out under vacuum conditions.
Hydrocracker (HC or HCU)	A process plant where long chain hydrocarbon molecules are broken into high value short chain molecules using a catalyst to add hydrogen molecules.
Hydroskimmer	A refinery using primary distillation and treatment processes.
In-Situ	A term meaning "at site;" usually used in the context of doing work without moving the work piece to a workshop.
Inspection	Those activities carried out to determine whether an asset is at its required level of functionality and integrity and the rate of change (if any) in these levels.
Instrument Protective Systems	These instruments protect equipment from high-consequence failures by tripping them when pre-set limits are exceeded.
Key performance Indicator	Business ratio or measure that is significant to the business.
Lagging Indicators	Performance measurements of past events/activities.
Leading Indicators	Performance measurements of past activities meant to promote beneficial effects in future and thus enhance business performance.
Life Cycle Costs	The total cost of ownership of equipment, taking into account the costs of acquisition, personnel training, operation, main tenance, modification, and disposal. It is used to decide between alternative options on offer.
Liquefied Natural Gas	Methane, liquefied at -260°F.
Liquefied Petroleum Gas	A mixture of propane and butane, liquefied under pressure at ambient temperature.
Maintainability	The ability of an item, under stated conditions of use, to be retained in or restored to a state in which it can perform its required functions, when maintenance is performed under stated conditions and using

prescribed procedures and resources. It is usually characterised by the time required to locate, diagnose, and rectify a fault.

Maintenance The combination of all technical and administrative actions intended to retain an item in or restore it to a state in which it can perform its required function.

Maintenance Costs Total maintenance costs including capital replacement items, averaged over two years for routine maintenance and over a complete cycle for turnaround maintenance.

Maintenance Index Maintenance costs divided by Equivalent Distillation Capacity. See Equivalent Distillation Capacity and Complexity above.

Maintenance Strategy Framework of actions to prevent or mitigate the consequences of failure in order to meet business objectives. The strategy may be defined at a number of levels (i.e., corporate, system, equipment, or failure modes).

Man-Way An opening in a vessel through which people can enter it.

Mean Time Between Failures (MTBF) For repairable systems, a measure of average operating performance,obtained by dividing the cumulative time in service (hours, cycles, miles, or other equivalent units) by the cumulative number of failures.

Mean Time To Failures (MTTF) For non-repairable systems, a measure of average operating performance obtained by dividing the cumulative time in service (hours, cycles, miles, or other equivalent units) by the cumulative number of failures.

Mean Time To Restore (MTTR) A measure of average maintenance performance, obtained by dividing the cumulative time for a number of consecutive repairs on a given repairable item by the cumulative number of failures of the item. The term restore means the time the defect was reported to the time the equipment was ready to restart and operate satisfactorily.

Mechanical Seal A device to prevent leakage of fluids from inside the equipment to the outside at shaft (or piston rod) entry points.

Method Study The systematic recording and critical examination of existing and proposed methods of doing work, as a means of developing and applying easier and more effective methods of reducing effort and costs.

MIG Also called Gas Metal Arc Welding, a process in which metal wire is fed continuously from a spool and shielded at the arc by an inert gas, usually Argon. MIG stands for Metal Inert Gas welding and is used for welding difficult metals, e.g., aluminum.

Modification An alteration made to a physical item, procedure, or software, usually resulting in an improvement in performance and carried out after a design review.

Monte Carlo Simulation — A mathematical model used to predict the performance of a complex system.

Net Positive Suction Head — The difference between the suction pressure of a pump and the vapor pressure of the fluid, measured at the impeller inlet.

Network Planning — See Critical Path planning.

Non Routine Maintenance — Any maintenance work which is not undertaken on a periodic time basis.

Normalizing Factor — A factor used to enable like-for-like comparisons.

On Stream Factor — 100% minus percentage of all downtimes (maintenance and others).

Operational Integrity — The continuing ability of a facility to produce as designed and forecast.

OREDA — Offshore REliability DAta, a reliability database collected by an association of offshore oil and gas companies and used in risk analysis,mathematical modeling, selection of maintenance strategies, etc. Data collection methodology is an ISO standard 14 224 1999.

Outage — The state of an item being unable to perform its required function.

Overhaul — A comprehensive examination and restoration of an item, or a major part of it, to an acceptable condition.

Overtime — Work done outside normal working hours; it usually involves payment at premium rates.

Paradigm — A model or theory; the mental map people use to assess a situation

Partial Closure Tests — When total closure of executive elements resulting in a shutdown is technically or economically undesirable, the movement of the executive element is physically restrained. Such tests prove that these elements would have closed in a real emergency. In the case of blow-down valves, the equivalent is partial opening tests.

Performance Indicator — A variable, derived from one or more measurable parameters, which when compared with a target level or trend, provides an indication of the degree of control being exercised over a process.

Planned Maintenance — The maintenance organized and carried out with forethought, control, and the use of records, to a pre-determined plan.

Platformer (or reformer) (PFU)	A process plant using a platinum catalyst to improve the octane number of gasoline streams.
Power Factor Correction	Use of devices to align current and voltage waveforms and vectors to maximize power per transmitted ampere, thus reducing power losses.
Pressure Relief Valve	A safety device used to discharge excess pressure automatically from a vessel or pipeline.
Preventive Maintenance	The maintenance carried out at pre-determined intervals or corresponding to prescribed criteria and intended to reduce the probability of failure or loss of performance of an item.
Project	A one-off finite piece of work with fixed start and end points and a clear objective.
Prototyping	Creation of a working information system by the use of high level soft ware modeling languages. This is useful where clarification of needs is a must before the specification is finalized.
Redundancy	The spare capacity that exists in a given system that enables it to tolerate failure of individual equipment items without total loss of system function, over a period of time during which the defective item can be restored.
Regulator	A government appointed, but usually independent, authority who regulates the affairs of businesses where market forces are insufficient to protect public interest.
Reliability	The probability that an item or system will fulfill its function when required to do so, under given conditions.
Reliability Block Diagram	This diagram shows the elements of a system connected in series or parallel as appropriate and allowing the use of Boolean logic to predict failure effects.
Reliability Centered Maintenance	(RCM) SAE 1011/1012; A process used to determine the maintenance requirements of any physical asset in its operating context.
Repair	To restore an item to an acceptable condition by the adjustment, renewal, replacement, or mending of misaligned, worn, damaged, or corroded parts.
Replacement Value	Investment needed to replace a production plant (asset) in its same location. See asset replacement value above.
Resource Leveling	Tasks that make up a project are shared evenly between all members (the fixed resources) of the team. Tasks are delayed until a resource becomes available. Also known as resource-limited scheduling.

Risk	The combined effect of the probability of occurrence of an undesirable event and the magnitude of the event.
Risk Based Inspection (RBI)	A structured and auditable method for establishing the appropriate inspection strategy and frequency for static mechanical equipment, e.g. vessels, piping in their operating context. Many variants but all based on API 580.
Root Cause	(RCA) A structured and systematic evidence-based process for determining the true causes of failures.
Routine Maintenance	Maintenance work of a repetitive nature which is undertaken on a periodic time (or equivalent) basis.
Routine Maintenance Index	Routine maintenance costs averaged over two years divided by Equivalent Distillation Capacity. See Equivalent Distillation Capacity and Complexity above.
Safety	Freedom from conditions that can cause death, injury, occupational illness, or damage to asset value or the environment.
SBM or Single Buoy Mooring	A system where a ship carrying liquid cargo is tethered to a buoy through which the cargo is discharged.
Schedule 80	Pipe specification (API) for thick-wall pipe.
Shutdown Duration	Time from feed out to product back on specification.
Shutdown Interval	Time measured from product back on specification in previous shutdown to feed out on next shutdown.
Shutdown Maintenance	Maintenance which can only be carried out when the item, system, or plant is out of service.
Shutdown or Turnaround	A term designating a complete stoppage of production in a plant, system, or sub-system to enable planned or unplanned maintenance work to be carried out. Planned shutdowns are usually periods of significant inspection and maintenance activity, carried out periodically. Shutdowns to reconfirm technical integrity are called major shutdowns in Europe and turnarounds in North America.
Spade	A metal plate inserted between a pair of pipe flanges to provide positive isolation.
Spectacle Blind	A special spade wilth two "eyes," one solid plate and the other a hole of pipe bore size. Such a blind can be in one of two positions, open or closed.
Stakeholder	Parties who contribute to or have an interest in an activity or a project.

String Diagram	A process used to identify the density of traffic of parts or people's movements, enabing improvements to be made.
Stuffing Box	A part of a rotating or reciprocating machine used for sealing the shaft or piston, thus preventing leakage of fluid.
Sub-surface Safety Valve	A safety valve placed at the foot of the oil or gas well to isolate the reservoir.
Synchronize Generators	Before a generator is connected to live busbars certain conditions must be met: 1) The magnitude of the generator voltage must be equal to the busbar voltage. 2) The generator voltage must be in phase with the busbar voltage. 3) The frequencies must match. The process of obtaining the above conditions is known as synchronizing.
Technical Integrity	Absence, during specified operation of a facility, of foreseeable risk of failure endangering safety of personnel, environment, or asset value.
Test Interval	The elapsed time between the initiation of identical tests on an item to evaluate its state or condition. Inverse of test frequency.
Thermal Cracking Unit (TCU)	A process unit where long-chain hydrocarbon molecules are broken into high-value short-chain molecules using heat to remove carbon molecules.
Tiered Quantity Discount	Discounts offered at increasing (or decreasing) rates for equal progressive volumes.
Tool-Box Talks	Short discussion between worker and supervisor at the start of work to familiarize the worker with the safety precautions and work procedures.
Toroidal	Shaped like a doughnut.
Total Equivalent Distillation Capacity (EDC) of a refinery	Sum of the EDCs of the individual units in a refinery. This EDC is used as a divisor to normalize aspects such as costs, personnel numbers, etc., for benchmarking.
Trays	Metal plates with holes or special devices (called bubble-caps), placed at equal spacing along the height of a distillation column, to enable mixing of the upward-flowing gases and downward-flowing liquids.
Trip	Operation of a protective device causing a plant (or part of it) to cease production.
Trunnions	An arrangement where a supported axle runs through the equipment allowing it to tilt through a small angle, up to 90°.

Tungsten Inert Gas Welding (TIG)	Tungsten Inert Gas welding, a non-consumable electrode welding process using metal filler wire and under an inert gas for shielding the arc; often used for welding stainless steel.
Turnaround	A term used in North America meaning planned shutdown. See Shutdown above.
Turnaround Maintenance Index	Annualized turnaround costs divided by EDC.
Uptime	The period of time during which an item is in a condition to perform its intended function.
Utilization%	100 x Actual annual intake in bbl divided by (365x annual design capacity in bbl/day)
Weibull	A commonly-used failure distribution model.
Well Head Valve	An isolation valve at the entry to the Production Facility.
Work Order	Work that has been approved for scheduling and execution. Materials, tools, and equipment can then be ordered and labor availability determined.

Index